Lecture Notes in Computer Science 2092

Edited by G. Goos, J. Hartmanis and J. van Leeuwen

Springer
Berlin
Heidelberg
New York
Barcelona
Hong Kong
London
Milan
Paris
Singapore
Tokyo

Lars Wolf   David Hutchison
Ralf Steinmetz (Eds.)

Quality of Service –
IWQoS 2001

9th International Workshop
Karlsruhe, Germany, June 6-8, 2001
Proceedings

Springer

Series Editors

Gerhard Goos, Karlsruhe University, Germany
Juris Hartmanis, Cornell University, NY, USA
Jan van Leeuwen, Utrecht University, The Netherlands

Volume Editors

Lars Wolf
University of Karlsruhe, Faculty of Informatics and Computing Center
Institute of Telematics, Zirkel 2, 76128 Karlsruhe, Germany
E-mail: Lars.Wolf@uni-karlsruhe.de

David Hutchison
Lancaster University, Faculty of Applied Sciences, Computing Department
Lancaster, LA1 4YR, UK
E-mail: dh@comp.lancs.ac.uk

Ralf Steinmetz
GMD IPSI, German National Research Center for Information Technology
Integrated Publication and Information Systems
Dolivostr. 15, 64293 Darmstadt, Germany
E-mail: Ralf.Steinmetz@darmstadt.gmd.de

Cataloging-in-Publication Data applied for

Die Deutsche Bibliothek - CIP-Einheitsaufnahme

Quality of service : 9th international workshop ; proceedings / OWQoS
2001, Karlsruhe, Germany, June 6 - 8, 2001. Lars Wolf ... (ed.). -
Berlin ; Heidelberg ; New York ; Barcelona ; Hong Kong ; London ;
Milan ; Paris ; Singapore ; Tokyo : Springer, 2001
    (Lecture notes in computer science ; Vol. 2092)
    ISBN 3-540-42217-X

CR Subject Classification (1998): C.2, D.4.4, H.3.5-7, H.4, H.5.1, K.4.4, K.6.5

ISSN 0302-9743
ISBN 3-540-42217-X Springer-Verlag Berlin Heidelberg New York

Springer-Verlag Berlin Heidelberg New York
a member of BertelsmannSpringer Science+Business Media GmbH

http://www.springer.de

© Springer-Verlag Berlin Heidelberg 2001
Printed in Germany

Typesetting: Camera-ready by author
Printed on acid-free paper   SPIN: 10839396   06/3142   5 4 3 2 1 0

Preface

Welcome to IWQoS 2001 in Karlsruhe!

Quality of Service is a very active research field, especially in the networking community. Research in this area has been going on for some time, with results getting into development and finally reaching the stage of products. Trends in research as well as a reality check will be the purpose of this Ninth International Workshop on Quality of Service.

IWQoS is a very successful series of workshops and has established itself as one of the premier forums for the presentation and discussion of new research and ideas on QoS. The importance of this workshop series is also reflected in the large number of excellent submissions. Nearly 150 papers from all continents were submitted to the workshop, about a fifth of these being short papers. The program committee were very pleased with the quality of the submissions and had the difficult task of selecting the relatively small number of papers which could be accepted for IWQoS 2001. Due to the tough competition, many very good papers had to be rejected.

The accepted papers included in these proceedings can be neatly structured into sessions and we have a very interesting workshop program which covers the following areas: Provisioning and Pricing; Systems QoS; Routing; TCP Related; Aggregation and Active Networks Based QoS; Wireless and Mobile; Scheduling and Dropping; and Scheduling and Admission Control. Finally, we have a session in which the accepted short papers will be presented. The contributed workshop program is complemented by a strong invited program with presentations on *"Quality of Service – 20 years old and ready to get a job?"* and *"Automated, dynamic traffic engineering in multi-service IP networks"*, and by a panel discussion on *"How will media distribution work in the Internet?"*.

While IWQoS is now a very well established event, we nevertheless want it to be a lively workshop with lots of interesting discussions. Thus, we would like to encourage all participants to take an active part during the discussions of the presented work as well as during the breaks and social events.

Here we would also like to thank our sponsors and patrons, namely SAP AG / CEC Karlsruhe – Corporate Research, Enterasys Networks, Ericsson Eurolab, Siemens, IBM, NENTEC, and Gunther-Schroff-Stiftung, who helped to make the workshop possible. Moreover, IWQoS 2001 is supported by technical co-sponsorship in-cooperation with the IEEE Communications Society, ACM SIGCOMM, and IFIP WG 6.1.

We heartily thank the program committee, and also all the expert reviewers, for their efforts and hard work especially in the reviewing and selection process – a large number of reviews had to be prepared in a very short time!

Finally, many thanks to the local organizers, especially Klaus Wehrle and Marc Bechler, but also all the other people helping with the workshop organization.

June 2001   Lars Wolf, David Hutchison, and Ralf Steinmetz

Workshop Co-chairs

Lars Wolf	University of Karlsruhe
David Hutchison	Lancaster University
Ralf Steinmetz	GMD IPSI and Darmstadt University of Technology

IWQoS Steering Committee

Jon Crowcroft	University College London
Rich Friedrich	HP Labs
Edward Knightly	Rice University
Peter Steenkiste	Carnegie Mellon University
Hui Zhang	Carnegie Mellon University

Program Committee

Nina Bhatti	Nokia
Gordon Blair	Lancaster University
Jose Brustoloni	Bell Labs
Andrew Campbell	Columbia University
Georg Carle	GMD FOKUS
Jon Crowcroft	University College London
Bruce Davie	Cisco
Hermann de Meer	University College London
Jan de Meer	GMD FOKUS
Serge Fdida	University Paris 6 – LIP6
Rich Friedrich	HP Labs
David Hutchison	Lancaster University
Kevin Jeffay	University of North Carolina
Edward Knightly	Rice University
Jim Kurose	University of Massachusetts
Jorg Liebeherr	University of Virginia
Qingming Ma	Cisco
Laurent Mathy	Lancaster University
Klara Nahrstedt	University of Illinois
Andrew Odlyzko	AT&T Research
Jim Roberts	France Telecom
Jens Schmitt	Darmstadt University
Yuval Shavitt	Tel Aviv University
Cormac Sreenan	University College Cork
Peter Steenkiste	Carnegie Mellon University
Ralf Steinmetz	GMD IPSI and Darmstadt University of Technology
Burkhard Stiller	ETH Zurich
Ion Stoica	Carnegie Mellon University
Lars Wolf	University of Karlsruhe
John Wroclawski	MIT

Local Organization

Lars Wolf   University of Karlsruhe
Klaus Wehrle   University of Karlsruhe
Marc Bechler   University of Karlsruhe

Reviewers

Ralf Ackermann, Darmstadt Univ.
Pascal Anelli, Univ. Paris 6
Marc Bechler, Univ. of Karlsruhe
Nabil Benameur, France Telecom
Nicole Berier, Darmstadt Univ.
Nina Bhatti, Nokia
Gordon Blair, Lancaster Univ.
Roland Bless, Univ. of Karlsruhe
Thomas Bonald, France Telecom
Jose Brustoloni, Bell Labs
Andrew Campbell, Columbia Univ.
Roberto Canonico, Univ. Napoli
Georg Carle, GMD FOKUS
Coskun Cetinkaya, Rice Univ.
Anna Charny, Cisco
Nicolas Christin, Univ. of Virginia
Luis Costa, Univ. Paris 6
Jon Crowcroft, Univ. College London
Vasilios Darlagiannis, Darmstadt Univ.
Bruce Davie, Cisco
Hermann de Meer, Univ. College London
Jan de Meer, GMD FOKUS
Martin Dunmore, Lancaster Univ.
Larry Dunn, Cisco
Hasan, ETH Zurich
Christopher Edwards, Lancaster Univ.
Thomas Erlebach, ETH Zurich
Serge Fdida, Univ. Paris 6
Ulrich Fiedler, ETH Zurich
Clarence Filsfils, Cisco
Anne Fladenmuller, Univ. Paris 6
Placi Flury, ETH Zurich
Rich Friedrich, HP Labs
Huirong Fu, Rice Univ.
Violeta Gambiroza, Rice Univ.
Carsten Griwodz, Univ. of Oslo
Stefan Gruhl, Rice Univ.
Robert Haas, IBM Zurich
Oliver Heckmann, Darmstadt Univ.
Andreas Hoffmann, GMD FOKUS
Polly Huang, ETH Zurich
David Hutchison, Lancaster Univ.
Kevin Jeffay, Univ. of North Carolina
Verena Kahmann, Univ. of Karlsruhe
Vikram Kanodia, Rice Univ.
Martin Karsten, Darmstadt Univ.
Edward Knightly, Rice Univ.
Jim Kurose, Univ. of Massachusetts

Aleksandar Kuzmanovic, Rice Univ.
Gwendal Le Grand, Univ. Paris 6
Chengzhi Li, Rice Univ.
Yuhong Li, Univ. of Karlsruhe
Jorg Liebeherr, Univ. of Virginia
Yonghe Liu, Rice Univ.
Qingming Ma, Cisco
Laurent Mathy, Lancaster Univ.
Klara Nahrstedt, Univ. of Illinois
Andrew Odlyzko, AT&T Research
Philippe Olivier, France Telecom
Sara Oueslati-Boulahia, France Telecom
Philippe Owezarski, LAAS-CNRS
Dimitris Pezaros, Lancaster Univ.
Roman Pletka, IBM Zurich
Alexandre Proutiere, France Telecom
Supranamaya Ranjan, Rice Univ.
Christoph Reichert, GMD FOKUS
Axel Rennoch, GMD FOKUS
Hartmut Ritter, Univ. of Karlsruhe
James Roberts, France Telecom
Rudolf Roth, GMD FOKUS
Bahareh Sadeghi, Rice Univ.
Kave Salamatian, Univ. Paris 6
Henning Sanneck, Siemens AG
Susana Sargento, Rice Univ.
Jens Schmitt, Darmstadt Univ.
Yuval Shavitt, Tel Aviv Univ.
Promethee Spathis, Univ. Paris 6
Cormac Sreenan, Univ. College Cork
Peter Steenkiste, Carnegie Mellon Univ.
Ralf Steinmetz, GMD IPSI & Darmstadt U.
Burkhard Stiller, ETH Zurich
Ion Stoica, Carnegie Mellon Univ.
Kim Thai, Univ. Paris 6
Rolland Vida, Univ. Paris 6
Klaus Wehrle, Univ. of Karlsruhe
Dorota Witaszek, GMD FOKUS
Lars Wolf, Univ. of Karlsruhe
Thomas Wolfram, GMD FOKUS
John Wroclawski, MIT
Ping Yuan, Rice Univ.
Sebastian Zander, GMD FOKUS
Remi Zara, Univ. Paris 6
Michael Zink, Darmstadt Univ.
Artur Ziviani, Univ. Paris 6
Tanja Zseby, GMD FOKUS

Supporting/Sponsoring Societies

IEEE Communications Society
    Technical Committees on Computer Communications (TCCC) and Internet (ITC)
ACM SIGCOMM
IFIP WG6.1

Supporting/Sponsoring Companies and Foundations

CEC Karlsruhe – Corporate Research

Gunther-Schroff-Stiftung

Table of Contents

Routing

TCP Related

Wireless and Mobile

Short Paper Session

Aggregation and Active Networks Based QoS

Scheduling and Dropping

Scheduling and Admission Control

Author Index

Panel Discussion: How Will Media Distribution Work in the Internet?

Andrew Campbell[1], Carsten Griwodz[2], Joerg Liebeherr[3], Dwight Makaroff[4], Andreas Mauthe[5], Giorgio Ventre[6], and Michael Zink[7]

[1] Columbia University
[2] University of Oslo
[3] University of Virginia
[4] University of Ottawa
[5] tecmath AG
[6] University of Napoli
[7] Darmstadt University of Technology

Abstract. The panelists discuss the directions that future QoS research should to support distributed multimedia applications in the Internet, in particular applications that rely on audio and video streaming.

While real-time streaming services that are already deployed attract only a minor share of the overall network traffic and have only a minor impact on the network, this is in part due to the distribution infrastructures that are already in place to reduce network and server loads.

However, it is questionable whether these infrastructures for non-interactive applications, that rely largely on the regional over-provisioning of resources, will also be capable of supporting future multimedia applications. The panelists discuss their views of the current and future demand for more sophisticated QoS mechanisms and the resulting research issues.

1   Introduction

Researchers in distributed multimedia applications in the Internet, and in audio and video streaming in particular, have investigated the provision of services with a well-defined quality, and a considerable share has gone into the development of approaches for generic QoS provision. The commercial applications in the streaming area make a strict distinction between application areas such as conferencing and on-demand streaming, reflecting the equally strict distinction between the business and consumer markets. For high-quality delivery over long distances, the costumer demand for quality guarantees is pressing, but an application of QoS techniques that have been developed so far is not necessarily the solution.

In the multimedia server business, scalable and performance-optimized servers are currently preferred over such that guarantee fine-grained QoS per session. We can observe a concentration of real-time delivery development in a few companies and an obvious interest of infrastructure providers to integrate storage and distribution services into their portfolio. Their nationwide and sometimes global

L. Wolf, D. Hutchison, and R. Steinmetz (Eds.): IWQoS 2001, LNCS 2092, pp. 3–5, 2001.

distribution infrastructures rely largely on the transfer of data into regional centers which allow to fulfill demands regionally. Researchers in the network resource management are increasingly paying attention to reservation schemes that work on aggregated flows rather than individual reservations. Applications are supposed to cope with variations in service quality that result from such infrastructures by adaptation. However, services that are extremely attractive to consumers for a short time are not served well by these approaches, as demonstrated by live high-quality broadcasts, that have so far been rare and unrewarding events.

Depending on the point of view, the lack of control over the Internet infrastructure or the lack of open standards for distributed multimedia applications impede the deployment of commercial services to the general public.

The network orientation of this year's IWQoS asks for a comparison of networking researchers' and other distributed systems researchers' views of the future developments.

The panelists discuss their assumptions about future QoS requirements and the question whether current networking research topics are closer to real-world applications in today's Internet than tomorrow's. Giorgio Ventre, Jörg Liebeherr and Andrew Campbell address the development of QoS research from the networking point of view while their Dwight Makaroff, Andreas Mauthe and Michael Zink provide their views on the distributed applications' future needs for QoS and they identify the existing and upcoming research issues. They demonstrate the influence that modified networking developments –such as the unlikeliness of negotiable per-flow guarantees in the foreseeable future– have on the goals of research and development of distribution systems, and identify the specific QoS requirements for audio and video distribution that can not be discarded or evaded.

A preview of their points of view is given in the following.

Dwight Makaroff: Is research in Network Quality of Service going to ever be applied to the Internet, given that no commercial force in the Internet seems willing to enable any protocols or mechanisms that will be enforceable. Even though CISCO puts QoS capabilities in their routers and Windows will eventually have RSVP, if routers in the middle are going to turn off these features, all the QoS work just can't be tested or implemented in the large scale.

It may well be the case that there is currently (and will continue to be) so much excess capacity in the backbone of the Internet that all we need is fiber to the home and we won't need QoS in the network. All the packets will get there just fine. So, in this case, QoS work is not needed. Since "fiber is cheap", individual users will be able to purchase/lease/lay down their own fiber or reserve wavelengths on a fiber.

What remains to be solved? Server issues will still be a problem, because the scalability of the content provider will continue to be difficult. One reason for the scalability problem is the need to maintain consistency in a replicated environment, caching issues, pushing the content out further to the edges of the backbone. This must all work seamlessly in a heterogeneous environment with multiple carriers and pricing policies.

Additionally, the display devices will continue to have differing bandwidth and resolution capabilities. Thus, applications will have to be adaptable, either at the server end, or at some intermediate location to avoid overwhelming the device that does not have the processing power to intelligently discard the excess media data. These resources will be at the control of the intra-domain (in the lab, in the house, in the building) sphere to properly adjust the bandwidth for individual devices.

Andreas Mauthe: The 'ubiquitous' Internet has penetrated almost all communication sectors. Without the Internet the World Wide Web would not have been possible and most LAN implementations use Internet protocol. Professional media applications at productions houses and broadcasters more and more use IT networks for the transmission of their data (in-house and in the wide area). In this context there are two major problems, lack bandwidth and lack of sufficient QoS support. The data rates of broadcast or production quality video are between 8Mb/s and 270 Mb/s. In a streaming environment these data rates have to be guaranteed. At present the reaction to these problems are either dedicated networks or an adaptation at the application level. A fully integrated communication platform catering for the needs of these applications are not in sight.

Michael Zink: It seems that streaming applications will not be able to make use of Network QoS in the near future because QoS features are not enabled in today's Internet. Caused by the lack of this functionality streaming applications need to make use of other mechanisms to support high quality even in best effort networks. One possible solution is a distribution infrastructure for streaming applications that will overcome some of the existing shortcomings. Scalability issues on media servers can be solved by load reduction through caches. The reliability of the system can be increased by an increased fault tolerance against server failures, and in addition the amount of long distance distribution will be reduced.

In an Internet without any QoS mechanism, streaming must be TCP-friendly to adopt the "social" rules implied by TCP's cooperative resource management model. TCP-friendly streaming mechanisms are also needed in the case that Network QoS is realized by reservation schemes that are based on aggregated flows (e.g DiffServ). Such aggregated flows that provide QoS guarantees are the minimum requirement for an infrastructure that reasonably supports wide area distribution of continuous media. It is rather questionable whether the other extreme, individual reservations, will ever be affordable for an end user.

Invited Talk:
Automated, Dynamic Traffic Engineering in Multi-service IP Networks

Joseph Sventek

Agilent Laboratories Scotland
Communication Solutions Department
joe_sventek@agilent.com

Abstract. Since the initial conversion of the ARPAnet from the NCP-family of protocols to the initial TCP/IP-family of protocols in the early 1980s, and especially since the advent of tools (such as browsers) for easily accessing content in the early 1990s, the Internet has experienced (and continues to experience) meteoric growth along any dimension that one cares to measure. This growth has led to the creation of new business segments (e.g. Network Element Manufactures and Internet Service Providers), as well as to partitioning of the operators providing IP services (access/metro/core).

Many of the core IP operators also provide long-distance telephony services. Since the majority of their sunk costs are concerned with laying and maintaining optical fibre (as well as the equipment to light up the fibre plant), these operators would like to carry both classes of traffic (data and voice) over a single core network. Additionally, the meteoric growth in number of users forces these carriers, and their NEM suppliers, to continue to look for ways to obtain additional capacity from the installed fibre and network element plants.

The net result of these pressures is that core IP networks have become increasingly complex. In particular, traditionally form of management products and processes are becoming less effective in supporting the necessary service provisioning, operation, and restoration.

I will describe some of the research that Agilent Laboratories is pursuing to enable automated, dynamic traffic engineering in multi-service IP networks.

L. Wolf, D. Hutchison, and R. Steinmetz (Eds.): IWQoS 2001, LNCS 2092, pp. 6–6, 2001.

Dynamic Core Provisioning for Quantitative Differentiated Service

Raymond R.-F. Liao and Andrew T. Campbell

Dept. of Electrical Engineering, Columbia University, New York City, USA
{liao,campbell}@comet.columbia.edu

Abstract. Efficient network provisioning mechanisms supporting service differentiation and automatic capacity dimensioning are important for the realization of a differentiated service Internet. In this paper, we extend our prior work on edge provisioning [7] to interior nodes and core networks including algorithms for: (i) dynamic node provisioning and (ii) dynamic core provisioning. The dynamic node provisioning algorithm prevents transient violations of service level agreements by self-adjusting per-scheduler service weights and packet dropping thresholds at core routers, reporting persistent service level violations to the core provisioning algorithm. The dynamic core provisioning algorithm dimensions traffic aggregates at the network ingress taking into account fairness issues not only across different traffic aggregates, but also within the same aggregate whose packets take different routes in a core IP network. We demonstrate through analysis and simulation that our model is capable of delivering capacity provisioning in an efficient manner providing quantitative delay-bounds with differentiated loss across per-aggregate service classes.

1 Introduction

Efficient capacity provisioning for the Differentiated Services (DiffServ) Internet appears more challenging than in circuit-based networks such as ATM and MPLS for two reasons. First, there is a lack of detailed control information (e.g., per-flow states) and supporting mechanisms (e.g., per-flow queueing) in the network. Second, there is a need to provide increased levels of service differentiation over a single global IP infrastructure. In traditional telecommunication networks, where traffic characteristics are well understood and well controlled, long-term capacity planning can be effectively applied. We argue, however, that in a DiffServ Internet more dynamic forms of control will be required to compensate for coarser-grained state information and the lack of network controllability, if service differentiation is to be realistically delivered.

There exists a trade-off intrinsic to the DiffServ service model (i.e., qualitative v.s. quantitative control). DiffServ aims to simplify the resource management problem thereby gaining architectural scalability through provisioning the network on a per-aggregate basis, which results in some level of service differentiation between service classes that is *qualitative* in nature. Although under

L. Wolf, D. Hutchison, and R. Steinmetz (Eds.): IWQoS 2001, LNCS 2092, pp. 9–26, 2001.

normal conditions, the combination of DiffServ router mechanisms and edge regulations of service level agreements (SLA) could plausibly be sufficient for service differentiation in an over-provisioned Internet backbone, network practitioners have to use *quantitative* provisioning rules to automatically re-dimension a network that experiences persistent congestion or device failures while attempting to maintain service differentiation. Therefore, a key challenge for the emerging DiffServ Internet is to develop solutions that can deliver suitable network control granularity with scalable and efficient network state management.

In this paper, we propose an approach to provisioning quantitative differential services within a service provider's network (i.e., the intra-domain aspect of the provisioning problem). Our SLA provides quantitative per-class delay guarantees with differentiated loss bounds across core IP networks. We introduce a distributed *node provisioning algorithm* that works with class-based weighted fair (WFQ) schedulers and queue management schemes. This algorithm prevents transient service level violations by adjusting the service weights for different classes after detecting the onset of SLA violations. The algorithm uses a simple but effective analytic formula to predict persistent SLA violations from measurement data and reports to our network *core provisioning algorithm*, which in turn coordinates rate regulation at the ingress network edge (based on our prior work of edge provisioning [7]).

In addition to delivering a quantitative SLA, another challenge facing DiffServ provisioning is rate control of any traffic aggregate comprising of flows exiting at different network egress points. This problem occurs when ingress rate-control can only be exerted on a per traffic aggregate basis (i.e., at the root of a traffic aggregate's point-to-multipoint distribution tree). In this case, any rate reduction penalizes traffic flowing along branches of the tree that are not congested. We call such a penalty, *branch-penalty*. One could argue for breaking down a customer's traffic aggregate into per ingress-egress pairs and provisioning in a similar way as MPLS tunnels. Such an approach, however, would not scale as the network grows because adding an egress point to the network would require reconfiguration of all ingress rate-controllers at all customer sites. We implement a suite of policies in our core provisioning algorithm to address the provisioning issues that arises when supporting point-to-multipoint traffic aggregates. Our solution includes a policy that minimizes branch-penalty, delivers fairness with equal reduction across traffic aggregates, or extends max-min fairness for point-to-multipoint traffic aggregates.

Node and core provisioning algorithms operate on a medium time scale, as illustrated in Figure 1. As can be seen in the figure, packet scheduling and flow control operate on fast time scales (i.e., sub-second time scale); admission control and dynamic provisioning operate on medium time scales in the range of seconds to minutes; and traffic engineering, including rerouting and capacity planning, operate on slower time scales on the order of hours to months. Significant progress has been made in the area of scheduling and flow control, (e.g., dynamic packet state and its derivatives [11]). In the area of traffic engineering, solutions for circuit-based networks has been widely investigated in literature [10]. There has been recent progress on the application of these techniques to IP routed networks [5]. In contrast, in the area of dynamic provisioning, most research effort

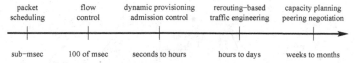

Fig. 1. Network Provisioning Time Scale.

has been focused on admission control issues such as endpoint-based admission control [3]. However, these algorithms do not provide fast mechanisms that are capable of reacting to sudden traffic pattern changes. The dynamic provisioning algorithms introduced in this paper are complementary to scheduling and admission control algorithms. These dynamic algorithms are capable of quickly restoring service differentiation under severely congested and device failure conditions. Our method bears similarity to the work on edge-to-edge flow control [1] but differs in that we provide a solution for point-to-multipoint traffic aggregates rather than point-to-point ones. In addition, our emphasis is on the delivery of multiple levels of service differentiation.

This paper is structured as follows. In Section 2, we introduce a dynamic provisioning architecture and service model. Following this, in Section 3, we present our dynamic node provisioning mechanism. In Section 4, we present a core provisioning algorithm. In Section 5, we discuss our simulation results demonstrating that the proposed algorithms are capable of supporting the dynamic provisioning of SLAs with guaranteed delay, differential loss and bandwidth prioritization across per-aggregate service classes. We also verify the effect of rate allocation policies on traffic aggregates. Finally, in Section 6, we present some concluding remarks.

2 Dynamic Network Provisioning Model

2.1 Architecture

We assume a DiffServ framework where edge traffic conditioners perform traffic policing/shaping. Nodes within the core network use a class-based weighted fair (WFQ) scheduler and various queue management schemes for dropping packets that overflow queue thresholds.

The dynamic capacity provisioning architecture illustrated in Figure 2 comprises dynamic core and node provisioning modules for bandwidth brokers and core routers, respectively, as well as the edge provisioning modules that are located at access and peering routers. The edge provisioning module [7] performs ingress link sharing at access routers, and egress capacity dimensioning at peering routers.

2.2 Control Messaging

Dynamic core provisioning sets appropriate ingress traffic conditioners located at access routers by utilizing a *core traffic load matrix* to apply rate-reduction (via a *Regulate_Ingress Down* signal) at ingress conditioners, as shown in Figure 2. Ingress conditioners are periodically invoked (via the *Regulate_Ingress*

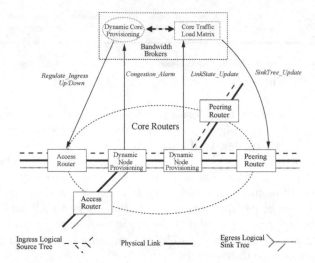

Fig. 2. Dynamic Capacity Provisioning Model.

Up signal) over longer restoration time scales to increase bandwidth allocation restoring the max-min bandwidth allocation when resources become available. The core traffic load matrix maintains network state information. The matrix is periodically updated (via *Link_State* signal) with the measured per-class link load. In addition, when there is a significant change in rate allocation at ingress access routers, a core bandwidth broker uses a *SinkTree_Update* signal to notify egress dimensioning modules at peering routers when renegotiating bandwidth with peering networks, as shown in Figure 2. We use the term "sink-tree" to refer to the topological relationship between a single egress link (representing the root of a sink-tree) and two or more ingress links (representing the leaves of a sink-tree) that contribute traffic to the egress point.

Dynamic core provisioning is triggered by *dynamic node provisioning* (via a *Congestion_Alarm* signal as illustrated in Figure 2) when a node persistently experiences congestion for a particular service class. This is typically the result of some local threshold being violated. Dynamic node provisioning adjusts service weights of per-class weighted schedulers and queue dropping thresholds at local core routers with the goal of maintaining delay bounds and differential loss and bandwidth priority assurances.

2.3   Service Model

Our SLA comprises:

- a *delay guarantee*: where any packet delivered through the core network (not including the shaping delay of edge traffic conditioners) has a delay bound of D_i for network service class i;
- a *differentiated loss assurance*: where network service classes are loss differentiated, that is, for traffic routed through the same path in a core network,

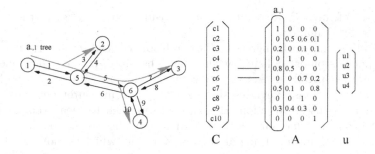

Fig. 3. Example of a Network Topology and Its Traffic Matrix.

the long-term average loss rate experienced by class i, $\bar{P}_{loss}(i)$ is no larger than $P^*_{loss}(i)$. The thresholds $\{P^*_{loss}(i)\}$ are differentiated by $P^*_{loss}(i-1) = \alpha_i P^*_{loss}(i), \alpha_i < 1$; and

- a *bandwidth allocation priority*: where the traffic of class j never affects the bandwidth/buffer allocation of class $i, i < j$.

We argue that such a service is suitable for TCP applications that need packet loss as an indicator for flow control while guaranteed delay performance can support real-time applications. We define a service model for the core network that includes a number of algorithms. A node provisioning algorithm enforces delay guarantees by dropping packets and adjusting service weights accordingly. A core provisioning algorithm maintains the dropping-rate differentiation by dimensioning the network ingress bandwidth. Edge provisioning modules [7] perform rate regulation based on utility functions. Even though these algorithms are not the only solution to supporting the proposed SLA, their design is tailored toward delivering quantitative differentiation in the SLA with minimum complexity.

In the remaining part of this paper, we focus on core network provisioning algorithms which are complementary components to the edge algorithms of our dynamic provisioning architecture shown in Figure 2.

2.4 Core Traffic Load Matrix

We consider a core network with a set $\mathcal{L} \triangleq \{1, 2, \cdots, L\}$ of link identifiers of unidirectional links. Let c_l be the finite capacity of link $l, l \in \mathcal{L}$.

A core network traffic load distribution consists of a matrix $\mathbf{A} = \{a_{l,i}\}$ that models per-DiffServ-aggregate traffic distribution on links $l \in \mathcal{L}$, where $a_{l,i}$ indicates the fraction of traffic from traffic aggregate i passing link l. Let the link load vector be $\mathbf{c}$ and ingress traffic vector be $\mathbf{u}$, whose coefficient u_i denotes a traffic aggregate of one service class at one ingress point. A network customer may contribute traffic to multiple u_i for multiple service classes and at multiple network access points.

The constraint of link capacity leads to: $\mathbf{Au} \leq \mathbf{c}$. Figure 3 illustrates an example network topology and its corresponding traffic matrix.

The construction of matrix $\mathbf{A}$ is based on the measurement of its column vectors $\mathbf{a}_{.,i}$, each represents the traffic distribution of an ingress aggregate u_i

over the set of links $\mathcal{L}$. In addition, the measurement of u_i gives the trend of external traffic demands.

In a DiffServ network, ingress traffic conditioners need to perform per-profile (usually per customer) policing or shaping. Therefore, traffic conditioners can also provide per-profile packet counting measurements without any additional operational cost. This alleviates the need to place measurement mechanisms at customer premises. We adopt this simple approach to measurement that is proposed in [5] and measure both u_i and $\mathbf{a}_{.,i}$ at the ingress points of a core network, rather than measuring at the egress points which is more challenging. The external traffic demands u_i is simply measured by packet counting at profile meters using ingress traffic conditioners. The traffic vector $\mathbf{a}_{.,i}$ is *inferred* from the flow-level packet statistics collected at a profile meter. Some additional packet probing (e.g., traceroute) methods can be used to improve the measurement accuracy of intra-domain traffic matrix.

3   Dynamic Node Provisioning

3.1   Algorithm Control Logic

The design of the node provisioning algorithm follows the typical logic of measurement based closed-loop control. The algorithm is responsible for two tasks: (i) to predict SLA violations from traffic measurements; and (ii) to respond to potential violations with local reconfigurations. If violations are severe and persistent, then reports are sent to the core provisioning modules to regulate ingress conditioners, as shown in Figure 2.

The calculation of rate adjustment is based on an M/M/1/K model where K represents the current packet-dropping threshold. The Poisson hypothesis on arrival process and service time is validated in [9] for mean delay and loss calculation under exponential and bursty inputs. We argue that because the overall network control is an iterative closed-loop control system, the impact of modeling inaccuracy would increase the convergence time but does not affect the steady state operating point.

Since the node provisioning algorithm enforces delay guarantees by dropping packets, the packet-dropping threshold K needs to be set proportionally to the maximum delay value D_{max}, i.e., $K+1 = D_{max} * \mu$, where μ denotes the service rate. In addition, we denote traffic intensity $\rho = \lambda/\mu$, and λ is the mean traffic rate.

3.2   Invocation Condition

The provisioning algorithm is invoked by the detection of overload and underload conditions. Given a packet loss bound P^*_{loss} for one traffic aggregate class, the dynamic node provisioning algorithm's goal is to ensure that the measured average packet loss rate $\bar{P}_{loss}$ is below P^*_{loss}. When $\bar{P}_{loss} > \gamma_a P^*_{loss}$, the buffer for this class is considered overload, and when $\bar{P}_{loss} < \gamma_a P^*_{loss}$, the buffer is considered underload. Here $0 < \gamma_b < \gamma_a < 1$.

When P_{loss}^* is small, solely counting rare packet loss events can introduce a large bias. Therefore, the algorithm uses the average queue length N_q to improve the measurement accuracy. Since the average queue length N_q is represented as:

$$N_q = \frac{\rho}{1-\rho}(\rho - (K+1)P_{loss}), \tag{1}$$

Given P_{loss} and K, we need to formulate ρ in order to calculate N_q.

Proposition 1. *Given the packet loss rate of a M/M/1/K queue as P_{loss}, the corresponding traffic intensity ρ is bounded as: $\rho_a \le \rho \le \rho_b$, where $\rho_b = f(\kappa_{inf})$ and $\rho_a = f(\kappa_{sup})$.*

The detailed derivation of $f(x)$, κ_{inf} and κ_{sup} , as well as the error analysis of this approximation method are given in [8] due to lack of space here.

With the upper loss threshold $\gamma_a P_{loss}^*$, we calculate the corresponding upper threshold on traffic intensity ρ^{sup} with the value of ρ_b in Proposition (1), and subsequently N_q^{sup}, the upper threshold on the average queue length, from Equation (1). Similarly, with $\gamma_b P_{loss}^*$, we can calculate the lower threshold ρ^{inf} using ρ_a in Proposition (1), and then N_q^{inf}.

3.3 Target Control Value and Feedback Signal

We use the target traffic intensity $\tilde{\rho}$ as our target control value. It is calculated as:

$$\tilde{\rho} = \left(\rho^{sup} + \rho^{inf}\right)/2, \tag{2}$$

Subsequently, we use $\bar{\rho}$, the measured traffic intensity as the feedback signal. By definition, $\bar{\rho} = \bar{\lambda}/\bar{\mu}$. However, directly measuring $\bar{\mu}$ is not easy because it requires measuring the time a packet at the head of queue waits for packets from other queues to complete service. We use an indirect approach to solve this: when a queue is overloaded, measuring $\bar{\mu}$ can be simply done by counting the packet departure rate r_{depart}, and $\bar{\rho} = \bar{\lambda}/r_{depart}$. When a queue is underloaded ($\bar{\lambda} = r_{depart}$), we use Equation (1) to calculate $\bar{\rho}$ using the average queue length $\bar{N}_q$ and packet loss $\bar{P}_{loss}$. Therefore, we have:

$$\bar{\rho} = \begin{cases} \bar{\lambda}/r_{depart} & \bar{\lambda}/r_{depart} > 1 \\ (\sqrt{(\bar{N}_q - (K+1)\bar{P}_{loss})^2 + 4\bar{N}_q} - (\bar{N}_q - (K+1)\bar{P}_{loss}))/2 \; otherwise \end{cases} \tag{3}$$

The performance of the proposed node algorithm depends on the measurement of queue length $\bar{N}_q$, packet loss $\bar{P}_{loss}$, arrival rate $\bar{\lambda}$ and departure rate r_{depart} for each class. We use the same form of exponentially weighted moving average function proposed in [11] to smooth the measurement samples.

3.4 Control Actions

The control conditions that invoke changes to the traffic intensity $\bar{\rho}(i)$ are as follows:

1. If $\bar{N}_q(i) > N_q^{sup}(i)$, reduce traffic intensity to $\tilde{\rho}(i)$ by either increasing service weights or reducing arrival rate by applying multiplicative factor β_i; and
2. If $\bar{N}_q(i) < N_q^{inf}(i)$, increase traffic intensity to $\tilde{\rho}(i)$ by either decreasing service weights or increasing arrival rate by multiplying β_i.

In both cases, the control factor β_i is: $\beta_i = \tilde{\rho}(i)/\bar{\rho}(i)$.

Reducing arrival rate is achieved by signaling (via the Regulate_Down signal) the core provisioning algorithm (discussed in Section 4) to reduce the allocated bandwidth at the appropriate edge traffic conditioners. Similarly, an increasing arrival rate is signaled (via the Link_State signal) to dynamic core provisioning, which increases the allocated bandwidth at the edge traffic conditioners.

For simplicity, we introduce a strict priority in the service weight allocation procedure, i.e., higher priority classes can "steal" service weights from lower priority classes until the service weight of a lower priority class reaches its minimum (w_i^{min}). In addition, we always change local service weights first before sending a Congestion_Alarm signal to the core provisioning module to reduce the arrival rate which would require a network-wide adjustment of ingress traffic conditioners at edge nodes. An increase in the arrival rate is deferred to a periodic network-wide rate re-alignment algorithm which operates over longer time scales. In other words, the control system's response to rate reduction is immediate, while, on the other hand, its response to rate increase to improve utilization is delayed to limit any oscillation in rate allocation.

The details of the algorithm are given in [8].

4   Dynamic Core Provisioning

Our core provisioning algorithm has two functions: to reduce edge bandwidth immediately after receiving a *Congestion_Alarm* signal from a node provisioning module, and to provide periodic bandwidth re-alignment to establish a modified max-min bandwidth allocation for traffic aggregates. We will focus on the first function and discuss the latter function in Section 4.2.

4.1   Edge Rate Reduction Policy

Given the measured traffic load matrix $\mathbf{A}$ and the required bandwidth reduction $\{-c_l^\delta(i)\}$ at link l for class i, the allocation procedure *Regulate_Ingress_Down()* needs to find the edge bandwidth reduction vector $-\mathbf{u}^\delta = -[\mathbf{u}^\delta(1) \vdots \mathbf{u}^\delta(2) \vdots \cdots \vdots \mathbf{u}^\delta(J)]^T$ such that: $\mathbf{a}_{l,.}(j) * \mathbf{u}^\delta(j) = c_l^\delta(j)$, where $0 \le u_i^\delta \le u_i$.

When $\mathbf{a}_{l,.}$ has more than one nonzero coefficients, there is an infinite number of solutions satisfying the above equation. We will choose one based on optimization policies such as fairness, minimizing the impact on other traffic and a combination of both. For clarity, we will drop the class (j) notation since the operations are the same for all classes.

The policies for edge rate reduction may be optimize for two quite different objectives.

Equal Reduction. Equal reduction minimizes the variance of rate reduction among various traffic aggregates, i.e., $\min_i \left\{ \sum_{i=1}^{n} \left(u_i^\delta - \frac{\sum_{i=1}^{n} u_i^\delta}{n} \right)^2 \right\}$ with constraints $0 \le u_i^\delta \le u_i$ and $\sum_{i=1}^{n} a_{l,i} u_i^\delta = c_l^\delta$. Using Kuhn-Tucker condition [6], we have:

Proposition 2. *The solution to the problem of minimizing the variance of rate reductions comprises three parts:*

$$\forall i \ \text{ with } a_{l,i} = 0, \quad \text{we have } u_i^\delta = 0; \tag{4}$$

then for notation simplicity, we re-number the remaining indeces with positive $a_{l,i}$ as $1, 2, \cdots, n$; and

$$u_{\sigma(1)}^\delta = u_{\sigma(1)}, \cdots, u_{\sigma(k-1)}^\delta = u_{\sigma(k-1)}; \quad and \tag{5}$$

$$u_{\sigma(k)}^\delta = \cdots = u_{\sigma(n)}^\delta = \frac{c_l^\delta - \sum_{i=1}^{k-1} a_{l,\sigma(i)} u_{\sigma(i)}}{\sum_{i=k}^{n} a_{l,\sigma(i)}}, \tag{6}$$

where $\{\sigma(1), \sigma(2), \cdots, \sigma(n)\}$ is a permutation of $\{1, 2, \cdots, n\}$ such that $u_{\sigma(i)}^\delta$ is sorted in increasing order, and k is chosen such that: $c_{eq}(k-1) < c_l^\delta \le c_{eq}(k)$, where $c_{eq}(k) = \sum_{i=1}^{k} a_{l,\sigma(i)} u_{\sigma(i)} + u_{\sigma(k)} \sum_{i=k+1}^{n} a_{l,\sigma(i)}$.

Remark: Equal reduction gives each traffic aggregate the same amount of rate reduction until the rate of a traffic aggregate reaches zero.

Minimal Branch-Penalty Reduction. A concern that is unique to DiffServ provisioning is to minimize the penalty on traffic belonging to the same regulated traffic aggregate that passes through non-congested branches of the routing tree. We call this effect the "branch-penalty", which is caused by policing/shaping traffic aggregates at an ingress router. For example, in Figure 3, if link 7 is congested, the traffic aggregate #1 is reduced before entering link 1. Hence penalizing a portion of traffic aggregate #1 that passes through link 3 and 9.

The total amount of branch-penalty is $\sum_{i=1}^{n}(1 - a_{l,i})u_i^\delta$ since $(1 - a_{l,i})$ is the proportion of traffic not passing through the congested link. Because of the constraint that $\sum_{i=1}^{n} a_{l,i} u_i^\delta = c_l^\delta$, we have $\sum_{i=1}^{n}(1 - a_{l,i})u_i^\delta = \sum_{i=1}^{n} u_i^\delta - c_l^\delta$. Therefore, minimizing the branch-penalty is equivalent to minimizing the total bandwidth reduction, that is: $\min \sum_{i=1}^{n}(1 - a_{l,i})u_i^\delta \iff \min \sum_{i=1}^{n} u_i^\delta$, with constraints $0 \le u_i^\delta \le u_i$ and $\sum_{i=1}^{n} a_{l,i} u_i^\delta = c_l^\delta$.

Proposition 3. *The solution to the minimizing branch-penalty problem comprises three parts:*

$$u_{\sigma(1)}^\delta = u_{\sigma(1)}, \cdots, u_{\sigma(k-1)}^\delta = u_{\sigma(k-1)}; \tag{7}$$

$$u_{\sigma(k)}^\delta = c_l^\delta - \sum_{i=1}^{k-1} a_{l,\sigma(i)} u_{\sigma(i)}; \quad and \tag{8}$$

$$u^\delta_{\sigma(k)} = \cdots = u^\delta_{\sigma(n)} = 0, \tag{9}$$

where $\{\sigma(1), \sigma(2), \cdots, \sigma(n)\}$ is a permutation of $\{1, 2, \cdots, n\}$ such that $a_{l,\sigma(i)}$ is sorted in decreasing order, and k is chosen such that: $c_{br}(k-1) < c^\delta_l \le c_{br}(k)$, where $c_{br}(k) = \sum_{i=1}^{k} a_{l,\sigma(i)} u_{\sigma(i)}$.

A straightforward proof by contradiction is omitted due to lack of space. See [8] for details.

Remark: The solution is to sequentially reduce the u_i with the largest $a_{l,i}$ to zero, and then move on to the u_i with the second largest $a_{l,i}$ until the sum of reductions amounts to c^δ_l.

Remark: A variation of the minimal branch-penalty solution is to sort based on $a_{l,\sigma(i)} u^\delta_{\sigma(i)}$ rather than $a_{l,\sigma(i)}$. In this case, the solution minimizes the number of traffic aggregates affected by the rate reduction procedure.

Penrose-Moore Inverse Reduction. It is clear that equal reduction and minimizing branch-penalty have conflicting objectives. Equal reduction attempts to provide the same amount of reduction to all traffic aggregates. In contrast, minimal branch-penalty reduction always depletes the bandwidth associated with the traffic aggregate with the largest portion of traffic passing through the congested link. To balance these two competing optimization objectives, we propose a new policy that minimizes the Euclidean distance of the rate reduction vector u^δ: $\min\left\{\sum_{i=1}^{n}(u^\delta_i)^2\right\}$, with constraints $0 \le u^\delta_i \le u_i$ and $\sum_{i=1}^{n} a_{l,i} u^\delta_i = c^\delta_l$.

Similar to the solution of the minimizing variance problem in the equal reduction case, we have:

Proposition 4. *The solution to the problem of minimizing the Euclidean distance of the rate reduction vector comprises three parts:*

$$\forall i \ \text{ with } a_{l,i} = 0, \ \text{ we have } u^\delta_i = 0; \tag{10}$$

then for notation simplicity, we re-number the remaining indeces with positive $a_{l,i}$ as $1, 2, \cdots, n$; and

$$u^\delta_{\sigma(1)} = u_{\sigma(1)}, \cdots, u^\delta_{\sigma(k-1)} = u_{\sigma(k-1)}; \ \text{and} \tag{11}$$

$$\frac{u^\delta_{\sigma(k)}}{a_{l,\sigma(k)}} = \cdots = \frac{u^\delta_{\sigma(n)}}{a_{l,\sigma(n)}} = \frac{c^\delta_l - \sum_{i=1}^{k-1} a_{l,\sigma(i)} u_{\sigma(i)}}{\sum_{i=k}^{n} a^2_{l,\sigma(i)}}, \tag{12}$$

where $\{\sigma(1), \sigma(2), \cdots, \sigma(n)\}$ is a permutation of $\{1, 2, \cdots, n\}$ such that $u^\delta_{\sigma(i)}/a_{l,\sigma(i)}$ is sorted in increasing order, and k is chosen such that: $c_{pm}(k-1) < c^\delta_l \le c_{pm}(k)$, where $c_{pm}(k) = \sum_{i=1}^{k} a_{l,\sigma(i)} u_{\sigma(i)} + (u_{\sigma(k)}/a_{l,\sigma(k)}) \sum_{i=k+1}^{n} a^2_{l,\sigma(i)}$.

Remark: Equation (12) is equivalent to the Penrose-Moore (P-M) matrix inverse [4], in the form of $[u^\delta_{\sigma(k)} \ u^\delta_{\sigma(k+1)} \cdots u^\delta_{\sigma(n)}]^T = [a_{l,\sigma(k)} \ a_{l,\sigma(k+1)} \cdots a_{l,\sigma(n)}]^+ \ast$

$(c_l^\delta - \sum_{i=1}^{k-1} a_{l,\sigma(i)} u_{\sigma(i)})$, where $[\cdots]^+$ is the P-M matrix inverse. In particular, for an $n \times 1$ vector $\mathbf{a}_{l,\cdot}$, the P-M inverse is a $1 \times n$ vector $\mathbf{a}_{l,\cdot}^+$ where $a_{l,i}^+ = a_{l,i}/(\sum_{i=1}^n a_{l,i}^2)$.

We name this policy as the P-M inverse reduction because of the property of P-M matrix inverse. The P-M matrix inverse always exists and is unique, and gives the least Euclidean distance among all possible solution satisfying the optimization constraint.

Proposition 5. *The performance of the P-M inverse reduction lies between the equal reduction and minimal branch-penalty reduction. In terms of fairness, it is better than the minimal branch-penalty reduction and in terms of minimizing branch-penalty, it is better than the equal reduction.*

Proof: By simple manipulation, the minimization objective of P-M inverse is equivalent to the following: $\min \left\{ \sum_{i=1}^n \left(u_i^\delta - (\sum_{i=1}^n u_i^\delta)/n \right)^2 + \left(\sum_{i=1}^n u_i^\delta \right)^2 /n \right\}$. The first part of this formula is the optimization objective of the equal reduction policy. The second part of the above formula is scaled from the optimization objective of the minimizing branch penalty policy by squaring and division to be comparable to the objective function of equal reduction.

That is, the P-M inverse method minimizes the sum of the objective functions minimized by the equal reduction and minimal branch penalty methods, respectively. Therefore, the P-M inverse method balances the trade-off between equal reduction and minimal branch penalty.   □

Remark: It is noted that the P-M inverse reduction policy is not the only method that balances the optimization objectives of fairness and minimizing branch penalty. However, we choose it because of its clear geometric meaning (i.e., minimizing the Euclidean distance) and its simple closed-form formula.

4.2 Edge Rate Alignment

Unlike edge rate reduction, which is triggered locally by a link scheduler that needs to limit the impact on ingress traffic aggregates, the design goal for the periodic rate alignment algorithm is to re-align the bandwidth distribution across the network for various classes of traffic aggregates and to re-establish the ideal max-min fairness property.

However, we need to extend the max-min fair allocation algorithm given in [2] to reflect the point-to-multipoint topology of a DiffServ traffic aggregate. Let $\mathcal{L}^u$ denote the set of links that are not saturated and $\mathcal{P}$ be the set of ingress aggregates that are *not bottlenecked*, (i.e., have no branch of traffic passing a saturated link). Then the procedure is given as follows:

(1) identify the most loaded link l in the set of non-saturated links:
     $l = \arg\min_{j \in \mathcal{L}^u} \left\{ x_j = (c_j - \texttt{allocated\_capacity})/\sum_{i \in \mathcal{P}} a_{j,i} \right\}$
(2) increase allocation to all ingress aggregates in $\mathcal{P}$ by x_l and
     update the allocated_capacity for links in $\mathcal{L}^u$
(3) remove ingress aggregates passing l from $\mathcal{P}$ and

```
        remove link l from L^u
(4)  if P is empty, then stop; else go to (1)
```

Our modification of step (1) changes the calculation of remaining capacity from $(c_l - allocated\_capacity)/\|\mathcal{P}\|$ to $(c_l - allocated\_capacity)/\sum_{i \in \mathcal{P}} a_{l,i}$.

Remark: The convergence speed of max-min allocation for point-to-multipoint traffic aggregates is faster than for point-to-point session because it is more likely that two traffic aggregates have traffic over the same congested link. In the extreme case, when all the traffic aggregates have portions of traffic over all the congested links, there is only one *bottlenecked* link. In this case, the algorithm takes one round to finish, and the allocation effect is equivalent to the equal reduction (in this case, "equal allocation") method.

5   Simulation Results

5.1   Simulation Setup

We evaluate our algorithms by simulation using the ns-2 simulator with the DiffServ module provided by Sean Murphy. Unless otherwise stated, we use the default values in the standard ns-2 release for the simulation parameters.

The ns-2 DiffServ module uses the Weighted-Round-Robin scheduler which is a variant of WFQ. Three different buffer management algorithms are used for different DiffServ classes; these are, tail-dropping for the Expedited Forwarding (EF) Per-Hob-Behavior (PHB), RED-with-In-Out for the Assured Forward (AF) PHB Group, and Random Early Detection for the best-effort (BE) traffic class. In our simulation, we consider four classes: EF, AF1, AF2, and BE. The order above represents the priority for allocation of service weight and bandwidth, but does not reflect packet scheduling. The initial service weights for the four class queues are 30, 20, 10 and 10 with a fixed total of 100. The minimum service weight for EF, AF1 is 10, and for AF2 it is 5. There is no minimum service weight for the BE class. The initial buffer size is 30 packets for the EF class queue, 100 packets each of the AF1 and AF2 class queues, respectively, and 200 packets for the BE class queue. When a RED or RIO queue's buffer size is changed, the thresholds are changed proportionally.

A combination of TCP, Constant-Bit-Rate (CBR) and Pareto On-Off traffic sources are used in the simulation. We use CBR sources for EF traffic, and a combination of infinite FTP with TCPReno and Pareto bursty sources for others. The mean packet size of the EF traffic is 210 bytes, and 1000 bytes for others. The traffic conditioners are configured with one profile for each traffic source.

The measurement window τ_l for packet loss is set to 30 seconds for the EF class and to 10 seconds for other classes. The measurement interval τ for exponential weighted moving average is set to 500 ms. The multiplicative factor for upper and lower queue length thresholds are set to $\gamma_a = 0.75$ and $\gamma_b = 0.3$.

The simulation network comprises eight nodes with traffic conditioners at the edge, as shown in Figure 4. The backbone links are configured with 6 Mb/s capacity with a propagation delay of 1 ms. The three backbone links (C1, C2 and C3) highlighted in the figure are overloaded in various test cases to represent

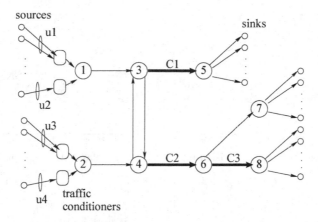

Fig. 4. Simulated Network Topology.

the focus of our traffic overload study. The access links leading to the congested link have 5 Mb/s with a 0.1 ms propagation delay.

5.2 Dynamic Node Provisioning

Service Differentiation Effect. To evaluate the effect of dynamic node provisioning on our network service model (defined in Section 2.3), we compare the results where the algorithm is enabled and disabled. We set the per-node delay bound to be 100 ms for the EF class, 500 ms for the AF1 class and 1 s for the AF2 class. The packet loss threshold that invokes edge rate reduction is set to 5×10^{-6} for the EF class, 5×10^{-4} for the AF1 class and 10^{-3} for the BE class.

In this experiment, we use 2 CBR sources for the EF class, and 3 FTP and 3 Pareto sources each for the AF1 and AF2 classes, respectively. In addition, 2 FTP and 2 Pareto sources are used for the BE class. The dynamic node provisioning algorithm's update_interval is set to 500 ms.

The delay comparison shown in Figure 5(a) and 5(b) illustrates the gain in delay differentiation when the node provisioning algorithm is enabled. Unlike 5(a) where both AF1 and AF2 delays are in the same range, in 5(b), the AF1 class has a quite flat delay of 100 ms. In addition, the EF class delay is better in 5(b) than in 5(a). The initial large delay for the AF2 class shown in 5(b) reflects the difficulty of performing node provisioning without sufficient initial measurement data.

In the packet loss comparison, the lack of loss differentiation is also evident in Figure 5(c), while in 5(d) with node provisioning enabled, only AF2 experiences loss.

We also use simulation to investigate the appropriate time scale for the update_interval when invoking the node provisioning algorithm. Our results show that a short update_intervals (100 msec) can affect system stability. We also note that increasing the update_interval to 5 second also risks violating the packet loss threshold. This comparison leads us to use an update_interval of 500 ms for the remaining simulations.

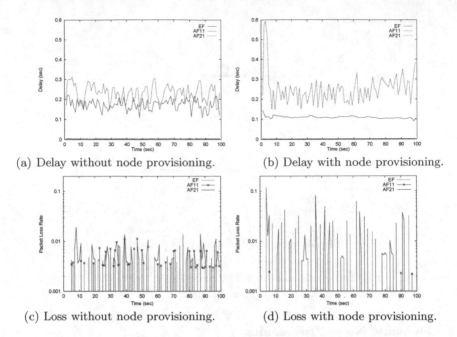

(a) Delay without node provisioning.   (b) Delay with node provisioning.

(c) Loss without node provisioning.   (d) Loss with node provisioning.

Fig. 5. Node Provisioning Service Differentiation Effect.

Sensitivity to Traffic Source Characteristics. Since our node provisioning algorithm uses a Markovian traffic model to derive the target control value $\tilde{\rho}$, we further test our scheme with TCP and self-similar bursty sources to demonstrate the algorithm's insensitivity to non-Markovian traffic sources. We first run the simulation with only CBR and FTP sources.

As shown in 6(a), the service weights remain unchanged after an initial increase from an arbitrarily chosen initial value. This indicates that the provisioning algorithm quickly finds a stable operating point for each queue, interoperating well with the TCP flow control mechanism. In the next experiment, we increase the traffic dynamics by adding eight Pareto sources with 500 Kb/s peak rate and 10s On-Off interval. We run the simulation for 1000s. Figure 6(b) shows that the dynamics of the service weight allocation is well differentiated against different service classes.

In addition, the delay plots given in [8] also show that all the delay results are well below the per-node delay bound of 100ms, 500ms and 1s for the EF, AF11 and AF21 classes, respectively. These results verify that even though we use a simple and concise analytical model, we are able to provide performance differentiation for non-Markovian long-range dependent traffic.

5.3   Dynamic Core Provisioning

Responsiveness to Network Dynamics. We use a combination of CBR and FTP sources to study the effect of our dynamic core provisioning algorithm (i.e., the P-M Inverse method for rate reduction and max-min fair for rate alignment).

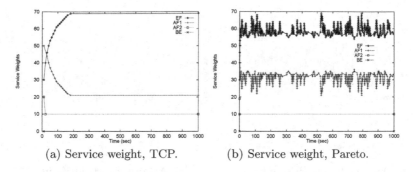

(a) Service weight, TCP.   (b) Service weight, Pareto.

Fig. 6. Node Provisioning Sensitivity to Bursty Traffic.

The update_interval for edge rate reduction is set to 1s. Periodic edge rate alignment is invoked every 60s. We use CBR and FTP sources for EF and AF1 traffic aggregates, respectively. Each traffic class comprises four traffic aggregates entering the network in the same manner as shown in Figure 4. A large number (50) of FTP sessions are used in each AF1 aggregate to simulate a continuously bursty traffic demand. The distribution of the AF1 traffic across the network is the same as shown in Table 1.

The number of CBR flows in each aggregate varies to simulate the effect of varying bandwidth availability for the AF1 class (which could be caused in reality by changes in traffic load, route, and/or network topology). The changes of available bandwidth for AF1 class includes: at time 400s into the trace, $C2$ (the available bandwidth at link 2) is reduced to 2Mb/s; at 500s into the trace, $C3$ is reduced to 0.5 Mb/s; and at 700s into the trace, $C3$ is increased to 3 Mb/s. In addition, at time 800s into the trace, we simulate the effect of a route change, specifically, all packets from traffic aggregate $u1$ and $u3$ to node 5 are rerouted to node 8.

Figure 7 illustrates the allocation and delay results for the four AF1 aggregates. We observe that not every injected change of bandwidth availability triggers an edge rate reduction, however, in such a case it does cause changes in packet delay. Since the measured delay is within the performance bound, the node provisioning algorithm does not generate Congestion_Alarm signals to the core provisioning module. Hence, rate reduction is not invoked. In most cases, edge rate alignment does not take effect either because the node provisioning algorithm does not report the needs for an edge rate increase. Both phenomena demonstrate the robustness of our control system.

The system correctly responds to route changes because the core provisioning algorithm continuously measures the traffic load matrix. As shown in Figure 7(a) and (b), after time 800s into the trace, the allocation of $u1$ and $u3$ at link C1 drops to zero, while the corresponding allocation at link C2 increases to accommodate the surging traffic demand.

Effect of Rate Control Policy. In this section, we use test case examples to verify the effect of different rate control policies in our core provisioning

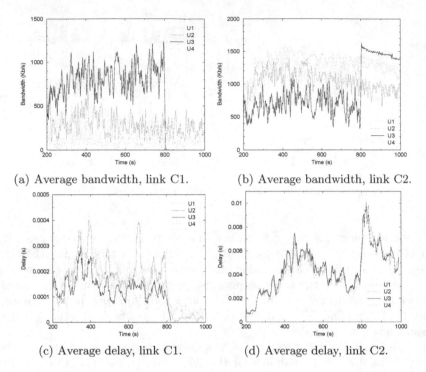

(a) Average bandwidth, link C1.   (b) Average bandwidth, link C2.

(c) Average delay, link C1.   (d) Average delay, link C2.

Fig. 7. Core Provisioning for AF1 Aggregates: Bandwidth Allocation and Delay (Averaged over 10s).

algorithm. We only use CBR traffic sources in the following tests to focus on the effect of these policies.

Table 1 gives the initial traffic distribution of the four EF aggregates comprising only CBR flows in the simulation network, as shown in Figure 4. For clarity, we only show the distribution over the three highlighted links (C1, C2 and C3). The first three data-rows form the traffic load matrix $\mathbf{A}$, and the last data-row is the input vector $\mathbf{u}$.

In Figure 8, we compare the metrics for equal reduction, minimal branch-penalty and the P-M inverse reduction under ten randomly generated test cases. Each test case starts with the same initial load condition, as given in Table 1. The change is introduced by reducing the capacity of one backbone link to cause congestion which subsequently triggers rate reduction.

Figure 8(a) shows the fairness metric: the variance of rate reduction vector $\mathbf{u}^{\delta}$. The equal reduction policy always generates the smallest variance, in most of the cases the variances are zero, and the non-zero variance cases are caused by the boundary conditions where some of the traffic aggregates have their rates reduced to zero. Here we observe that the P-M inverse method always gives a variance value between those of equal reduction and minimizing branch penalty. Similarly, Figure 8(b) illustrates the branch penalty metric: $\sum_i (1 - a_{l,i}) u_i^{\delta}$. In this case, the minimizing branch penalty method consistently has the lowest branch penalty values, followed by the P-M inverse method.

Table 1. Traffic Distribution Matrix.

Bottleneck	Traffic Aggregates			
Link	U_1	U_2	U_3	U_4
C_1	0.20	0.25	0.57	0.10
C_2	0.80	0.75	0.43	0.90
C_3	0.40	0.50	0.15	0.80
Load (Mb/s)	1.0	0.8	1.4	2.0

The results support our assertion that the P-M Inverse method balances the trade-off between equal reduction and minimal branch penalty.

6   Conclusion

We have argued that dynamic provisioning is superior to static provisioning for DiffServ because it affords network mechanisms the flexibility to regulate edge traffic maintaining service differentiation under persistent congestion and device failure conditions when observed in the core network. Our core provisioning algorithm is designed to address the unique difficulty of provisioning DiffServ traffic aggregates (i.e., rate-control can only be exerted at the root of any traffic distribution tree). We proposed the P-M Inverse method for edge rate reduction which balances the trade-off between fairness and minimizing the branch-penalty. We extended max-min fair allocation for edge rate alignment and demonstrated its convergence property. Our node provisioning algorithm prevents transient service level violations by dynamically adjusting the service weights of a weighted fair queueing scheduler. The algorithm is measurement-based and uses a simple but effective analytic formula for closed-loop control. Collectively, these algorithms contribute toward a more quantitative differentiated service Internet, supporting per-class delay guarantees with differentiated loss bounds across core IP networks.

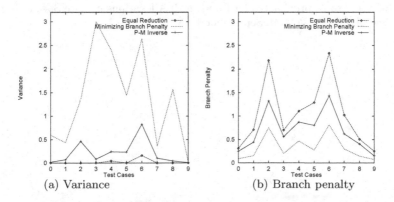

(a) Variance   (b) Branch penalty

Fig. 8. Reduction Policy Comparison (Ten Independent Tests).

Acknowledgments

This work is funded in part by a grant from the Intel Corporation for "Signaling Engine for Programmable IXA Networks".

References

1. C. Albuquerque, B. J. Vickers, and T. Suda. Network Border Patrol. In *Proc. IEEE INFOCOM*, Tel Aviv, Israel, March 2000.
2. D. Bertsekas and R. Gallager. *Data Networks*. Prentice-Hall, Englewood Cliffs, NJ, 1992.
3. L. Breslau, E. Knightly, S. Shenker, I. Stoica, and H. Zhang. Endpoint Admission Control: Architectural Issues and Perormance. In *Proc. ACM SIGCOMM*, Stockholm, Sweden, September 2000.
4. S.L. Campbell and C.D. Meyer, Jr. *Generalized Inverses of Linear Transformations*. Pitman, London, UK, 1979.
5. A. Feldmann, A. Greenberg, C. Lund, N. Reingold, J. Rexford, and F. True. Deriving Traffic Demands for Operational IP Networks: Methodology and Expeience. In *Proc. ACM SIGCOMM*, September 2000.
6. H. W. Kuhn and A. W. Tucker. Non-linear programming. In *Proc. 2nd Berkeley Symp. on Mathematical Statistics and Probability*, pages 481–492. Univ. Calif. Press, 1961.
7. R. R.-F. Liao and A. T. Campbell. Dynamic Edge Provisioning for Core Networks. In *Proc. IEEE/IFIP Int'l Workshop on Quality of Service*, June 2000.
8. R. R.-F. Liao and A. T. Campbell. Dynamic Core Provisioning for Quantitative Differentiated Service. Technical report, Dept. of Electrical Engineering, Columbia University, April 2001.
   http://www.comet.columbia.edu/~liao/publications/coreProv.ps.
9. M. May, J. Bolot, A. Jean-Marie, and C. Diot. Simple performance models of Differentiated services schemes for the Internet. In *Proc. IEEE INFOCOM*, New York City, March 1999.
10. D. Mitra, J.A. Morrison, and K. G. Ramakrishnan. Virtual Private Networks: Joint Resource Allocation and Routing Design. In *Proc. IEEE INFOCOM*, New York City, March 1999.
11. I. Stoica, S. Shenker, and H. Zhang. Core-Stateless Fair Queueing: A Scalable Architecture to Approximate Fair Bandwidth Allocations in High Speed Networks. In *Proc. ACM SIGCOMM*, September 1998.

Towards Provisioning Diffserv Intra-Nets

Ulrich Fiedler, Polly Huang, and Bernhard Plattner

Compute Engineering and Networks Laboratory
Swiss Federal Institute of Technology
ETH-Zentrum, Gloriastrasse 35
CH-8092 Zurich, Switzerland
{fiedler,huang,plattner}@tik.ee.ethz.ch

Abstract. The question of our study is how to provision a diffserv (differentiated service) intra-net serving three classes of traffic, i.e., voice, real-time data (e.g. stock quotes), and best-effort data. Each class of traffic requires a different level of QoS (Quality of Service) guarantee. For VoIP the primary QoS requirements are delay and loss; for real-time data response-time. Given a network configuration and anticipated workload of a business intra-net, we use ns-2 simulations to determine the minimum capacity requirements that dominate total cost of the intra-net. To ensure that it is worthwhile converging different traffic classes or deploying diffserv, we cautiously examine capacity requirements in three sets of experiments: three traffic classes in i) three dedicated networks, ii) one network without diffserv support , and iii) one network with diffserv support. We find that for the business intra-net of our study, integration without diffserv may need considerable over-provisioning depending on the fraction of real-time data in the network. In addition, we observe significant capacity savings in the diffserv case; thus conclude that deploying diffserv is advantageous. The relations we find give rise to, as far as we know, the first rule of thumb on provisioning a diffserv network for increasing real-time data.

1 Introduction

Integration of services for voice, real-time data, and best-effort data on IP intra-nets is of special interest, as it has the potential to save network capacity that dominates costs [1]. Moreover, Bajaj et al. [2] suggested that when network utilization is high, having multiple service levels for real-time traffic may reduce the capacity requirement. While Bajaj et al. examined voice and low quality video in a diffserv network, we are interested in voice and emerging transaction-oriented applications. Such applications are web-based and include support for response-time-critical quotes, and transactions on securities, stocks, and bonds. We expect to derive guidelines to provision an IP intra-net with multiple levels of services for such applications.

Recently Kim et al. evaluated provisioning for voice over IP (VoIP) in a diffserv network [3]. They found a roughly linear relationship between the number of VoIP connections and the capacity requirement. The multiplexing gain was in

L. Wolf, D. Hutchison, and R. Steinmetz (Eds.): IWQoS 2001, LNCS 2092, pp. 27–43, 2001.
© Springer-Verlag Berlin Heidelberg 2001

agreement with the telecommunication experience. Although they investigated the effect of competing web traffic on VoIP in diffserv networks, they did not measure performance for any type of traffic other than VoIP.

It is not clear yet how to provision for web-like data traffic. This provisioning problem is of special interest because in the future a portion of this data traffic may be real-time. This work addresses the problem of provisioning web traffic with response-time requirement on diffserv networks that carry, in addition to real-time data, VoIP and other best effort data. Our goal is to measure with simulations the benefit of integration and differentiation of services. Therefore we try to answer, which of the following three cases requires lower capacity.

1. Keeping voice, real-time data, and best-effort data on dedicated IP networks
2. Integrating voice, real-time data, and best-effort data in a conventional IP network without diffserv support
3. Integrating voice, real-time data, and best-effort data in a network with diffserv support for the three classes of traffic

To determine the minimal capacity requirement, we measure quality of VoIP and real-time data. For VoIP, similar to Kim et al. [3], we measure end-to-end delay and loss. For real-time data traffic, we measure response-times of web down-loads.

We observe that case two needs significant over-provisioning compared to case one when the fraction of real-time data is small to medium. If all data traffic is real-time, the capacity requirements for case one and two are not very different. For case three, we find a lower capacity requirement than that of case one. This capacity requirement shows an approximately linear relationship to the fraction of real-time data; more real-time data require more capacity. Compared to dedicated networks, the percentage of capacity savings for integration with diffserv remains approximately constant for a small to medium fraction of real-time data. These results give a guideline on how to provision diffserv networks for web traffic with response-time requirements.

The rest of this report is organized as follows: Section 2 describes the simulation environment, topology, source model and explains measurement metrics for QoS requirements to assess capacity provisioning. In Section 3 we explore capacity provisioning for voice, real-time and best-effort data services on dedicated networks. In Section 4 we present results on provisioning a conventional IP network without diffserv support for these services. In Section 5 we investigate provisioning and capacity savings with diffserv support. In Section 6 we discuss our results, limitations, and further work.

2   Setup

This section introduces the simulation environment, topology, source model, and measurement metrics for QoS requirements to assess capacity provisioning.

2.1   Simulation Environment

We use the ns-2 [4] as our simulation environment. We have augmented ns-2 with diffserv additions by Murphy [5], and scripts that explicitly model the interactions of HTTP/1.1 [6]. We have customized routines for collecting performance statistics.

The diffserv additions model diffserv functionality. Diffserv [7,8,9] provides different levels of service by aggregating flows with similar QoS requirements. At the network edges, packets are classified and marked with *code-points*. Inside the network, packets are forwarded solely depending on their code-points. We consider three levels of service based on different forwarding mechanisms as specified by the IETF.

- *expedited forwarding (EF)*
  This forwarding mechanism implements a virtual wire [10]. It is intended for delay sensitive applications such as interactive voice or video. Wroclawski and Charny [11] propose to implement guaranteed service[12] based on EF. We use it to forward voice traffic.
- *assured forwarding (AF)*
  This forwarding mechanism is intended for data that should not be delayed by network congestion [13]. Wroclawski and Charny [11] propose to implement controlled-load service[14] based on AF. We use AF to differentiate real-time data traffic from best-effort data traffic.
- *best effort (BE)*
  This forwarding mechanism implements best-effort service. We use it to forward best-effort data traffic.

The diffserv package implements three queues to buffer the packets of these three service classes. The queues are serviced with a deficit round-robin scheduler. We do no conditioning at ingress and configure a single level of drop precedences for AF, i.e. we mark all AF traffic as AF11.

2.2   Topology

For our simulation we choose a simple dumbbell topology as depicted in Figure 2. This topology may e.g. represent the bottleneck link in an intra-net, on which capacity is scarce (see Figure 1). It may also be viewed in a more general sense, since many have argued that there is always a single bottleneck link on any network path [2], which justifies our choice to start experiments with such a network topology. We distribute sources as depicted in Figure 2. We position 50 phones and 50 web clients at the right side of the bottleneck and 50 phones and 5 web servers at left side. Access links have 10Mb in capacity and 0.1ms in delay. The bottleneck link has a propagation delay of 10ms to model the worst case for a connection going through all of Europe. The variables in this simulation are the capacity of the bottleneck link and its queue size. If not explicitly mentioned, queue size is set to 20KB.

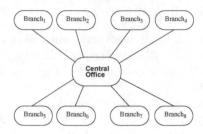

Fig. 1. Intra-Net Topology.

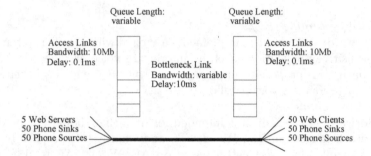

Fig. 2. Simulation Topology.

2.3   Source Model

We model voice traffic as VoIP. A telephone is a source/sink pair that produces CBR traffic at call times. We assume no compression or silence suppression; the source thus produces a 64Kb stream. As packetization for VoIP is not standardized, we assume that each packet contains 10ms of speech. We think that this is a realistic tradeoff between packetization delay and overhead due to IP and UDP headers. The net CBR rate of a phone is thus 86.4Kb.

To model call duration and inter-call time we use parameters as depicted in Table 1. In residential environments, call durations are simply exponentially distributed with a mean of 3 minutes [15]. However, from our partners we know that a bi-modal distribution for call durations, representing long and short calls, better suits the situation of a busy hour in a large bank. Both long and short calls, as well as the inter-call time, were represented by exponential distributions. This model offers a mean load of 800Kb on the bottleneck link given that 50 phone pairs interact.

We model data traffic as web traffic. We explicitly model the traffic generated by request/reply interactions of HTTP/1.1 between web server and clients. This modeling of HTTP/1.1 interactions includes features like persistent connections and pipelining. We keep TCP connections at servers open for further down-

Table 1. User/Session Related Parameters to Generate Voice Traffic.

Parameter	Distribution	Average
Call duration	Bimodal	long call: 20% Short Call: 80%
Long Call	Exponential	8 min
Short Call	Exponential	3 min
Inter-Call time	Exponential	15 min

loads and eliminate the need for a client to wait for the server's response before sending further requests. We model a requested web page as an index page plus a number of embedded objects. To generate the corresponding traffic, we need the distributions for the following entities:

1. size of requested index objects
2. number of embedded objects
3. size of embedded objects
4. server selection
5. think time between two successive down-loads.

Table 2. User/Session Related Parameters to Generate Web Traffic.

Parameter	Distribution	Average	Shape
Size of Index Objects	Pareto	8000 B	1.2
Size of Embedded Objects	Pareto	4000 B	1.1
Number of Embedded Objects.	Pareto	20	1.5
Think Time	Pareto	30 sec	2.5

We use parameters as depicted in Table 2. Distributions of object sizes and number of embedded objects were determined with web crawling in a bank's intra-net [16]. Measured distributions basically agree with those of Barford et al. [17,18]. Our model does not address the matching problem, i.e. which request goes to which server. Instead we randomly chose the server for each object. Server latencies were not modeled, as this exceeds the scope of this work. This model produces a mean load of 850Kb on the bottleneck link in direction from servers to clients, given that five web servers interact with 50 web clients on a dedicated network with 2.4Mb provisioned at the bottleneck link. In experiments with integrated networks, we measure slightly more data traffic than voice traffic on the bottleneck link given the number of sources and topology in Figure 2.

2.4   QoS Requirements

To determine the capacity requirement for a given network, we need to define QoS requirements for each type of traffic. We chose these requirements as realistically as possible. For this reason we were in contact with a large Swiss bank. Thus, the QoS requirements for voice and real-time data are derived from the practice.

For VoIP transmission, we measure end-to-end delay and packet loss. We define three requirements: one for the end-to-end delay and the other two for packet loss. End-to-end packet delay for VoIP consists of coding and packetization delay as well as network delay. Coding and packetization delays can be up to 30ms. Steinmetz and Nahrstedt [19] state that one-way lip-to-ear delay should not exceed 100–200ms based on testing with human experts. In conventional telephony networks, users are even accustomed to much lower delays. For these reasons we define 50ms as an upper limit for end-to-end network delay. The impact of loss to VoIP depends on the speech coder. We assume the ITU-T G.711 coder [20], which is widely deployed in conventional telephones in Europe. To quantify the impact of packet loss to packetized speech coded with this coder, we group successive losses of VoIP packets into *outages*. The notion of outages comes from Paxson who investigated on end-to-end Internet packet dynamics [21]. Such outages are usually perceived as a crackling sound when played out at the receiver's side. We think that number and duration of outages may characterize the impact of loss reasonably well for this coder in situations of low or moderate loss. Since there is no standard on how to assess human perception of VoIP transmission with this coder, we conducted trials in our lab, in which we mapped pre-recorded speech on outage patterns of our simulations. From these trials, we define the requirements on outages for VoIP as follows: The number of outages must not exceed 5 per minute. Outage duration must not exceed 50ms. This approach is similar to the one used by Kim et al. [3].

For real-time data we measure the fraction of response-times of web page down-loads that are below five seconds. We define this fraction as the *five seconds response time quantile*. We define the *response-time* as the time that has elapsed between sending out the first TCP *syn* packet and receiving the last TCP packet containing relevant data. To control QoS for this type of real-time data traffic, banks measure the percentage of data that is received in five seconds. Banks require this quantile to exceed some limit in the very high nineties. After discussion with partners, we set this limit to 99%.

3   Dedicated Networks

We investigate the provisioning for dedicated networks for VoIP and web traffic to be able to assess the benefit in terms of capacity savings when integrating these networks. We determine provisioning by assessing QoS for each type of traffic.

3.1   Voice

For VoIP traffic we measure the capacity needed to accommodate the traffic generated by 50 phones. We measure end-to-end delay, and number and duration of outages at increasing link capacity.

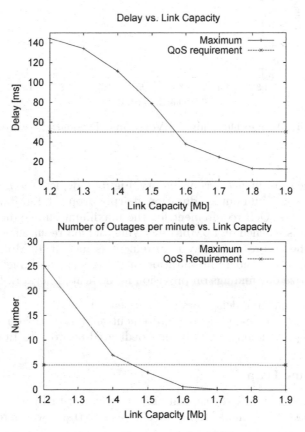

Fig. 3. Packet Delay and Outage Frequency for VoIP on a Dedicated Network.

Figure 3 (top) depicts maximum end-to-end delay of VoIP packets. We find that the maximum delay decreases from 145ms at 1.2Mb to 11ms at 1.9Mb. The QoS requirement for end-to-end packet delay, to be less than 50ms, is reached at 1.6Mb. For capacities larger than 1.8Mb, the end-to-end delay equals the physical propagation delay. This means that queues in the system are empty at this capacity. Figure 3 (bottom) demonstrates the number of outages for VoIP on a dedicated network decreases from 25 at 1.2Mb to zero at 1.8Mb. For larger capacities the number of outages remains zero. The QoS requirement for the number of outages, to be less than five per minute, is met at 1.5Mb. At 1.8Mb

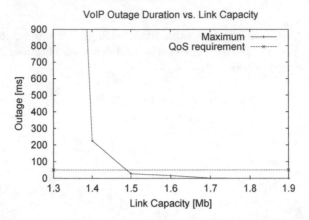

Fig. 4. Outage Duration for VoIP on a Dedicated Network.

outages disappear as queues become empty. Figure 4 depicts the maximum outage durations. Maximum outage duration sharply drops at 1.4Mb and becomes zero at 1.8Mb. The QoS requirement for the maximum outage duration, to be less than 50ms, is reached at 1.5Mb. Reviewing all QoS requirements at a time, we find that the end-to-end delay requirement is met at 1.6Mb, whereas the requirements on the number and duration of outages are both met at 1.5Mb.

We summarize our findings on provisioning dedicated network for VoIP:

- Outages and queuing delay totally disappear at 1.8Mb.
- The stringent end-to-end delay requirement of 50ms is the dominant factor that limits provisioning for VoIP in a dedicated network to 1.6Mb.

3.2   Real-Time Data

In this section, we measure the capacity requirement for real-time data traffic on a dedicated network. We model real-time data as web traffic with a response-time requirement.

We start with all data traffic in our experiment being real-time, i.e. all 50 web clients requesting web pages with a response-time requirement. Figure 5 depicts the fraction of web pages that can be down-loaded within five seconds at increasing capacities. This fraction grows from 93.5% at 1.2Mb to 99.3% at 2.6Mb. The curve is concave from above, which means that additional capacity has a stronger impact on response-times at lower capacities and a smaller impact at higher capacities. The QoS requirement for real-time data, that 99% of the web pages can be down-loaded in less than five seconds, is met at 2.4Mb.

These response-time quantiles are measured at a queue size of 20KB at both ends of the bottleneck link. Since queue size 20KB is already larger than the bandwidth delay product, which is required to make TCP perform at its maximum, varying queue size has not much impact on response-times.

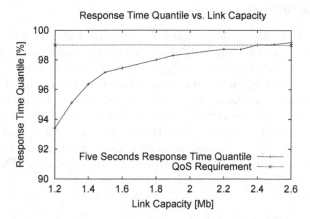

Fig. 5. Five Seconds Response Time Quantile for Web Traffic on a Dedicated Network.

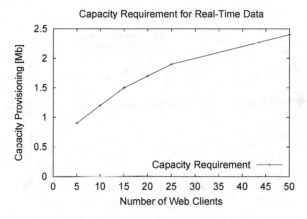

Fig. 6. Multiplexing for Real-Time Data Traffic.

As we want to experiment with various fractions of data traffic having a real-time requirement, we investigate the capacity requirement of an increasing number of web clients requesting pages with a response-time requirement. Figure 6 shows that the capacity requirement grows less and less with increasing number of clients. It grows from 0.9Mb for 5 web clients to 2.4Mb for 50 web clients.

We summarize our findings on provisioning dedicated networks for real-time data:

– The capacity requirement for 50 web clients requesting web pages with response-time requirement is met at 2.4Mb.

3.3   Three Dedicated Networks

We determine the capacity requirement for a set of dedicated networks as follows. A dedicated network for VoIP has a capacity requirement of 1.6Mb. To accommodate the real-time data traffic on a dedicated network, we need between zero and 2.4Mb depending on the number of clients requesting web pages with response-time requirement. The remaining best-effort traffic on a dedicated network needs marginal capacity as it has no QoS requirements to meet. We assume that this capacity for best-effort traffic on a best-effort network is zero, which is in favor of the dedicated networks case. Summing up these requirements, we conclude that dedicated networks need between 1.6Mb and 4.0Mb provisioning depending on the fraction of real-time traffic.

4   Integrated Networks

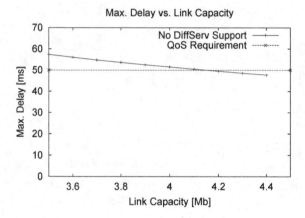

Fig. 7. Maximal End-to-End Delay without Diffserv Support.

In this section, we investigate capacity requirements for an integrated network without diffserv support. A network without diffserv support cannot differentiate between best-effort and real-time data. Therefore, we have to over-provision the network such that VoIP meets its QoS requirements and all web traffic, best-effort and real-time, meet the response-time requirement.

For VoIP packets we measure a maximum end-to-end delay of 58ms at 3.5Mb that linearly decreases with increasing capacity to 45ms at 4.4Mb (Figure 7). In contrast to the dedicated networks case, these values remain considerably high. To measure the impact of loss for VoIP, we depict number and duration of outages at increasing capacity. Figure 8 (top) shows that outages start at 5.5 per minute for 3.5Mb and slowly decrease to 2.8 per minute at 4.5Mb. The QoS

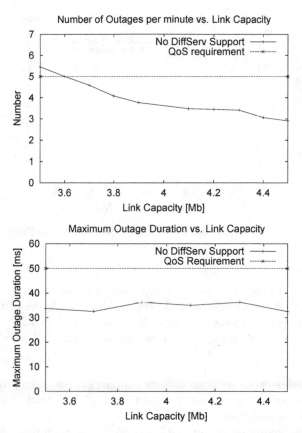

Fig. 8. Number of Outages and Outage Duration without DiffServ Support.

requirement of a maximum of five outages per minute is met at 3.6Mb. Outage durations (see Figure 8, bottom) show no clear trend. They fluctuate between 30ms and 40ms in the studied interval between 3.5Mb and 4.5Mb. This is below the QoS requirement of 50ms. Other than in the dedicated networks case, there is no capacity at which VoIP delay sharply drops and outages disappear. We presume that this effect comes from the bursty nature of competing web traffic, which cannot be alleviated in the studied range of capacities.

For data traffic, we monitor the response-time of all web pages, as a network without diffserv support cannot differentiate between real-time and best-effort data. Figure 9 depicts the fraction of all down-loads below five seconds. This fraction increases from 99.2% for 3.6Mb to 99.6% for 4.5Mb. The QoS requirement, that 99% of the web pages can be down-loaded in five seconds, is fulfilled for all studied capacities. We presume that the reason for this excellent performance for real-time data is the unlimited use of the total capacity at the bottleneck link.

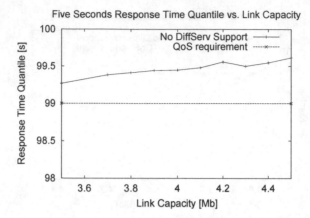

Fig. 9. Five Seconds Response Time Quantile without DiffServ Support.

To determine the capacity requirement for a network without diffserv support, we review QoS requirements for VoIP and data traffic. VoIP's delay is the dominant requirement that sets the capacity requirement to 4.2Mb.

We summarize our results for provisioning an integrated network without diffserv support:

– An integrated network without diffserv support cannot differentiate services. Therefore we have to over-provision such that all data traffic meets the QoS requirement for real-time data.
– The stringent requirement on VoIP's delay is the dominant factor setting the capacity requirement for an integrated network without diffserv support to 4.2Mb.

5   DiffServ Support

In this section we investigate provisioning an integrated network for voice, real-time, and best-effort data with diffserv support. To differentiate three levels of service, we mark the traffic as follows:

1. Voice traffic as expedited forwarding (EF).
2. Real-time data traffic as assured forwarding (AF).
3. The remaining traffic as best effort (BE).

After some first trials, in which we have observed larger capacity requirements for a diffserv network than for a network without diffserv support, we realized a performance problem caused by the deficit round-robin (DRR) scheduler in Murphy's diffserv package. The DRR scheduler implements a queue for each level of service and assigns a fixed number of byte-credits to each queue on a

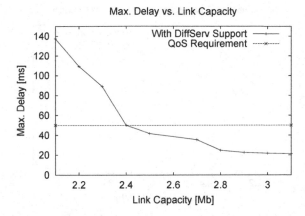

Fig. 10. Maximal End-to-End Delay with DiffServ Support.

per-round basis. These byte-credits can then be used to forward packets of back-logged queues. If a queue is not back-logged, the byte-credits can be unlimitedly accumulated to be used when packets arrive on this queue, which then causes high queuing delays for packets in other queues. We have found in our simulations that bursty real-time traffic has led the AF queue to accumulate enough byte-credits to cause large delay and outages for VoIP traffic in EF. The performance problem is now also reported on the web page of the diffserv package[5]. We thus revised the scheduler and modified it into a weighted round-robin (WRR) based on packet counts. This WRR scheduler has the advantage that we can give tight delay bounds for every queue. With this modification, we were able to measure significantly lower capacity requirements than in the integration without diffserv case.

Before varying the fraction of data traffic that is real-time, we have thoroughly studied a setup in which 20% of the web clients have requested real-time data and 80% of the web clients have requested best-effort data. We have found that the capacity requirement is minimized when service rates for the queues in the WRR scheduler are configured such that QoS for voice and real-time data is simultaneously at the performance limit. We show figures that depict QoS metrics vs. capacity for such a configuration. For VoIP, Figure 10 shows that end-to-end delay lowers from 140ms at 2.1Mb to 20ms at 3.1Mb. The QoS requirement of 50ms is fulfilled at 2.4Mb. Figure 11 (top) shows that the number of outages for VoIP drops from eight per minute at 2.1Mb to zero at 2.8Mb and remains zero for capacities larger than 2.8Mb. Figure 11 (bottom) shows that the maximal outage duration for VoIP drops from 15ms at 2.1Mb to zero at 2.8Mb and remains zero for capacities larger than 2.8Mb. The QoS requirement for outage durations, not to exceed 50ms, is met for all capacities in the interval studied. We see that all QoS metrics for VoIP in an integrated network with diffserv support evolve similar to the corresponding QoS metrics in a ded-

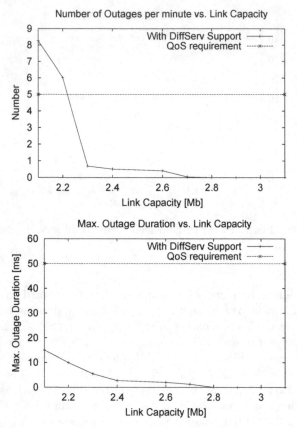

Fig. 11. Number of Outages and Outage Duration with DiffServ Support.

icated network. Delay, and the number and duration of outages sharply drop at some sufficient capacity around 2.4Mb. The delay requirement determines the provisioning from the VoIP side.

For real-time data, the fraction of web pages that can be down-loaded in less than five seconds increases from 98.7% at 2.1Mb to 99.7% at 3.1Mb (Figure 12. The 99% QoS requirement, determining provisioning from the real-time data side, is met at 2.3Mb. For best-effort data, the fraction of pages with down-loads times of less than five seconds increases from around 80% at 2.1Mb (not shown) to 97.6% at 3.1Mb. This is a clear service differentiation between real-time traffic and best-effort traffic.

Next, we vary the fraction of data traffic that is real-time to investigate the relationship between provisioning requirements and the fraction of real-time data traffic. We find an approximately linear correlation for small to medium fractions of data traffic being real-time; more real-time traffic needs more capacity (Figure 13). Compared to the dedicated networks case, the savings in capacity are

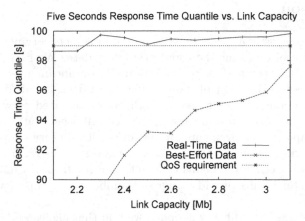

Fig. 12. Five Seconds Response Time Quantiles with DiffServ Support.

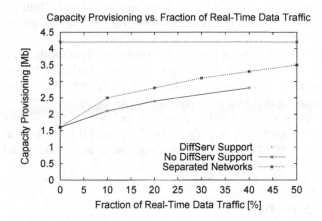

Fig. 13. Provisioning Requirement versus Fraction of Real-Time Data Traffic.

approximately 15% for the WRR scheduler with our source model that produces slightly more data traffic than voice traffic. First measurements with a weighted fair queuing scheduler seem to show slightly higher savings. In addition, we see that a network without diffserv support needs up to 60% over-provisioning.

We additionally experimented with conditioning the real-time data traffic with a token bucket and found that configuration of rate and depth had no significant effect on provisioning.

6   Conclusion

In this study, we investigate QoS provisioning of intra-nets serving three classes of traffic, i.e., voice, real-time data and best-effort data. We find that integration on a network without diffserv support needs significant over-provisioning if a small to medium fraction of data traffic is real-time. If all data traffic is real-time, provisioning for dedicated networks and integrated networks with or without diffserv support is not much different. For all other traffic compositions we find lower capacity requirements for networks with diffserv support than for dedicated networks. Compared to dedicated networks, the percentage of savings for integration with diffserv remains constant for a small to medium fraction of real-time data. Our results provide a rule of thumb towards capacity planing for real-time data.

For the networks with diffserv support, we find that choice and implementation of the scheduling algorithm had significant impact on capacity requirements. We are aware that the WRR scheduler, which we used in Section 5 does not lead to the optimal performance. However, we can give tight delay bounds for this scheduler, which we think is crucial as delay seems to be the limiting factor. To clarify this issue, we are currently experimenting with a weighted fair queuing scheduler[22].

In addition, we plan to further vary traffic compositions, particularly to study variations of offered load for voice and data, and investigate more complex topologies. To generalize our results for the Internet, we plan to investigate capacity provisioning on randomly generated topologies with Zipf's law connectivity [23]. Finally, we would like to stress that, in addition to the results reported in this paper, our simulation package is publicly available. It will enable the research community, network service providers, banks and other large enterprises to derive provisioning guidelines for their networks.

Acknowledgements

We would like to thank many people for helpful discussions particularly Marcel Dasen, Thomas Erlebach, George Fankhauser, Felix Gaertner, Matthias Gries, Albert Kuendig, Sean Murphy, and Burkhard Stiller.

References

1. A. M. Odlyzko, "The internet and other networks: Utilization rates and their implications," *Information Economics and Policy*, vol. 12, pp. 341–365, 2000.
2. S. Bajaj et. al., "Is service priority useful in networks?," in *Proceedings of the ACM Sigmetrics '98*, Madison, Wisconsin USA, June 1998.
3. G. Kim et. al., "Qos provisioning for voip in bandwidth broker architecture: A simulation approach," in *Proceedings of the communication networks and distributed systems modeling and simulation conference (CNDS)'01*, Phoenix, Arizona, USA, Jan. 2001.

4. L. Breslau et. al., "Advances in network simulations," *IEEE Computer*, May 2000.
5. S. Murphy, "Diffserv package for ns-2," available at
   `http://www.teltec.dcu.ie/~murphys/ns-work/diffserv/index.html`.
6. R. Fielding et. al., "Hypertext transfer protocol — http/1.1," RFC 2616, Internet Request For Comments, June 1999.
7. S. Blake et. al., "An architecture for differentiated services," RFC 2475, Internet Request For Comments, Dec. 1998.
8. J.Wroclaski D. Clark, "An approach to service allocation in the internet," IETF Draft, July 1997.
9. L. Zhang K. Nichols, V. Jacobson, "A two-bit differentiated services architecture for the internet," IETF Draft, Apr. 1999.
10. V. Jacobson et. al., "An Expedited Forwarding PHB," RFC 2598, Internet Request For Comments, June 1999.
11. A. Charny J. Wroclawski, "Integrated service mappings for differentiated services networks," IETF Draft, Feb. 2001.
12. S. Shenker et al., "Specification of guaranteed quality of service," RFC 2212, Internet Request For Comments, Sept. 1997.
13. J. Heinanen et. al., "Assured Forwarding PHB Group," RFC 2597, Internet Request For Comments, June 1999.
14. J. Wroclawski, "Specification of the controlled-load network element service," RFC 2211, Internet Request For Comments, Sept. 1997.
15. Siemens, *Telephone traffic theory tables and charts*, Siemens Aktiengesellschaft, Berlin - Muenchen, Germany, 1981, 3rd edition.
16. M. Dasen, "Up-to-date information in web accessible information resources," Ph.d. thesis – tik-schriftenreihe nr. 42, ETH Zurich, June 2001.
17. P. Barford et. al., "Changes in web client access patterns: Characteristics and caching implications," *World Wide Web, Special Issue on Characterization and Performance Evaluation*, vol. 2, no. 2, pp. 15–28, 1999.
18. P. Barford and M. E. Crovella, "Generating representative web workloads for network and server performance evaluation," in *Proc. of Performance'98/ACM SIGMETRICS'98*, 1997.
19. R. Steinmetz and K. Nahrstedt, *Multimedia: Computing, Communications & Applications*, Prentice-Hall, 1995.
20. International Telecommunication Union Telecom Standardization Sector, "Itu-t g.711 - pulse code modulation *pcm* of voice frequencies," available at `http://www.itu.int`, Nov. 1988.
21. V. Paxson, "End-to-end internet packet dynamics," *IEEE/ACM Transactions on Networking*, vol. 7, no. 3, pp. 277–292, June 1999.
22. C. Partridge, "Weighted fair queueing," in *Gigabit Networking*. 1994, p. 276, Addison-Wesley Publishing.
23. Faloutsos et. al., "On power-law relationships of the internet topology," in *Proceedings of ACM SIGCOMM*, Stockholm, Sweden, Aug. 1999.

Analysis of Paris Metro Pricing Strategy for QoS with a Single Service Provider

Ravi Jain, Tracy Mullen, and Robert Hausman

Telcordia Technologies*
445 South Street, Morristown, NJ 07960, USA
{rjain,mullen,rhausman}@research.telcordia.com

Abstract. As the diversity of Internet applications increases, so does the need for a variety of quality-of-service (QoS) levels on the network. The Paris Metro Pricing (PMP) strategy uses pricing as a tool to implement network resource allocation for QoS assurance; PMP is simple, self-regulating, and does not require significant communications or bandwidth overhead. In this paper, we develop an analytic model for PMP. We first assume that the network service provider is a single constrained monopolist and users must participate in the network; we model the resultant consumer behavior and the provider's profit. We then relax the restriction that users must join the network, allowing them to opt-out, and derive the critical QoS thresholds for a profit-maximizing service provider. Our results show that PMP in a single-provider scenario can be profitable to the provider, both when users must use the system and when they may opt out.

1 Introduction

There has been tremendous interest in engineering the Internet to provide Quality of Service (QoS) guarantees. Most proposals for providing QoS in networks focus on the details of allocating, controlling, measuring and policing bandwidth usage. Many of the Internet resource reservation schemes such as RSVP [2] initially proposed for the Internet have also been adapted and proposed for wireless access networks (e.g. [3,4,5] and references therein); so have simpler alternatives like the philosophy behind the Differentiated Services (Diff-Serv) [6] approach (e.g. [7,8] and references therein).

In this paper we consider the use of market-based mechanisms, and in particular, pricing, for resource allocation to provide differentiated services. Odlyzko [9] has recently proposed a simple approach, called Paris Metro Pricing (PMP), for providing differentiated QoS service on the Internet. The PMP proposal was inspired by the Paris Metro system, which until about 15 years ago operated 1st and 2nd class cars that contained seats identical in number and quality, but with 1st class tickets costing twice as much as 2nd class. The 1st class cars were thus less congested, since only those passengers who cared about getting a seat, fresher air, etc., paid the premium. The system was self-regulating: when the 1st

---

* ©Telcordia Technologies, Inc. All Rights Reserved.

L. Wolf, D. Hutchison, and R. Steinmetz (Eds.): IWQoS 2001, LNCS 2092, pp. 44–58, 2001.
© Springer-Verlag Berlin Heidelberg 2001

class cars became too crowded passengers decided they were not worth the premium and traveled 2nd class, eventually thus restoring the quality differential.[1]

As Odlyzko [9] points out, pricing is a crude tool, since different applications have different requirements for bandwidth, latency, jitter, etc. PMP is similar to Diff-Serv in that no hard QoS guarantees are provided, only expected levels of service, but unlike Diff-Serv, it integrates economics and pricing with traffic management. In this sense the present paper is interdisciplinary in nature. The appeal of PMP lies in its simplicity and low overhead, not only in terms of technical aspects of marking and measuring packets, etc., but also in terms of charging and pricing, which can represent very significant overheads in commercial networks like the Public Switched Telephone Network (PSTN). We believe that the simplicity and low overhead of PMP make it particularly attractive for wireless access networks, given the constrained bandwidth of wireless networks and the resource limitations of mobile devices. However this model can be applied to any network, not just wireless ones.

In Sec. 2 we develop a simplified model of PMP using tools drawn from economic analysis to investigate how well it provides differentiated QoS. Gibbens et al. [1] have previously developed an analytical model for PMP, and we base our work on this model. We investigated the Gibbens model and found that it did not correctly take into account the number of users accessing the network. We have developed a modified analytical model that does so; although the modification is subtle, it has significant impact. We validated our model by simulation. We describe our analytical model and simulation results in Sec. 2. The simulation shows that when the system is bootstrapped, with high and low-QoS channels being initially empty, and users subsequently arrive and make choices in accordance with their willness to pay for QoS, the system does eventually converge, and it converges to the equilibrium predicted by the analytic model. Thus the simulation model confirms that users under PMP obtain differentiated services, and users who are willing to pay more do in fact enjoy a less congested channel.

We next turn our attention to whether a service provider offering network access in an enterprise environment will in fact enhance profits by offering PMP. Clearly this question is crucial if PMP is to be viable. Note that in an enterprise situation the service provider may be an external company (as is increasingly the case where support services are outsourced) or an in-house operation (where internal charging mechanisms are used to maintain accountability and reduce waste.)

In Sec. 3 we consider the situation where all users use the network, using either the low-priced or high-priced channel, but the service provider is constrained to provide some minimum level of capacity at the low price (a common type of clause in service contracts.) We then consider the question: can the provider enhance profits by introducing a premium channel of high QoS and using PMP? We show by means of enhancements to our analytical model that in fact he can.

In Sec. 4 we turn to the situation where users may in fact opt out of using the network altogether if they feel it does not have sufficient utility (e.g., the price

---

[1] The Bombay commuter trains, affectionately known as "locals", still operate under this principle, so the proposal should perhaps be called Bombay Local Pricing (BLP).

is too high or the quality too low). We develop our analytical model further and show that once again, for this situation, the service provider can enhance profits by splitting the available bandwidth into a high priced and a low-priced channel and using PMP.

Thus this paper develops an analytical model for the viability of PMP and of user behavior, validates this model by simulation, and shows by further development of the model that PMP is desirable from a service provider's point of view for the situations we have discussed. Clearly there are numerous further issues that can and need to be studied further. In Sec. 5 we briefly address these in our conclusion.

2   The Basic PMP Model

In this section, we develop the basic PMP model. Our derivation follows the approach used in Gibbens [1] but with some key differences described below.

Let the entire bandwidth (or capacity) available in a given cell of the network be C Mbps, and when using PMP this is divided into a low-priced channel with capacity $C_L = \alpha C$, and a high-priced channel with capacity $C_H = (1 - \alpha)C$. For a particular system, the possible values of α may need to be fixed, but in general they may be any value in $[0, 1]$. For simplicity in this paper we assume *subscription pricing*, i.e., the user pays a flat rate per unit time for accessing the network, rather than *usage pricing*, where the user would pay based on the amount of data transmitted. Thus let the price of the low-priced channel be P_L per unit time and of the high-priced channel be $P_H = RP_L$, $R \geq 1$, where R is the ratio of P_H/P_L.

A user specifies, for each job, the strength of her preference for high QoS, θ, where θ is a normalized value in the range $[0, 1]$. For a wireless environment, it is quite feasible that a portable terminal, such as a two-way pager or a screen cellular phone with a Wireless Application Protocol (WAP) browser, might allow the user to specify QoS in this way. Alternatively, the user's terminal (e.g. a PDA) might assign default values for desired QoS for different types of jobs (ftp, e-mail, voice, etc.) or even different destinations (e.g. based on a profile) to reduce the burden on the user. Let the QoS for each user's jobs be drawn from a probability distribution function, $f(\theta)$, with a cumulative distribution function $F(\theta)$.

At the moment that a new user arrives to the cell, let the total number of jobs already in the low and high channels be J_L and J_H respectively. Then we assume that obtained QoS in the low and high channels respectively is $Q_H = C_H/J_H = \frac{(1-\alpha)C}{J_H}$ and $Q_L = C_L/J_L = \frac{\alpha C}{J_L}$.

We will find it more convenient to use the *congestion* in each channel, where

$$K_H = \frac{1}{Q_H} = \frac{J_H}{(1 - \alpha)C} \qquad (1)$$

$$K_L = \frac{1}{Q_L} = \frac{J_L}{\alpha C} \qquad (2)$$

We assume that all users join exactly one channel (we will relax this assumption in Sec. 4), and on doing so a user receives a benefit, or, in economics parlance [10,11], a *utility*, U. The user receives some benefit V simply for obtaining access to the network, but for any given channel $c \in \{H, L\}$, this is offset by the amount of congestion K_c the user's job experiences, and the price P_c the user has to pay. The utility to the user for using channel c is then set to

$$U(\theta, c) = V - w\theta K_c - P_c \qquad (3)$$

where w is a scaling factor that converts the user's dislike of congestion into units of \$ per unit time. Fig. 1 shows how utility varies with the user's desire for high QoS, θ, for a job and the actual QoS, Q_c, obtained on a given channel. For any given θ, increasing QoS beyond a certain point has diminishing returns, while as θ increases, the utility of using the channel drops more sharply when the obtained QoS is low. This utility function thus captures some of the essential relevant features of the system while remaining tractable.

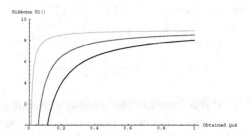

Fig. 1. Utility as a function of obtained QoS Q_c for $\theta = 1, 0.5, 0.1$ (black, grey, light grey), and $V = 10, w = 1, P_c = 1, C = 1$.

2.1   User Job Allocation

With the basic model above, we now consider how users react to PMP, i.e., how users decide which channel they should use for their jobs. We use (slightly strengthened forms of) two properties from [1] and provide proofs (omitted in [1]).

Property 1: When the system is at equilibrium, the high-priced channel has lower congestion, i.e., $P_H \geq P_L$ iff $K_H \leq K_L$.

Proof. Consider the QoS, θ, for a job where the consumer is indifferent between using the high and low quality channels. For this θ:

$$U(\theta, H) = U(\theta, L)$$
$$V - wK_H\theta - P_H = V - wK_L\theta - P_L$$

$$\text{Since } P_H \geq P_L$$
$$K_H \leq K_L$$

□

Property 2: When the system is at equilibrium, users who dislike congestion more will join the higher-priced channel, i.e., Given $P_H \geq P_L$, $\exists \theta^*$ such that $\forall \theta$, $\theta \geq \theta^*$ iff $U(\theta, H) \geq U(\theta, L)$.
Proof.

$$U(\theta, H) \geq U(\theta, L)$$
$$V - wK_H\theta - P_H \geq V - wK_L\theta - P_L$$
$$\theta \geq \frac{P_H - P_L}{w(K_L - K_H)}$$
$$= \theta^*$$

□

Property 2 implies that a user with a job of desired QoS $\theta = \theta^*$ is indifferent between the two channels. Thus, $U(\theta^*, H) = U(\theta^*, L)$, i.e.,

$$V - wK_H\theta^* - P_H = V - wK_L\theta^* - P_L \tag{4}$$

$$\theta^* = \frac{P_L(R-1)}{w(J_L/C_L - J_H/C_H)} \tag{5}$$

Recall that the jobs in the system are distributed with cdf $F(\theta)$. Thus if there are J users in the system, the number of jobs in the low-priced channel is $J_L = F(\theta^*)J$ and in the high-priced channel is $J_H = (1 - F(\theta^*))J$. Hence,

$$\theta^* = \frac{P_L(R-1)}{wJ(F(\theta^*)/C_L - (1 - F(\theta^*))/C_H)}$$
$$= \frac{\alpha(1-\alpha)C^2 P_L(R-1)}{wJC(F(\theta^*) - \alpha)} \tag{6}$$

We note here that Eq. 6 differs in subtle but important ways from [1]. Firstly, the formulation is more general since it allows for an arbitrary distribution of jobs, $F(\theta)$. More importantly, however, it takes into account the number of jobs J present in the system at equilibrium, a crucial factor overlooked in [1], leading to the following observation.

Observation 1: The threshold value θ^* beyond which, at equilibrium, users will choose to utilize the high-priced channel decreases as the number of jobs in the system increases. When the desired QoS for the users' jobs is drawn from the uniform distribution, θ^* approaches α as the number of jobs in the system increases.

Proof. When θ is uniformly distributed we have $F(\theta) = \theta$, so using Eq. 6,

$$wJC\theta^{*^2} - wJ\alpha C\theta^* - \alpha(1-\alpha)C^2 P_L(R-1) = 0$$

Solving for θ^* and taking the positive root since $\theta \geq 0$,

$$\theta^* = \frac{\alpha}{2} + \frac{1}{2}\sqrt{\alpha^2 + 4\alpha(1-\alpha)P_L(R-1)\frac{C}{wJ}} \qquad (7)$$

□

Observation 1 points out that when the system is at equilibrium but overloaded, jobs simply get assigned roughly in accordance with the bandwidth allocated in each channel. We can see this from Eq. 7. As $J \to \infty$, the term $4\alpha(1-\alpha)P_L(R-1)\frac{C}{wJ} \to 0$, and therefore $\theta^* \to \alpha$. In practice some mechanism like admission control is required to prevent overload. For example, in a wireless network, system overload could be avoided either by the base station, which ignores requests for transmission that it cannot satisfy, or because contending mobile hosts are unable to gain access to the uplink signaling channel to transmit their requests to the base station.

When the system is not overloaded, however, the model indicates that PMP does result in users receiving differentiated services, i.e., users willing to pay the higher price access the channel with the lower congestion.

2.2   Model Validation by Simulation

To validate the model developed above we carried out a simulation experiment that simulates the behavior of individual users and verifies that Eq. 7 does in fact hold at equilibrium.

The model initially begins with a channel of capacity C divided into C_H and C_L, and no jobs. Users arrive sequentially, and each user has a job with desired QoS θ drawn from the uniform distribution. The user examines the current congestion levels of the high and low channels, and chooses the channel which currently provides her with the highest utility in accordance with Eq. 3.

In order to simulate the equilibrium situation, we use the following method to "prime the pump". Suppose we wish to study the equilibrium with J jobs. We let the simulation run for $N \gg J$ job arrivals. However, after J jobs, job number $J + i$ results in job i being removed from the system. Thus after N jobs have arrived, there are still exactly J jobs in the system, but the effect of individual jobs and the effect of starting from an empty channel has been smoothed out. (Note that we are interested in the equilibrium condition and not the dynamic behavior of the system in these experiments.)

In Fig. 2 we show the results of the simulation experiment for $J = 1000, N = 10000$, for the case where the channel is split equally ($\alpha = 0.5$) with the high-priced channel having a premium of $R = 1.25$. For this set of parameters, the theoretical equilibrium threshold $\theta^* = 0.604$. Three curves, each corresponding to a different experimental measure of θ^*, are shown in the Figure. The curve with the most volatility shows $\theta^*(i)$, the instantaneous value of θ^* as calculated for each arriving job $i, 1 \leq i \leq N$, using Eq. 5 and the current number of jobs in each channel. The curve with the least volatility shows θ_e^*, the equilibrium value of θ^* as calculated by each arriving user using Eq. 7 and the current number of total jobs in the system; this line decreases until $i = J$, after which point it becomes fixed at the theoretical value $\theta^* = 0.604$. Finally, the third line shows the instantaneous value $\phi(i)$ of the proportion of users in the low-priced channel; i.e., the instantaneous empirical value of $F(\theta^*)$.

Fig. 2 shows that as the simulation is run for a large number of jobs, the threshold value $\theta^*(i)$ observed in the simulation by an arriving job i does in fact converge to the theoretical value of θ^*, and the proportion of users in the low-priced channel also converges to θ^*. Thus the simulation has approached equilibrium and the theoretical model derived above is validated by the simulation. Other simulation runs confirm that both $\theta^*(i)$ and θ^* converge to lower values as J increases, as predicted from Eq. 7, as well as the behavior predicted for other values for the other parameters; we omit these plots for brevity.

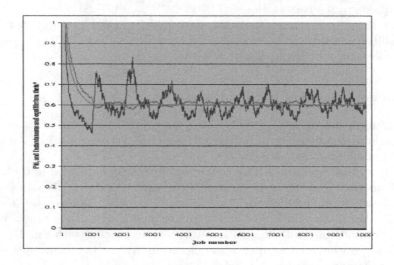

Fig. 2. Basic PMP Model: Instantaneous $\phi(i)$, instantaneous $\theta^*(i)$, and equilibrium θ^* from simulations with $J = 1000$, $\alpha = 0.5$, $R = 1.25$.

3   PMP for Profit

In this section, we consider the question of whether the network service provider would enhance profits by using PMP. Clearly if PMP is to be viable this is a key concern.

The Basic PMP model developed above considers price but not the notion of profit. In an enterprise environment it is quite feasible that revenue or profit might be a concern. For example, there is an increasing tendency in corporations and universities to outsource services that are not part of the core mission of the enterprise. Thus the enterprise would typically ask a third-party service provider to provide quotes or bid for a contract for network services. The service provider would then seek to maximize profit. Even when services are provided in-house, enterprises frequently use an internal system of charging employees (or their projects or organizations) to maintain accountability and discourage waste. Regardless of whether the service provider is in-house or an external company, in this situation typically the provider has a kind of *constrained monopoly*: there is no direct competition from other providers, but the service contract often places stipulations and constraints to provide certain minimum levels of service, prevent overcharging, etc.

In this section, we make the assumption that customers have no choice but to purchase service from the provider. (In the following section we will relax this assumption, allowing users to opt out of the service.) In this scenario, if the provider had no constraints and could fix both price and capacity, he would simply make the entire bandwidth a "high QoS" channel, setting $\alpha = 0$ and P_H arbitrarily high.

Thus here we consider a provider who is constrained to provide some minimum amount of capacity, say αC, for a fixed price, say $P_L = 1$. The question we consider in this section is: can the provider enhance profits by introducing a premium channel of high QoS and using PMP?

Let the revenue of the provider (or profit, assuming for simplicity that costs are set to zero) offering undifferentiated service be π_{UD} and under PMP be π_D. Then at equilibrium, if there are J jobs which under PMP are split into J_H and J_L jobs in the high and low-priced channels as before,

$$\pi_{UD} = P_L J \tag{8}$$

$$\pi_D = \pi_H + \pi_L = P_H J_H + P_L J_L$$
$$= J P_L (R - (R - 1) F(\theta^*)) \tag{9}$$

When the distribution of θ is uniform, Eq. 9 simplifies to

$$\pi_D = J P_L (R - (R - 1)\theta^*) \tag{10}$$

3.1   Maximizing Profit under PMP

We now find the price ratio that optimizes profit under PMP for a given α, $R^*(\alpha)$. We find

$$\frac{\partial \pi_D}{\partial R} = J P_L \left(1 - \theta^* - (R - 1)\frac{\partial \theta^*}{\partial R}\right) \tag{11}$$

For convenience, let $S = \frac{C}{wJ}$, so that S is the average capacity per job scaled by w. Then, from Eq. 7

$$\theta^* = \alpha/2 + 1/2\sqrt{\alpha^2 + 4\alpha(1 - \alpha)P_L(R - 1)S}$$
$$\frac{\partial \theta^*}{\partial R} = \frac{\alpha(1 - \alpha)P_L S}{\sqrt{\alpha^2 + 4\alpha(1 - \alpha)P_L(R - 1)S}}$$

Substituting in Eq. 11 and solving for $\partial \pi_D / \partial R = 0$:

$$R^*(\alpha) = \frac{2 - \alpha^2 - 2\alpha + (2 - \alpha)\sqrt{\alpha^2 - \alpha + 1}}{9\alpha(1 - \alpha)P_L S} + 1 \qquad (12)$$

We can show $\lim_{\alpha \to 0} R^*(\alpha) = \infty$ and $\lim_{\alpha \to 1} R^*(\alpha) = 1 + \frac{1}{2P_L S}$. In Figure 3, we plot $R^*(\alpha)$, with the parameters $P_L, S, J = 1$. Thus for any given α to which the provider is constrained by his service contract, he can offer PMP and choose R so as to maximize profit. As α gets smaller, the low-priced channel gets extremely congested and customers are willing to pay much higher premiums for the high QoS channel. (This underscores the need for minimum levels of capacity in the provider's service contract.)

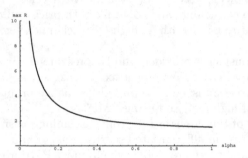

Fig. 3. The optimum price for the premium channel under PMP, $R^*(\alpha)$, given α, for $P_L, S = 1$.

Now Eq. 12 can be used with Eq. 9 to derive max π_D, the maximum profit for a given division of the channel.[2] Figure 4 plots max π_D and π_{UD}, showing that for any given $\alpha < 1$, the provider with a constrained monopoly has an incentive to offer a premium channel and increase his profits.

4   PMP with User Choice

So far we have assumed that all users have no choice but to join the network. In fact, even in an enterprise network with a provider having a constrained monopoly, this is often not the case. When prices are too high or quality is too low, users will simply find alternative methods to perform their tasks (or eventually cause the provider's service contract to be terminated.) Thus in this section we consider the realistic case where users can opt out and not use the network, i.e. not subscribe to any channel. The question we address is: in the situation where the provider is not only a constrained monopolist but users may

---

[2] We spare the reader this formula, particularly as it is difficult to format on the page.

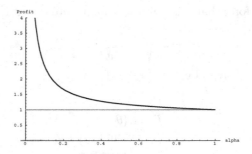

Fig. 4. Maximum profit under PMP, π_D, (black) and under undifferentiated QOS, π_{UD} (gray), for $P_L, S = 1$.

opt out of using his service, will it be beneficial for him to offer PMP? Clearly this question is critical if PMP is to be a viable option in this scenario.

It is in evaluating this option that the parameter V in the utility function of Eq. 3 becomes important. We interpret V as the value of the subscription to the user in the absence of any congestion. Thus, if the congestion and price components of the utility function more than cancel out the benefit V, the utility of the channel is negative and the user will not subscribe.

We approach this analysis by first studying the situation where the provider offers undifferentiated services to maximize profit, and then considering whether adding a premium channel would enhance profits.

4.1 Undifferentiated Service with User Choice

Observe that, unlike the situation studied in the previous section, when users have the choice to opt out the provider cannot simply provide an undifferentiated channel and set the price arbitrarily high. Thus we assume that if the provider has capacity C available, all of it is made available to users as an undifferentiated channel.

Let the price and congestion for this undifferentiated channel be P_L and K_L respectively. At equilibrium, a user with a job of desired QoS θ will choose to use the network only if the utility of subscribing is positive, that is, using Eq. 3, if

$$\theta < \frac{V - P_L}{wK_L} = \theta^* \tag{13}$$

At equilibrium, the proportion of users who subscribe will be $F(\theta^*)$ and so the number of subscribers will be $J_s = JF(\theta^*)$. Thus

$$\theta^* = \frac{(V - P_L)C}{wF(\theta^*)J}$$

If we assume, as before, that θ is uniformly distributed in $[0,1]$, $F(\theta) = \theta$ and so

$$\theta^* = \sqrt{\frac{(V - P_L)C}{wJ}} \tag{14}$$

Then the profit is given by:

$$\begin{aligned} \pi_{UD} &= P_L J_s \\ &= P_L J F(\theta^*) \\ &= P_L J \theta^*, \text{ if } F(\theta) = \theta \\ &= P_L J \sqrt{\frac{(V - P_L)C}{wJ}} \end{aligned} \tag{15}$$

To maximize profit, we set $\partial \pi_{UD} / \partial P_L = 0$, thus obtaining the optimal price P_L^* and thence the optimum profit π_{UD}^*. The resulting congestion at the optimum price is denoted K_L^*, and the corresponding optimal threshold value for θ is denoted τ. Thus,

$$P_L^* = \frac{2}{3}V \tag{16}$$

$$\pi_{UD}^* = \frac{2}{3}V\sqrt{\frac{VJC}{3w}} \tag{17}$$

$$\tau = \sqrt{\frac{VC}{3wJ}} \tag{18}$$

$$K^* = \sqrt{\frac{VJ}{3wC}} \tag{19}$$

Note that an increase in the number of users, J, will not affect the optimal price. However, it will result in a decrease in τ, the proportion of users that subscribe, but an increase in the absolute number of users who subscribe, i.e., $J\tau$, and so an increase in both the profit and the congestion.

4.2   Introducing a Premium Channel

We now consider the situation where the provider has offered a service at price P_L, and now wants to determine whether dividing the available capacity C by adding a premium channel and using PMP will enhance profits, given that users may opt out of the service altogether.

Under this scenario, a new user is faced with three options: 1) subscribe to the non-premium channel (LOW), 2) subscribe to the premium channel (HIGH) or 3) subscribe to neither channel (Opt out). To make this choice, the user considers three two-way comparisons:

1. *Which is better: LOW or Opt out?*. Based on the analysis in the previous section on undifferentiated service, i.e., Eq. 14, the user will select LOW if and only if

$$\theta < \theta_{LO} = \frac{V - P_L}{K_L w} \tag{20}$$

User chooses	Desired QoS	Condition
LOW:	$\theta \in [0, \min(\theta_{LO}, \theta_{LH})]$	$V > P_L$
HIGH:	$\theta \in [\theta_{LH}, \theta_{HO}]$	$\theta_{LH} < \theta_{HO}$, i.e., $V > P_H + \frac{(P_H - P_L)K_H}{K_L - K_H}$
Opt out:	$\theta \in [\max(\theta_{LO}, \theta_{HO}), 1]$	$V < \max(P_H + wK_H, P_L + wK_L)$

Fig. 5. User QoS Choices.

2. *Which is better: HIGH or Opt out?*. Again from Eq. 14 the user selects HIGH if and only if

$$\theta < \theta_{HO} = \frac{V - P_H}{K_H w} \tag{21}$$

3. *Which is better: LOW or HIGH?*. Based on the analysis in Section 2.1, i.e., Eq. 4, the user will select LOW if and only if

$$\theta < \theta_{LH} = \frac{P_H - P_L}{(K_L - K_H)w} \tag{22}$$

We can thus derive, for each value of θ for the user's job, which channel she will choose. Not all choices may exist, since the existence of a choice depends on how much value V the user places on access to the network. Figure 4.2 shows the user's QoS choices under different conditions.

In general, if V is high enough, no user will opt out. This scenario was covered in Section 2. Further, if V is low enough, every user will opt out. For intermediate values of V, the choices can be LOW only, LOW and Opt out only, or all three.

The other cases having been covered in earlier sections, we will focus here on the case in which all three choices are possible at equilibrium. For this to occur, the value of V must satisfy

$$P_H + \frac{(P_H - P_L)K_H}{K_L - K_H} < V < P_H + wK_H$$

To calculate the optimal profit for this situation, we must first find the critical QoS thresholds.

The Critical QoS Thresholds. At equilibrium, let ϕ_L be the proportion of the population in LOW and ϕ_H be that in HIGH. Then the congestions on the low and high channels are

$$K_L = \frac{\phi_L J}{\alpha C}$$

$$K_H = \frac{\phi_H J}{(1 - \alpha)C}$$

Using Eq. 20- 22:

$$\theta_{LO}^* = \frac{(V - P_L)\alpha C}{w\phi_L J}$$

$$\theta_{HO}^* = \frac{(V - P_H)(1 - \alpha)C}{w\phi_H J}$$

$$\theta_{LH}^* = \frac{(P_H - P_L)\alpha(1 - \alpha)C}{(\phi_L(1 - \alpha) - \phi_H\alpha)Jw}$$

For the case we are considering, the choice is given by

$$0 \le \theta \le \theta_{LH}^* \quad \text{LOW}$$
$$\theta_{LH}^* < \theta \le \theta_{HO}^* \quad \text{HIGH}$$
$$\theta_{HO}^* < \theta \le 1 \quad \text{Opt out}$$

Thus $\phi_L = F(\theta_{LH}^*)$ and $\phi_H = F(\theta_{HO}^*) - F(\theta_{LH}^*)$. This leads to

$$\theta_{LO}^* = \frac{(V - P_L)\alpha C}{wF(\theta_{LH}^*)J}$$

$$\theta_{HO}^* = \frac{(V - P_H)(1 - \alpha)C}{w(F(\theta_{HO}^*) - F(\theta_{LH}^*))J}$$

$$\theta_{LH}^* = \frac{(P_H - P_L)\alpha(1 - \alpha)C}{(F(\theta_{LH}^*)(1 - \alpha) - (F(\theta_{HO}^*) - F(\theta_{LH}^*))\alpha)Jw}$$

Note that in the current case, only θ_{HO}^* and θ_{LH}^* are of interest. Assuming that QoS is drawn from a uniform distribution, we have $F(\theta) = \theta$, so

$$\theta_{HO}^* = \frac{(V - P_H)(1 - \alpha)C}{w(\theta_{HO}^* - \theta_{LH}^*)J} \tag{23}$$

$$\theta_{LH}^* = \frac{(P_H - P_L)\alpha(1 - \alpha)C}{(\theta_{LH}^*(1 - \alpha) - (\theta_{HO}^* - \theta_{LH}^*)\alpha)Jw} \tag{24}$$

The simultaneous equations above can be solved for θ_{HO}^* and θ_{LH}^* (we omit the formulas for brevity). We plot the solutions assuming that the price of the LOW channel is set to the optimum value $P_L = \frac{2}{3}V$ found in the undifferentiated case.

Figure 6 displays the values of θ_{HO}^* and θ_{LH}^* as a function of P_H given $V = 1$, $P_L = 2/3$, $\alpha = 0.5$, and $w, S, J = 1$. When $P_H = P_L$, the subscribing users are split between LOW and HIGH in direct proportion to the capacity split (α and $1 - \alpha$). This situation corresponds to the undifferentiated case described above. As the value of P_H increases, some users move from HIGH to LOW while others, with higher values of θ, choose to opt out. When P_H reaches V, the utility associated with HIGH has a value of 0, even with no congestion. Thus, all users disappear from HIGH. At this point, $\theta_{LO}^*, \theta_{HO}^*$ and θ_{LH}^* all have the same value. Increases in P_H above this value are, of course, irrelevant.

Profit. The profit is modeled as:

$$\begin{aligned}
\pi_D &= P_L J_L + P_H J_H \\
&= P_L \phi_L J + P_H \phi_H J \\
&= J[P_L F(\theta_{LH}^*) + P_H(F(\theta_{HO}^*) - F(\theta_{LH}^*))] \\
&= J[P_L \theta_{LH}^* + P_H(\theta_{HO}^* - \theta_{LH}^*)], \text{ if } F(\theta) = \theta \tag{25}
\end{aligned}$$

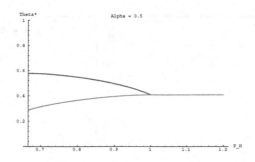

Fig. 6. How users choose channels when they may opt out: θ^*_{HO} (black) and θ^*_{LH} (gray) for $V = 1$, $P_L = 2/3$, $\alpha = 0.5$, and $w, S, J = 1$.

Figure 7 displays four plots of the profit as a function of P_H given $V = 1$, $P_L = 2/3$, $w, S, J = 1$. The first three show profit when α is set to 0.5, 0.75, and 0.99 respectively. The bottom right-hand graph shows the three plots superimposed. When $P_H = P_L$, the situation is identical to the undifferentiated case, and is as specified by Eq. 17. As P_H increases from P_L, the profit increases as well. Thus, there is an incentive for the provider to differentiate. As α increases, the maximum profit is diminished due to increased congestion in the HIGH channel. All three plots seem to be maximized at roughly the same value of $P_H = 0.75$, that is, the profit-maximizing value of P_H seems very insensitive to α.

In summary, we show that even when users have the ability to opt out of using the service, the provider has an incentive to offer differentiated service using PMP.

5 Conclusions

In this paper, we developed results and demonstrated the profit incentive for providers to use PMP in a variety of scenarios. In particular, we analyzed consumer behavior under PMP both when users must use a network service, and when they have the ability to opt-out. We also evaluated the profitability of using PMP when there is a single provider of network service. While our analytic model assumes consumers interact under equilibrium conditions, we developed a simulation that verifies that users behavior does indeed converge to the analytic equilibrium results.

Future work, based on this initial model, includes developing models with more than one service provider where each provider may offer different bundles of QoS and channel capacities. We also intend to investigate different architectural and protocol models for applying this work in a wireless environment.

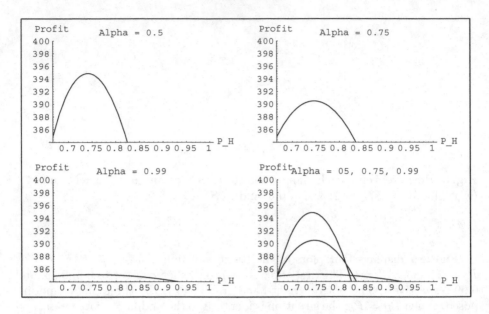

Fig. 7. Profit as a Function of P_H.

References

1. R. Gibbens, R. Mason, and R. Steinberg, "Internet service classes under competition," Tech. Rep., Univ. of Cambridge, Sep. 1999.
2. A. Mankin et al, "RSVP version 1: Applicability statement, some guidelines on deployment," Internet RfC, Sep. 1997.
3. A. K. Talukdar, B. R. Badrinath, and A. Acharya, "MRSVP: A reservation protocol for an Integrated Services packet network with mobile hosts," Tech. Rep. TR-337, Rutgers Univ., Dept. Comp. Sci., 1998.
4. S. Choi and K. G. Shin, "Comparison of connection admission-control schemes in the presence of handoffs in cellular systems," in *Proc. MobiCom*, Oct. 1998, pp. 264–275.
5. J.-C. Chen, K. M. Sivalingam, and R. Acharya, "Comparative analysis of wireless atm channel access protocols supporting multimedia traffic," *ACM MONET*, 1989.
6. R. Blake et al., "An architecture for differentiated services," Internet Draft, Oct. 1998.
7. P. Ramanathan and P. Agrawal, "Adapting packet fair queuing algorithms to wireless networks," in *Proc. MobiCom*, 1998, pp. 1–9.
8. S. Lu, T. nandagopal, and V. Bhargavan, "Fair scheduling in wireless packet networks," in *Proc. MobiCom*, 1998, pp. 10–20.
9. Andrew M. Odlyzko, "Paris metro pricing for the internet," in *ACM Conference on Electronic Commerce (EC'99)*, 1999, pp. 140–147.
10. H. Varian, *Intermediate Microeconomics: A Modern Approach*, W. W. Norton, 1993.
11. J. Tirole, *The Theory of Industrial Organization*, MIT Press, 1988.

Why Value Is Everything: A User-Centered Approach to Internet Quality of Service and Pricing

Anna Bouch and M. Angela Sasse

Department of Computer Science, University College London
Gower Street, London WC1E 6BT, UK
{A.Bouch,A.Sasse}@cs.ucl.ac.uk

Abstract. To create acceptable levels of Quality of Service (QoS), designers need to be able to predict users' behaviour in response to different levels of QoS. However, predicting behaviour requires an understanding of users' requirements for specific tasks and contexts. This paper reports qualitative and experimental research that demonstrates that future network service must be based on an old principle: service and its associate cost must represent value in terms of the contribution it makes to customers' goals. Human Computer Interaction (HCI) methods can be applied to identify users' goals and associated QoS requirements. Firstly, we used a qualitative approach to establish the mental concepts that users apply when assessing network services and charges. The subsequent experimental study shows that users' require certain types of feedback at the user interface to predict future levels of quality. Price alone cannot be used to regulate demand for QoS.

1 Introduction

The number of Internet users is expected to triple between 1998 and 2002 [1], largely because of new applications (such as videoconferencing) and new services (such as e-commerce). This shift in usage imposes higher Quality of Service (QoS) requirements at different levels of granularity. It also means that the traditional Internet way of managing quality (best-effort) has to be replaced by a more service-oriented approach, and that service providers need to find a way of creating revenue.

Traffic produced from different applications can be characterized through an associated payment [2],[3],[4]. For example, high-volume video may be prioritized by associating it with a high-price. The majority of pricing schemes are based on the assumption that the amount and type of quality that is detectable within the network is identical to the quality that will be paid for by users. However, the design of socio-technical environments, such as the evolving Internet, cannot be solely be based on technical considerations. By definition, integration of the requirements of users and technology has to take place.

There are several stakeholders in the design of Internet services: server designers, network providers, advertisers, companies whose products are sold on-line, and consumers themselves. While a vision of the future Internet offers the potential to break traditional barriers in communications and commerce, the current level of

L. Wolf, D. Hutchison, and R. Steinmetz (Eds.) : IWQoS 2001, LNCS 2092, pp. 59-72, 2001.
© Springer-Verlag Berlin Heidelberg 2001

service does not satisfy the requirements of many users [5],[1]. Failure to understand users' QoS requirements may affect users' conception of a company's stature and commercial viability which, in turn, affects the business interests of service providers and advertisers [6]. The future Internet will have more users and support a greater diversity of Internet applications. It has the potential to change the way that consumers interact with companies. Research is needed to identify users' requirements for QoS and the schemes that are used to charge them for it. Only through such identification will it be possible to achieve the customer satisfaction that leads to the success of any commercial system.

The aim of this research is to define users' requirements for network QoS and charging, and investigate the relationship between users' assessment of QoS in different contexts of use, and the mental models that motivate that assessment. The ultimate aim of the research is to provide models that can be used to predict users' demand for QoS in priced situations. The work reported establishes users' requirements, and what network service designers need to know to make design decisions that support users' goals. The developed models should aid the integration of users' requirements for QoS into systems design, and therefore have immediate practical benefits for systems designers and users themselves. The study reported in this paper was structured in 2 stages:

1) Construct conceptual models that motivate users' perceptions of network QoS and charging mechanisms.
2) Capture users' evaluations of objective QoS in different contexts.

2   Previous Results

2.1   Predictability

The effect of applying flat and usage-based charging mechanisms to the telephone service was extensively studied in the late seventies [7]. It was shown that users reduced their usage of the network when faced with usage-based charging. Additionally, customers would choose the flat-rate option regardless of the amount of calls made. These results indicate that the predictability and simplicity of charges are criteria valued by customers, and because of this they will actually pay more than their usage levels require. In a study of American households, it was shown that families did not want to install phone lines because they were uncertain about the magnitude of bills incurred by usage [8]. These results suggest that an adequate feedback mechanism, informing customers of likely charges, might increase the acceptability of usage-based charges.

Quota charging has been suggested [9]: Quotas for certain important QoS drivers can be bought prior to network use. Users then have a (dynamic) choice whether to use their quota for any particular transaction. From a user's perspective, this scheme combines the predictability of bounded usage with the flexibility to request increased QoS when it is needed. Under the quota scheme, the average rate of service requests are optimal for every user considering the price and anticipated delay.

Predictability is one of the most fundamental QoS drivers from a user's perspective [10]. The concept *Risk Assessment* is directly linked with *Predictability* - a low-risk situation is one that is predictable. This suggests that users are prepared to accept charging mechanisms that are predictable. However, somewhat contrary to the findings in [7], more recent studies suggest that users prefer to be able to dynamically change the levels of QoS they receive in line with the value given to the task being performed [11]. Therefore, dynamic pricing needs to provide feedback on network congestion, which would enable users to predict the risk involved in making certain payments.

2.2   Risk

Users make a *Risk Assessment* about whether the QoS they will receive represents value for money. To assess *Risk*, users consider several sub-concepts; the relevance of the different sub-concepts depend on users' level of knowledge and experience. Users' tolerance of certain levels of QoS, and consequently how much QoS they are prepared to pay for, depends on their expectations [11]. This suggests that the network should provide feedback to users, allowing them to make accurate predictions of future quality.

2.3   Context of Interaction

Previous research has shown that users judge the acceptability of pricing schemes according to a variety of dimensions [3]. The salience of these dimensions is determined, not by the fact that they are technically implementable, but by their semantic value. In networked multimedia applications in particular, variations in quality at the network level are not directly linked to the subjective assessment of quality received by users [12]. This suggests that users will pay for the QoS they receive in terms of the media quality they require to complete their goal. For example, with real-time video tasks, participants in previous studies mentioned that the ability to manage the video image in terms of operations such as resizing was important [10]. Clearly, users' need to resize the video image depends on the value placed on what is seen in that image.

3   Qualitative Study

3.1   Method

The phenomena under investigation are complex. The approach of our study was therefore to combine qualitative and experimental research. Qualitative data is needed to investigate users' motivations and conceptual models. We used focus groups to collect data about the acceptability of different charging mechanisms in more detail with users who have the responsibility for payment of line and usage charges. We used grounded theory methods [13], to analyze focus group data. Grounded theory allows characterization of concepts extracted from conversations. The definitions of these concepts can be systematically tested for validity under experimental conditions.

Grounded theory has been successfully applied in recent HCI research to elicit user perceptions on issues such as security and privacy [14], [15]. To have practical benefits, it must also be shown that these models have direct impact on users behavior when interacting with a system. Experimental data provides this evidence.

Users. The same 30 participants took part in both the focus groups and experimental parts of the study. Participants were selected to have experience in using multimedia applications. The following criteria was also used to select participants:
1. Responsibility for the payment of subscription and/or usage charges for their network.
2. Use of that network for more than 2 hours per week.
3. Use of a mobile phone.

Focus Groups. The focus groups were designed to investigate:
1. How network QoS charging compares to other forms of charging (e.g. mobile phone usage).
2. Whether the method of charging preferred depends on the importance of the user's task.
3. The extent of users' satisfaction with method of charging for mobile phone usage.
4. Attitudes to budgeting, e.g. do participants think of their budget when making calls *vs.* the absolute per-minute charge?

3.2   Results

The main results of the focus groups show that absolute price alone is not a good predictor of users' subjective perception of quality. This means that a service provider cannot assume that charging twice as much for providing twice as much speed on a Web-page download doubles the subjective value of quality to users. This is because users' conceptions of quality are influenced by a number of contextual and social factors. Users made the distinction between the need for an awareness of the charge, and control of that charge, the latter being a chance for users to dynamically and directly reflect QoS needs within an interaction.

Peace of Mind. *Peace of Mind* was found to be the top-level goal in situations where users were aware of the charge being made for the interaction. It describes the state of mind arising from users' ability to commit to a call and not experience unacceptable changes in price or QoS, that would force a re-evaluation of that commitment.

Usage Trade-Offs. Where users are not able to control the QoS they receive in a dynamic fashion, they make usage trade-offs in terms of time. Users also make tradeoffs concerning the charging schemes that they use at particular times of day, and for particular tasks. For example:

'I'll think, how important is it for me to make this call now and if it's not important then I won't do it'.
'If I'm making a long-distance call I won't use a...land-line, I'll use a different scheme, that's cheaper'.

Commitment. Users make commitments to an interaction. They make an up-front assessment of the importance of an interaction, and make trade-offs in terms of whether that call is important enough to make at that time. This means that users do not want to constantly re-evaluate the value of the interaction. The key, therefore, is to provide initial feedback that gains users' confidence in their ability to control QoS. This enhances users' Commitment to the interaction. To illustrate:

'Understanding what's going on, what I'll get and if it will meet what I want...I suppose it's a peace of mind thing. If you tell me what it's going to be like, then I can go ahead and know I'll complete the call without hassle'.

Participants stressed the benefits of using pre-paid packages. A popular example was where the provider refunds the difference between expenditure on the chosen tariff and that on a cheaper tariff, for the users particular type of usage that month. This provides users with the confidence to commit to an interaction under a guarantee that that interaction represents value for money.

Trust. The notion of Trust plays an important role in influencing requirements for dynamic control of QoS and the schemes used to charge for that QoS. Trust is concerned both with users trusting themselves, and the service provider. Allowing users to dynamically control charges for QoS, or making users aware of their expenditure so that they must reassess their Commitment to the interaction encourages users to believe that there is a risk of making an inappropriate assessment of that quality. Users not trusting their ability to re-evaluate appropriately induces unacceptable cognitive load and a sense of anxiety:

'Knowing that all the time how much it's costing, makes be feel paranoid that I'll go over without knowing, I can't judge that right'.

Providing continuous charging feedback, or even making users aware that continuous charging feedback is possible also promotes a sense of mistrust in the service provider. Users feel that their usage is being monitored and this is a violation of their sense of privacy and control over the outcome of their expenditure.

Critical Periods. Users attach the concept of a Critical Period to the charging scheme. Users set the Critical Period by specifying a Critical Threshold (see below). This allows them to ascribe different values to interactions whilst retaining the Peace of Mind that they will not exceed an upper bound. The type of tariff determines the Critical Period. For example, the Critical Period for a pre-paid tariff on a mobile telephone is the call that might cause the allowance to run out. Users cannot commit to this call because of their inability to predict future levels of service.

Critical Thresholds. The Critical Threshold is an upper price bound specified by the user, although default settings could be applied to interactions associated with the same task. With data networks, users want to be able to predict if their particular transaction is likely to exceed the Critical Threshold that they have selected. They are then able to assess if a Commitment to the interaction should be made. For example:

'Yes, if I could say, "don't let me spend more that five pounds, and tell me if my (Web-page) selection will take me over it", yep, that would be excellent, I'd know then whether to go ahead'.

Like Critical Periods, the definition of a Critical Threshold is dependent on the type of tariff users have. For example, the Critical Threshold can be specified in terms of time:

'I hate the time when your allowances are about to run out and you might be cut off in an important call...I want it to say "ok, it's that amount of time left", then I know I can get through the whole thing'.

4   Experimental Study

The results from the focus groups showed that, from the user's point of view, there is no linear correlation between QoS requested and the price of that QoS. We devised an experiment to test the ability of the model constructed in focus groups to predict users' behavior whilst interacting with a priced, variable quality network. The experiment asks: *what type and frequency of network feedback to users require when interacting with a priced network?*

4.1   Method

Task. The experiment involved listening to a recording of a 10-minute interview in which an actor played the role of a candidate who was interviewed for a university place. During the experiment users were asked to manipulate audio quality (packet loss), using the *QUASS* slider [16]. To situate the task in a realistic context, video was streamed at maximum quality.

Participants were either told that tasks involve a measure of task completion (*measured* tasks), or that no measure is required (*unmeasured* tasks). This distinction is made to manipulate the importance of the task. In the measured scenario, participants were asked to answer specific questions concerning the candidate's responses in the interview. All participants took part in both the unmeasured and measured task, and the same material was used for both conditions. The order of administration of conditions is reversed for half of the participants in order to control for the effects of participants' varying expectations. To further encourage participants to select realistic levels of QoS, they were told that they would be able to keep a monetary equivalent of the budget they had left at the end of the experiment.

Tools. The experiment uses software that allows users to choose the levels audio quality they receive. The software allocates a budget to users prior to interaction. Participants saw the following interfaces:

1. INFORMATION: This is a panel of three menus. The menus are labeled *Current Values*, *Set Preferences* and *Future Predictions*. The options selectable from the menus are:

Under *Current Values*:

- Price: Shows how much is being paid for the quality requested via QUASS.
- Quality: Shows the current quality being received, as a percentage.
- Network State: Shows the current network state. This could be *Empty*, *Congested* or *Very Congested*. This information is color-coded.
- Current Budget: Shows how much money left to spend by showing the *Your Budget* display, described below.

Under *Set Preferences*:

- Keep current quality setting: Keeps the audio quality at the current level. If this is selected the QUASS tool disappears. This setting has to be cancelled to reconfigure the QUASS tool.
- Cancel quality setting: Brings back the QUASS tool to allow control of audio quality.
- Keep current price the maximum: Selecting this means that the participant won't pay more than is currently being paid.
- Cancel maximum price: Allows the price to fluctuate again.

Under *Future Predictions*:

- Quality: Shows a prediction of the quality likely in the near future.
- Network State: Shows what the network state is likely to be in the near future. This could be *Empty*, *Congested* or *Very Congested*. This information is color-coded.
- Battery Life: Shows a prediction of the number of minutes the participant can carry on before their money runs out, if the current level of quality continues to be requested. It does this by showing the *Your Budget* display, described below.
  1. Your Budget: What this shows depends on whether it is selected from the *Future Predictions* (Fig. 1) or the *Current Values* (Fig. 2) menu. On selecting from the *Future Predictions* menu the display will show if the

budget will run before the end of the experiment by showing *Not enough time.*

2.   Information Display: Shows the information selected by choosing from the menus in INFORMATION.

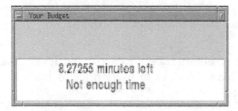

Fig. 1. Your Budget Showing Battery Life.

Fig. 2. Your Budget Interface Showing Current Budgett.

Experimental Hypotheses.

- *H1*: Users will be more likely to select a profile where QoS has priority over price when they are involved in an important task (the 'measured' task).
- *H2*: Users will be more likely to select a profile where price has priority over QoS when they are involved in a relatively unimportant task (the 'unmeasured' task).

The system used enables users to allow agents to control the levels of QoS received, via profiles. This allows users to specify whether a profile should prioritize delivering a specific level of QoS, or keep within a certain price, in situations where there is a perceived conflict between QoS and price dimensions. It is hypothesized that:

- *H3*: Users will tolerate lower levels of quality if they select a profile that controls for them.

According to focus group results, the degree to which users relinquish control over their QoS is affected by whether usage at the time is in a *Critical Period*:

- *H4*: Users will prioritize price over quality when their budget is relatively low.

Additionally, when QoS is priced, supplying feedback that allows users' to predict and therefore plan their spending affects their requests for QoS.

- *H5*: Feedback concerning future statistics (e.g. future network congestion) will be requested more frequently than feedback concerning current statistics.
- *H6*: Feedback concerning 'battery-life' (i.e. displaying how long people have left to interact before their budget runs out) will be the most frequently requested information.

4.2   Results

A strict linear correlation between the amount of quality delivered to users and the price of that quality cannot be assumed. Results show that a number of intervening factors influence whether users consider a certain price appropriate for a particular amount of audio quality. Table 1 presents a summary of results for each experimental hypothesis. These results are discussed below.

Table 1. Results for Each Experimental Hypothesis.

Hypothesis	Result
H1	High task importance = more *Quality* profile requests
H2	Low task importance = more *Price* profile requests
H3	Profile selected = lower quality tolerated
H4	Low budget = *Price* profile prioritized over *Quality* profile
H5	Future statistics selected more frequently than current statistics
H6	*Battery life* selected most frequently

Users' Tasks. Hypothesis *H1* was confirmed by the results of the study. Participants were more likely to select a profile where QoS has priority over price when they were involved in an important task. The frequency with which users selected a QoS profile over a price profile was compared between a measured and unmeasured task scenario.

- Users are more likely to select a profile where QoS has priority over price when they are involved in a relatively important task. This result is significant ($p<0.05$).

Using the same methods it was found that hypothesis *H2* was not confirmed by the results of the study. This means that:

- Users are not more likely to select a profile where price has priority over QoS when they are involved in a relatively unimportant task.

Fig. 3 shows the feedback required by participants for both measured and unmeasured tasks.

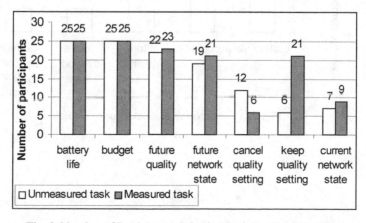

Fig. 3. Number of Participants Selecting Each Feedback Option.

The Levels of Control over QoS. Hypotheses *H3* and *H4* concerned the levels of quality users would accept depending on the amount of *Control* they were given over that quality. For each participant who selected a QoS profile, the lowest level of packet loss was compared to the lowest level of packet loss accepted under manual control. Statistical tests reveal that *H3* and *H4* can be confirmed by the results of the study. This means that:

- Users will tolerate lower levels of quality if they select a profile that controls for them. This result is statistically significant ($p < 0.05$).
- Users are more likely to prioritize price over quality when their budget is relatively low ($p < 0.05$).

Predictability. Hypotheses *H5* and *H6* were confirmed by the results of the study. This means that:

- Users request feedback concerning future statistics more frequently than feedback concerning current statistics.
- Users find information concerning Battery Life the most important type of feedback.

These results can be seen in Fig. 3. Fig. 4 shows the average number of times participants selected certain feedback options. Options that were selected only once have not been included in the analysis, because participants are likely to select options once through curiosity and not through a genuine requirement for feedback. Fig. 4 clearly shows the prevalence of predictive feedback. Statistical analysis shows that Battery Life is selected a significantly higher number of times than other options ($p < 0.05$).

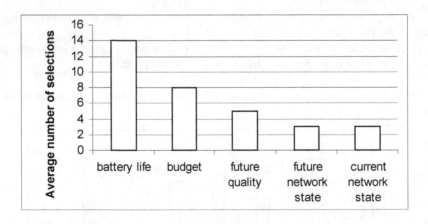

Fig. 4. Average Number of Selections per Option.

5   Discussion

Experimental results show that the concepts derived from grounded theory models have an impact on users' behaviour. Profile selection suggests that participants have an understanding that quality is linked to price. Selecting a profile allows users the *Peace of Mind* that they can concentrate on the task in hand without making re-evaluations of the value of received quality. The results show that users make conceptual links between price and quality depending on the context in which they are working. The effect of users' tasks on profile selection is further confirmed by the fact that, in the measured task, participants in this study were more likely to keep the profile selected than to cancel it, compared to those few participants who selected a profile in the unmeasured task. It is likely that participants chose QoS profiles in the measured task scenario to reduce the cognitive load of having to control the quality manually whilst attempting to hear crucial parts of the conversation:

'When I saw it's going to be worse I did try to not ask for as much, to try to get some leeway. But it was more important to hear it, to get the answers'.

Apart from when their budget was relatively full, very few participants selected a profile where price has precedence over quality (*H2*). These results show that the goal of users' interaction has priority over the price levied against that interaction. The confirmation of Hypothesis 4 (*H4*) shows that there is a complex relationship between users' tasks and the price they pay for the quality that helps them complete those tasks. The potential for price to correlate with QoS demands somewhat depends on whether users choose to use agents to control their QoS rather than to control it manually. This is because fluctuation in demands for QoS is greatly reduced if a profile is used. However, the results reported in this paper have shown that the importance of users' tasks varies according to their amount and rate of expenditure. This is shown by prioritizations of price over quality in a Critical Period, as predicted by *H4*. Results confirming the popularity of viewing a budget suggest that expenditure is correlated with QoS and not the absolute magnitude of price:

'The information that tells me about how much my spending is going down is the most useful...I don't understand about the charge for that bit or this bit, all that I think we really should know is how we're spending in total'.
'No, I don't feel good about seeing that blue representation of my money. How do I know the pace it's going down? I don't want to have to think so much about it'.

Users' evaluations of expenditure suggest that they make dynamic assessments concerning the value of an interaction, as a whole, against their assessment of its price, meaning that users' think about QoS as attached to an entire interaction. Charging schemes might therefore be designed to attach or estimate the cost of an interaction before that interaction starts, in order to provide accurate data to fit users' expectations. Users' assessments of value are therefore reflected by the Commitment they make to reach the goal of their task. This task is represented by the interaction as a whole.

6   Conclusions

The results reported in this paper have shown that:

1) QoS is correlated with expenditure and not absolute price at any one time.
2) Users value predictive feedback over feedback concerned with current network statistics.
3) The *Risk Assessment* associated with this expenditure depends on the user's goal of interaction.
4) Users have different requirements for feedback although default feedback profiles are feasible.

The work reported in this paper has shown that a vision of future network service must be based on an old principle, that economies are ultimately service-driven. As Peter Drucker (1999) points out:

'Quality in a product or service is not what the supplier puts in. It is what the customer gets out and is willing to pay for. [...] Customers pay only for what is of use to them and gives them value. Nothing else constitutes quality'.

The results reported in this paper have implications for network designers and HCI practitioners. With growing potential to support interactive applications, the real challenge for network designers does not solely lie in maximising utilisation of operations inside the network, but in ensuring that the service provided is both efficient and subjectively valuable to users. However, identifying a trend in behaviour and a cause for that behaviour is not the same thing as being able to predict it. Users' behaviour in response to levels of QoS needs to be predicted by defining the relationship, not only between conceptual factors, but between subjective perceptions and key magnitudes manipulated by the network infrastructure. The combination of grounded theory and experimental methods in this paper provide a framework for conducting HCI research that can, *a)* describe users' behaviour during interaction and, *b)* prescribe from conceptual models that represent the motivations for such behaviour. These models have shown that it is not enough to simply configure certain levels of quality to users, it is the interpretation of these figures that must be represented (i.e. the meaning of statistics to the user's current situation). For QoS to be acceptable, users must be able to make an informed decision about their requests, otherwise any valid link between demand and supply is diminished. By applying a combination of methods, this paper has shown what information users require to make such decisions.

We have shown that when users have to pay for the quality they receive, they require dynamic feedback concerning their expenditure. This requires a re-think about how network store and route information to the user; some kind of 'state' in routers is needed to maintain data of the quality likely to be received. Our results suggest how this should work. For example, we have been able to show that semi-intelligent QoS profiles can be implemented that receives pre-set QoS ranges from users and responds only when the quality drops below a certain range. This way, users would not have to constantly re-evaluate the cost of the quality.

Tolerance of QoS, as in the behavioural expression of any preference, is influenced by socio-technical systems that represent the context of that behaviour. These systems cannot be directly engineered for. What is possible is to recognise the role of such systems and to apply methods such as grounded theory to understand how their elements relate to each other. For example, this research has shown that users apply the principle of risk aversion when making decisions about the acceptability of a charging mechanism. This principle is based on an ability to predict future network conditions using feedback provided at the application level.

7   Further Work

There is a parallel in our society where technological and even political developments influence the way we interact with computers and the demand made for services [17]. The nature of demand for QoS is both dynamic and evolving. Further work that investigates network QoS and charging must reflect the evolving nature of that environment. The way users perceive network operations and the quality that they deliver depends on a number of non-static factors, such as the way QoS is marketed and the business model that lies behind the network. As the results reported in this paper represent users' perceptions of QoS they are subject to variation. The use of grounded theory has enabled the models to be explained at various levels of granularity. Thus, while top-level concepts are likely to continue to be relevant, continuing work is needed to test the concepts described in this paper for their relevance to an evolving environment. The aim of the experiments was to prove the efficacy of models constructed through grounded theory to predict users behaviour whilst interacting with networks. This has been shown. A next step would be an attempt to define objectively where users set an upper bound or Critical Threshold on the price they are willing to pay for a certain task, and how feedback showing them their expenditure could influence this figure.

Participants in this study show idiosyncratic behavior. This means that a participant who selected a certain option in the first condition they did was, in many cases, more likely to select the same option in the second condition. This means that users' feedback needs are likely to differ. Further work is needed to investigate the effects of demographics on users' requirements for QoS and the way they prefer to pay for it. Research is also needed to look at the effects of long term use of a priced quality system on users' feedback requests. It is likely that users' expectations of the levels of quality they receive will become more accurate through experience of typical levels. Requirements for feedback may then only be needed in Critical Periods.

References

1. Cullinane, P.: Ready, Set, Crash. Telephony **3** (1998) 3-13
2. Cocchi, R., Shenker, S., Estrin, D., Zhang, L.: Pricing in Computer Networks: Motivation, Formulation and Example. IEEE/ACM Transactions on Networking **1** (1993) 614-627

3. Mackie-Mason, J.K., Varian, H.R.: Economic FAQs about the internet. Journal of Economic Perspectives 8 (1994) 75-96
4. Clark, D.: Internet Cost Allocation and Pricing. In: McKnight, L.W., Bailey, J.P. (eds.): Internet Economics. MIT Press (1997)
5. Bruhns, J.: A Short Commentary on Usability in Northern Europe. Still a Long Way to Go. Available at http: //www.webword.com/moving/global0002.html (2001)
6. Bouch, A., Bhatti, N., Kuchinsky, A.J.: Quality is in the Eye of the Beholder: Meeting Users' Requirements for Internet Quality of Service. Proceedings of CHI'2000 (2000) 297-304
7. Cosgrove, J.G., Linhart, P.B.: Customer Choices Under Local Measured Telephone Service. Public Utilities Fortnightly, 30 (1979) 27-31.
8. Brittan, D.: Spending More and Enjoying it Less? Technical Review 100 (1997) 11-12
9. Bohn, R., Braun, H.W., Wolff, S.: Mitigating the Coming Internet Crunch: Multiple Service Levels via Precedence. Applied Network Research Technical Report, GA-A21530. University of California (1994)
10. Bouch, A., Sasse, M.A.: It Ain't What You Charge, it's the Way that You do it: A User Perspective of Network QoS and Pricing. Proceedings of IM'99 (Boston MA) (1999a) 639-655
11. Bouch, A., Sasse, M.A.: Network Quality of Service: What do Users Need?. Proceedings of IDC'1999. (Madrid, Spain) (1999b) 78-90
12. Watson, A.., Sasse, M. A.: Evaluating Audio and Video Quality in Low-cost Multimedia Conferencing Systems. Interacting with Computers 8 (1996) 255-275
13. Strauss, A., Corbin, J.: Basics of Qualitative Research: Grounded Theory Procedures and Techniques. 2nd edn. Sage (1997)
14. Adams, A., Sasse, M. A., Lunt, P.: (1997) Making Passwords Secure and Usable. Proceedings of HCI'97 (1997) 1-19
15. Adams, A., Sasse, M. A.: Privacy Issues in Ubiquitous Multimedia Environments: Wake Sleeping Dogs, or Let Them Lie? Proceedings of IFIP'99 (1999) 214-221
16. Bouch, A., Watson, A., Sasse, M.A.:QUASS-A Tool for Measuring the Subjective Quality of Real-time Multimedia Audio and Video. Proceedings HCI'98 (1998) (Sheffield, England) 94-96
17. Anagnostov, M.E., Cuthbert, L., Lyratzis, T.D., Pitts, J.M.: Economic Evaluation of a Mature ATM Network. Journal of Selected Areas in Communications 10 (1991) 1503-1509.
18. Drucker, P.: Management. Butterworth-Heinemann (1999)

Traveling to Rome: QoS Specifications for Automated Storage System Management

John Wilkes

Storage Systems Program, HP Laboratories,
1501 Page Mill Road, Palo Alto, CA 94304, USA
Tel: +1.650.857.3568   Fax: +1.650.857.5548
wilkes@hpl.hp.com

Abstract. The design and operation of very large-scale storage systems is an
area ripe for application of automated design and management techniques – and
at the heart of such techniques is the need to represent storage system QoS in
many guises: the goals (service level requirements) for the storage system,
predictions for the design that results, enforcement constraints for the runtime
system to guarantee, and observations made of the system as it runs. Rome is
the information model that the Storage Systems Program at HP Laboratories
has developed to address these needs. We use it as an "information bus" to tie
together our storage system design, configuration, and monitoring tools. In 5
years of development, Rome is now on its third iteration; this paper describes
its information model, with emphasis on the QoS-related components, and
presents some of the lessons we have learned over the years in using it.

1   Introduction

Designing, supporting, and managing storage systems is getting harder as they get
larger and more complicated. And they are getting larger very quickly: compound
annual growth rates of 150% in storage capacity are not unheard of. A data center of
the immediate future could easily contain hundreds or thousands of logical volumes
and file systems, hundreds of terabytes of disk drives, and handle tens to hundreds of
gigabytes per second (GB/s) of storage traffic. Data availability is crucial: if the
storage system goes down, so does the computer system that relies on it. Achieving
all this requires a great deal of complexity: multiple disk array types, different data
organizations, transactional support, automated fail-over schemes, and so on.

This complexity – compounded by the desire to avoid operator intervention
because of errors and the dearth of skilled system administrators – means that the
design and operation of large-scale storage systems is an area ripe for application of
automated design and management techniques. At the heart of such techniques is the
need to represent the QoS goals (service level requirements) of the storage system, the
design that results, and observations made of the system as it runs.

Even though many of the same approaches used in the large literature on QoS for
the networking domain (e.g., [1]) can be applied to storage systems, there are a
number of differences that make the mapping non-trivial: (1) the low-level storage
protocols (based on SCSI) are highly intolerant to packet loss, so dropping requests is
not a viable technique for handling overload or congestion; (2) there are very strong

L. Wolf, D. Hutchison, and R. Steinmetz (Eds.) : IWQoS 2001, LNCS 2092, pp. 75-91, 2001.

non-linearities in performance that result from the mechanical properties of disks drives and the use of large caches: it is easy to construct scenarios in which an inappropriate mix of I/O traffic to a disk drive can change the data transfer rate by a factor of 50; (3) there is no support for traffic shaping or QoS enforcement techniques in the storage system itself; (4) the cost of the storage system is often the dominant component of the overall system cost; and (5) dynamic quality adaptation at the application level is extremely rare. These factors mean that the primary approaches to providing QoS guarantees are provisioning and resource (re)-balancing, and that the only possible guarantees are probabilistic in nature. It also means that the portions of the QoS specifications for storage systems that describe I/O behavior have to be considerably richer than for many other domains, and the language used to describe them needs to be correspondingly more expressive than is usually the case for network traffic.

Our approach to the storage design problem has been to develop an architecture for a quality-of-service-based "attribute-managed" storage system [3]. This uses QoS goals to specify what is wanted to the storage system, which is then responsible for deciding how best to provide it – that is, the technically hard part of what we have built is an automated provisioning and load-balancing design tool that uses QoS goals as targets. We embed this design tool in a system that can automatically apply its decisions to a running system, monitor the result, and make better designs – all completely automatically.

In this scheme, the desired storage system goals are specified in terms appropriate to the client of the storage system (e.g., I/O requests per second, number of concurrent streams, access patterns, availability, etc.) and the storage system takes care of the details of deciding how many storage devices of what type to configure, how they should be connected, and how to lay out the data and balance the load across them.

We view this fundamentally as an optimization problem: something that computers are rather good at. Our contributions have been in defining the problem in a way that makes it tractable, applying our detailed knowledge of storage system components, developing techniques to understand and quantify the QoS specifications that such systems have to meet, and putting together design tools that can solve this problem, together with a complete suite of ancillary components to make it operational.

One of the most important – and certainly most central – of our contributions has been the Rome object model that is the subject of this paper. The Rome object (or information) model acts as an "information bus" to tie together our automated storage system design, configuration, and monitoring tools.

Rome is used to describe *everything* that we consider important about storage systems and their elements: the workloads presented to a storage system; the QoS goals of the system; the kinds of storage and network devices that storage systems can contain, and how they can be configured; how a specific storage system is configured; how the workloads are spread across those storage devices; how the storage components are connected together (e.g., through a FibreChannel or Ethernet-based SAN); end-user goals for the system and its behavior; both existing systems and potential "what if" designs that might better meet current or future needs; and information about the current state of a running system, and how it is behaving.

The tools we have developed either take in Rome description files, or emit them, or both. This makes it possible to compose these tools in many different ways, to achieve a wide range of different effects. It also makes it easier to write the tools:

each one need only concern itself with its portion of the problem, and can rely on other tools in the SSP suite to fill in the bits it doesn't deal with. In 5 years of development, Rome is now on its third iteration; this paper describes its information model, with emphasis on the QoS-related components, and presents some of the lessons we have learned over the years in using it.

2   The Flight fromTroy: Automated Storage System Designs

Our first activity was to survey the literature and develop a formulation of the design problem as a constraint-based knapsack (bin packing) problem [12]: storage devices were represented as knapsacks, with multiple constraint dimensions (capacity, throughput, bandwidth, utilization, etc.); storage loads as the things to pack into them. The design problem is to select the appropriate set of storage devices, and a packing of storage loads onto them, to minimize some objective function (typically system cost).

Our second action was to develop a storage system designer (called *Forum*) that would take in storage system QoS requirements (expressed as I/O workload demands) and a library of storage device descriptions (recording their performance capabilities), and explore the search space of designs or assignments: bindings of workload elements to storage devices. This paper focuses on how we represent the QoS requirements in this system and its successors.

It happened that we had earlier developed a technique to control a storage-system simulator using the Tcl language [6, 9, 17], so it was natural for us to apply this to Forum. The basic idea was very simple: make everything an object, and add attributes to those objects to describe additional properties. Tcl's nested lists made this easy to record in the obvious way (see Fig. 1).

In this architecture, a `store` is a container for data, such as a logical volume; a `stream` represents a dynamic access pattern to it – the important part of a QoS specification. More than one stream can target a single store. The store has two attributes: its size (`capacity`), and the fact that it is `boundTo` – in this case

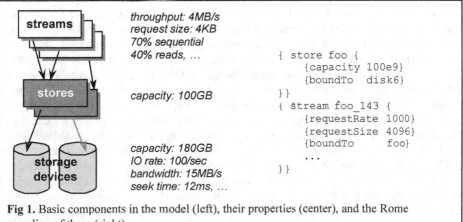

Fig 1. Basic components in the model (left), their properties (center), and the Rome encoding of these (right)

meaning *realized by* – a particular disk. Our Tcl-to-C++ interface made it easy to turn such Tcl statements into C++ objects. We chose to equate the statements with top-level objects, and the attributes with secondary objects hanging off the former, indexed by their names. An important behavior was that any attributes not recognized by a tool are merely passed through to the tool's output, but are otherwise ignored.

The basic object data structure with its attributes has served us very well. It made it easy to add new attributes – either to extend the standard set, or to support a tool-specific extension, or to try out a new idea. The "ignore things you don't understand rule" has made it easy to extend the set of attributes supported by some tools without affecting others: something that would have been very much harder if our primary interface had been an API instead of an object model.

We started with a very simple set of workload attributes: mean request rate (I/Os per second), mean request size (bytes), fraction of reads, and run length (the number of consecutive requests to sequential addresses) [12]. However, our prior work on understanding storage system behavior (e.g., [10, 11]) told us that these simple, fixed values would not be sufficient: storage system traffic is very bursty, and an understanding of this is vital to understanding its behavior (e.g., it is self-similar [7]). As a first step we augmented these simple mean values with variance, modeled on a normal distribution. Even though we suspected that this might not be all that good a fit to the real underlying process, it was compatible with our performance models, and therefore useful.

The QoS specification for a storage design could now be expressed in terms of a set of stores, each with zero or more streams directed to it, where the streams specified the required access patterns that were to be achieved. We found it helpful to distinguish between requirements and behaviors: *requirements* are the demands made on the system that its performance is to be measured against; *behaviors* are the actions taken by the load generator in asking for those requirements. For example, a requirement might be for a data rate of 1MB/s; a behavior might be to deliver this requirement as a stream of 1000 random-access 1KB reads per second – which would be a very different load on the storage system than a single 1MB request per second.

The obvious next question was: where should such data come from? Talking to customers and others trying to design storage systems, it became clear that few of them were comfortable providing estimates of the load currently imposed on their systems, let alone future loads. Despite this reluctance, it's important to point out that designers and customers are already forced to do just this by the existing manual methods of storage system design. These are still dominated by simple, rule of thumb "speeds and feeds" analyses: "this sounds a bit like the system we designed last week for _____ except that I think we should bump up the I/O rate by a bit (how about 20%?), and add a bit more random-access capacity in the disks for the indexing system." The main difference was in the degree of specificity required: we early on adopted the slogan that "the more you can tell us, the better a design job we can do". (For example, we once took advantage of anti-correlation data in the workload for a database benchmark to reduce the storage system cost by a factor of about 6.)

The two answers for where the data could come from were (1) measure an existing system, and, if necessary, extrapolate to a new one; (2) build up a library of prior workloads that we had met, and – just like our human designer counterparts – slowly accumulate wisdom about how workloads scaled, and capture these in tools that could emit an appropriately-scaled workload approximation given a small number of knobs (much like the workload scaling parameters in the TPC database benchmarks [14]). The latter proved relatively easy for some simple workloads, but the full generality of mapping upper-level application QoS specifications down to low-level storage system behavior remains a difficult research topic, an experience shared with others in different fields.

The measurement path proved much more conducive to automation. The HP-UX operating system, which we used as an experimental platform, contains a measurement system that we could use for this very purpose: it is able to emit a trace record for every single physical I/O request, at a negligible cost (a couple of percent processor utilization).   We wrote some tools to scan over such I/O traces, and generate the Tcl files describing the streams and stores.

One more component was needed: a representation of the performance characteristics of storage devices. For our first round, we restricted ourselves to disk drives, and were able to convert a subset of the data gathered for the simulation models used in the Pantheon simulator [17] into analytical models of disk drive's performance. The switch to analytical performance models inside the Forum design tool was required because the models can be called millions of times during the exploration of the design space, and simulations are simply too slow. To increase the flexibility of our models, we structured them as a set of composed components [13].

We completed the tool chain by developing an automatic configuration system called *Panopticon*: given a design of a storage system (i.e., a mapping of stores onto disk devices), it would construct the appropriate logical volumes to achieve this mapping, and we could then run a test on the resulting system.

2.1   Returning to the Imperial City: Supporting Disk Arrays

Bolstered by our success with the early Forum results, we thought it would be a simple transition to extend our performance models and design language to encompass disk arrays. We were wrong. Disk arrays themselves need to be configured, and frequently have a great deal of internal complexity. In fact, choosing how best to configure a disk array to support a given workload is itself a challenging, NP-hard design problem.

Our first approach was called Minerva; it used "divide and conquer" to break the dependencies. A Minerva run begins by estimating the amount and kind of disk arrays that would be needed to meet the workload's requirements, using some very simple performance and capacity models, and pre-configures this set as the device descriptions input to Forum, which then attempted to pack the workload on to these disk arrays. If the workload did not completely fit, the estimating process was repeated with the left-over workload; if it did, then the Forum solver was re-run with the objective of evening out the load imbalance across the configured hardware, rather than minimizing the system cost.

By this time, the range of workload parameters we had begun to record was becoming rather too large for an *ad hoc* tool, so we developed a flexible data-analysis framework, *Rubicon*, that used a structure rather like that of a packet filter to perform analysis of an I/O trace and generate workload information.

Separating the internal and external representations also proved vital: we learned the hard way to distinguish between the information model and the manifestation of it

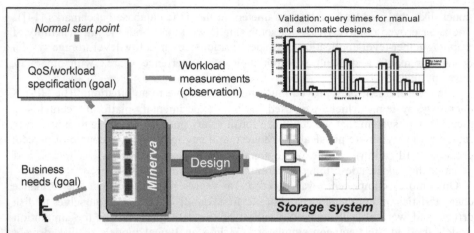

Fig. 2. Process flow for the storage-design process. The normal start point is at the top left; for the TPC-D-like validation we began with a manual system design whose performance was measured, and used to drive the redesign

in a tool as C++ objects: it is simply not possible to define a single representation of the important attributes of an object for all tools. (For example, performance data is completely ignored by the Panopticon configuration tool, so it would be counter-productive to insist that it use the same internal C++ data structures as Minerva.)

Using Minerva and its support tools, we were able to perform the following test (see Figure 2): we consulted a local database guru on how to configure the TPC-D benchmark [14] for our system; configured the system that way and ran the benchmark, measuring the I/O rates it achieved; then used these values as QoS goals input to the Minerva design tool, and asked it to match the performance QoS requirements at minimum cost. We used Panopticon to construct the resulting storage system design automatically (it had been extended to perform array configuration too); reloaded the database, and re-ran the benchmark. The performance result gave us query execution times within 2-3% of the original, but the Minerva-based design only needed 16 disk drives in a RAID-5 disk array configuration, whereas the manual design had required 30.

2.2   The Arrival of the Visigoths

As time went on, the internal structure of the Minerva design tool started to slow us down, and Eric Anderson developed a new storage system designer (*Ergastulum*), which explored the disk array design space at the same time as it assigned work to the array. This had a number of advantages, not the least of which was that it eliminated the rather cumbersome two-phase design process that Minerva had used.

Ergastulum happened to be much faster than Forum, and showed that parsing the old Tcl structures was becoming a significant overhead, so we switched to a hand-coded parser, and simplified the language design a little. Ergastulum also needed a way to express the range of disk array designs that it could explore, and, rather than hard-code these into its logic, we chose slightly to extend the flexibility of the language used to describe the storage devices in the device library. We called the result *Rome*.

3   Rome 2: The Renaissance – Rebuilding St Peters

After some time, the first version of Rome became a slightly messy hybrid, with some backward-compatible components, and some forward-looking elements. It was sufficient to prompt us to undertake an overhaul of the result. The result is the present version, known as Rome 2.

Rome 2 was required to be good at describing real-life storage systems – their workloads, configurations, and the components they contain (devices, hosts, software, networks, etc.); extensible to encompass new storage devices, workload attributes, and even new target domains, such as internet data center configurations; rich enough to represent many levels of complexity; and capable of being represented in ways that are easy to parse, generate, and use by computer tools, and in (possibly different) ways that are understandable by humans.

Rome 2 achieves these goals by separating its underlying *object model*; from the linear encodings of it, which are known as *Latin* and *Greek*. Latin is the "native" language of Rome, and is used to specify it in the descriptions that follow; it uses a Tcl-like syntax derived from our earlier experiences. (Figure 3 shows a simple QoS specification written in Latin.) Greek is an XML-based linear encoding derived from Latin.

3.1   The Rome Object Model

The Rome object model is built on an object-type inheritance hierarchy; it provides the underlying structure used by the Rome object model to describe things of interest to the Rome tools. *Objects* in Rome represent things like disk arrays, or part of a storage-system workload such as a stream. Each object is introduced by a single *declaration*, has a unique *name*, and has an associated set of *attributes*, which provide additional information about the object. Attributes are modeled as objects in their own right; some of them represent internal *components*, such as the I/O controllers on a disk array. Much of the Rome object model specification is concerned with describing the attributes that each object type can have associated with it.

The Rome object model is defined in two layers. The lower level is known as the Rome *shallow semantics*. This occupies a middle ground between the syntax of the representation language (e.g., Latin), and the deep semantics, which is an ever-extending set of object types, components, and attributes, with their associated meanings. Because the meanings of the shallow-semantics objects are reasonably well standardized across the tools, we can build software libraries to handle the common operations on these object types. Examples of such objects include approximate values (of which more below) and common properties such as cost.

The Rome *deep semantics* defines the meanings of the remaining object types and their attributes. Examples include workload measurements and performance requirements (i.e., QoS specifications); data mappings (such as RAID);

```
{ store store1 {                              # a 100GB store
  { capacity 100e9 }                          # mapped to a logical unit
  { boundTo array4.lu_3 }                     #  on an array (not shown)
}}

{ stream stream1 {                            # a stream
  { boundTo store1 }                          #  bound to that store
  { source  host_A1}                          # originating at this host
  { interArrivalTimeOpen {                    # inverse of request rate
      { datamodelNormal best {                # normal fit
          { mean 0.83e-3 }{ stddev 0.6e-3 }   # mean = 1200/sec
          { chiSquare 0.7 }                   # goodness of fit metric
      }}
      { datamodelExponential poor {           # exponential fit
          { mean 0.83e-3 } { chiSquare 0.2 }  # 1200/sec, less-good fit
      }}
  }}
  { requestSize {                             # a simple behavior
      { datamodelUniform {                    # uniform size in 4-12KiB,
          { mean 8192 }                       #  on 1024-byte boundaries
          { lbound 4096 } { ubound 12288 } { granularity 1024 }
      }}
  { responseTime {datamodelExponential {mean 50e-3}}}  # a goal
  { stream read {                             # just the read requests
      { filteredBy { opType read }}
      { interArrivalTimeOpen 1e-3 }           # 1000/sec
      { requestSize 9216 }                    # larger requests on avg.
  }}
  { stream write {
      { filteredBy { opType write }}
      { interArrivalTimeOpen 5e-3 }           # 200/sec
      { requestSize 4096 }
  }}
  { stream degraded {                         # something not right
      { filteredBy {{ outageDuration 3600 }   # 1 hour at a time
                    { outageFraction 0.002 }}} # 17 hours/year
      { interArrivalTimeOpen 1.67e-3 } # 600/sec
      { stream write {
          { filteredBy { opType write }}
          { interArrivalTimeOpen 0.1 } # 10/sec
      }}
  }}
  { stream broken {                           # completely stopped
      { filteredBy {{ outageDuration 300 }    # 5 min at a time
                    { outageFraction 0.00001 }}} # 5 min/year
      { interArrivalTimeOpen inf }            # nothing: 0/sec
  }}
}}
```

Fig. 3. A (much simplified) sample workload specification example. One store is accessed by one stream. In normal mode, it gets 1200 requests/sec; in "degraded" mode, it can limp along for an hour at a time at half that rate; and it can be "broken" (non-accessible) no more than 5 minutes a year ("five nines availability"). For simplicity, most of the datamodels shown are simple numeric values; in practice, distributions would normally be used.

storage devices; interconnect fabrics (including SANs); and host hardware and software. The interpretation of the deep-semantics varies considerably from one tool to another, so it is much harder to provide a shared code library for them that does much more than define the basic object type.

Rome represents the idea that many things are alike by giving the objects that represent those things a common *object type* (or `objectType`). Object types are first-class entities in Rome – that is, they are objects in their own right. An `objectType` declaration introduces (defines) a new Rome object type, after which objects and sub-objects (attributes) with that object type can be *declared*. Such declarations can occur at "the topmost level" (i.e., free-floating in the global namespace), or nested within another object.   That is, an `objectType` object is loosely equivalent to a programming language class definition. Rome object types form a single-inheritance `isA` hierarchy.

An `objectType` attribute in an `objectType` declaration defines a sub-object type that only applies in the context of objects of the enclosing `objectType`. Such components (and their types) can be arbitrarily nested.

```
objectType <qualname> {
  [ { isA            <qualname_objectType> } ]
    { objectType <name> { <obj-list_component>} }*
  [ { occurrenceCount <number> | <numeric-range>}} ]
    <obj-list_type-paramaters>
}
```

This description comes directly from the Rome BNF-like specification; <angle-brackets> enclose non-terminals in the grammar, [square brackets] denote optional components, and an asterisk (*) indicates elements that can be repeated zero or more times. It means that an object type declaration takes as first argument a qualified name such as `type.sub_type` (i.e., dot-separated nesting is allowed), and a list of attributes. One of these attributes may be an `isA` attribute, which takes as argument the qualified name of another `objectType`, and defines the inheritance hierarchy. One or more type-specific attributes exist, defined by their own `objectType` declarations. An `occurrenceCount` attribute may be present to bound the number of instances of this object type that should be instantiated for every instance of the enclosing object. And, finally, there can be a list of `objectType`-specific attribute values that act as defaults for objects of the newly-declared `objectType` (shown as <obj-list_type-paramaters>).

The set of object types recognized by a Rome tool varies from tool to tool; it is based purely on the name of the object's `objectType`. Some tools have various object types built in (for example, a storage-system design tool will probably know about storage devices that it can design for); and some will not (e.g., a pretty-printer). Unrecognized object types are quite acceptable: in such circumstances a tool should either ignore the unrecognized object, or pass it on to its output. However, a tool may demand that certain objects or attributes exist, and the `occurrenceCount` attribute in an `objectType` declaration can specify the allowed number of instances of an object.

3.2   Attribute Inheritance

Attribute inheritance is the process by which attributes are searched for in an object. If an attribute is present in the object, then it is used. If not, the attribute is looked for in the object's `objectType` declaration to see if a value is provided for it: that is, the search follows the `isA` type hierarchy, all the way up to the root, if necessary. Values provided closer to the point where the search originates take precedence. Each missing attribute is searched for independently.

This allows the `objectType` statement to store the attributes that all objects of that type have in common. Since adding an attribute or a component to an object does not change the type of the object itself, the number of base Rome object types is quite a bit smaller than in some other systems. Thus the Rome objectType `diskDrive_Quantum425S` is an instance of the `diskDrive` type, and includes values for the attribute parameters that don't vary across instances of Quantum 425S disk drives, such as capacity and performance parameters.

3.3   Approx Values and Datamodels

Rome treats data values that represent continuous, real-world values in a special way. It recognizes that such values are only approximate estimates of the underlying real-world process, and represents this by explicitly referring to such values as *approx values*. You can think of an *approx value* as the result of statistical sampling or characterization efforts on the real underlying process or value. Multiple independent measures or estimates can be provided; these are called *datamodels*. Datamodels can be simple or complex, ranging from a simple mean value up to complete distribution histograms of observed values. The currently supported set includes: Normal, Gamma, Exponential, Uniform, Constant, and Histogram.

There are three important properties of this idea. The first is that we use approx values almost everywhere that a value might be expected: this means that *everything* naturally becomes a statistical specification. The second is that approx values are used to express a range of allowed values, e.g., for a goal or a prediction. Most other QoS specification approaches that we are familiar with define a fixed, desired value, and then discuss what happens when variations from it occur. Our approach is to start by assuming the presence of variation, and then try to provision to support it. The third is that each datamodel has an associated random number generator that can produce values drawn from an equivalent distribution. This allows us, for example, to measure a trace, and then construct a similar one for replaying by simply generating I/O requests with similar distributions of inter-arrival time, request size, and so on.

Each datamodel type has its own particular set of parameters – as well as a set that can be applied to all datamodels. For example, a normal distribution datamodel fitted to the observed data would have mean and standard deviation parameters; it might also have data on the observed largest and smallest values observed, a count of the number of observations, how the data was gathered or filtered, and data about the underlying process that would otherwise be lost, such as the intrinsic minimum and maximum values.

Datamodels can represent truncated distributions: ones with hard upper and lower bounds, and ones with an intrinsic granularity, such as a block size.

A datamodel instance can have a name, so that there can be more than one of them: for example, two different datamodels of the same process, each of which can be provided with different goodness of fit measure to estimate how well it captures the underlying process. Note that a tool may choose to use a less-well fitting datamodel if it is unable to process the better mo del – for example, a tool using a queuing model may be restricted to exponential distributions, even though a Gamma model may be a better fit.

3.4   Storage Workloads

A *workload* for a storage system is made up from one or more related *workload elements* (streams and the stores they target, and other workloads) that are applied to a system all together, or not at all. A workload can be used to represent an application, or part of an application, or a group of applications. Workloads may physically contain workload elements, or merely group ones that are defined elsewhere together – or they can do both. Loops are not permitted; nesting is.

```
workload <qualname> {
    { store    <name> { <obj-list_store> } }*
    { stream   <name> { <obj-list_stream> } }*
    { contains <qualname_workload-element-list> }*
    { workload <obj-list workload> }*
}
```

We envisage that workloads will also capture relative importance of applications, and their security attributes, even though this doesn't yet occur.

3.5   Stores

A Rome store represents a container for file systems or database tables. A store can potentially handle many streams, and has only one intrinsic attribute: the capacity it provides to its clients, measured in bytes.

A store also demands backing space for its contents, and this is handled either by binding the store to a storage device (strictly, a logical unit on that device for block stores), or by mapping the store onto one or more lower-level stores through a `layout` attribute, which supports mappings such as mirroring, logical volume managers, and RAID data protection. These mappings may occur many times between the high-level logical volume seen by a database or file system, and the low-level disk mechanisms used to store their data.

3.6   Streams

A stream specifies the dynamic aspects of a workload imposed on a storage system. Each stream targets just one store. The stream attributes represent a combination of stream *requirements* and client *behaviors*. Requirements are goals that the storage system must meet (e.g., the request rate and request size); behaviors characterize the workload under which those requirements are to be provided (e.g., the request arrival process).

A stream can be looked at several different ways, and the specifications reflect this. The simplest is simply to record the desired, predicted, or observed access pattern.

streamType	block | NFS | CIFS | localUNIX | localWindowsNT	default is block
Identifies the type of operations that the stream supports		

boundTo	<qualname_store>	
Names the (one) store to which this stream is bound. A store can have multiple streams.		

source	<qualname_source>	
The name of the host system or device that generates the load represented by this stream.		

filteredBy	<obj-list_filterTypes>	
How (if at all) this substream was filtered down from the enclosing stream. The value is a list of parameters used for the filter, such as operation type, outage information, or phasing data.		

interArrivalTimeOpen | interArrivalTimeClosed	<approx_seconds>	default is 0
The time between requests issued to the storage system for this stream. If the arrival process is open, then this represents the rate at which requests will be generated regardless of the service time; if the arrival process is closed, it represents the "think time" between the completion of one request and the start of the next.		

numOutstanding	<approx_number>	
The number of requests outstanding at a time for this stream. In a goal, this attribute dominates a closed interarrival process specification, and may act as a limiter on the effective arrival rate.		

requestSize	<approx_bytes>	
The number of bytes read or written as a single request by this stream.		

runCount	<approx_I/O-count>	default is 1
A simple measure of spatial locality: the number of consecutive I/O requests that will be logically-consecutive addresses in the target store. There is no requirement that all the requests in a run have the same requestSize, nor need they all be reads or writes – this is solely a measure of the starting address of a set of consecutive requests.		

jumpDistance	<approx_bytes>	default is random uniform across store
A simple measure of spatial locality: the distance between the end of one request run and the beginning of the next request run in the I/O stream.		

responseTime	<approx_seconds>	
The time that a single I/O request takes to complete, including any queuing delays. A distribution can be used to specify the range of allowed values for a goal; if a single <numeric-value> is provided, it means that both the desired and maximum-allowed response time have the same value.		

onTogether	<qualname_stream-phase-approx_overlap-list>	default is independent
The fractions of total time that this rill is in the current phase at the same time as the other listed streams are in theirs. In practice, the list of other phases is likely to be sparse: the most important combinations are probably the "on together" and the "this on, other off". If no value is specified, the value to assume is that for independence: the product of the fractions of time that each rill is in the given phase.		

locationSkew	<approx_bytes>	default is no skew (uniform distribution)
A distribution to describe the access-location skew, in terms of byte offsets within the store for the beginning of independent runs of requests. (That is, if the run length is exactly 2, the locationSkew attribute is used to specify the start address of exactly half the I/Os.) The attribute value is usually expected to be a histogram, or other non-point distribution. The distribution represents the relative rate at which an I/O request in the stream commences at the given portion of the target store's address space; a point value causes all runs to begin at that precise address.		

Others include filtering the accesses by (for example) operation type, or on the permitted degraded modes of operation. We have found every one of the attributes described here to be necessary; doubtless, as we progress with our workload modeling, we will add to this list.

We used to use the request rate to specify the I/O request-arrival process, but Rome 2 changed this to one based solely on inter-arrival times, to avoid recurring difficulties associated with knowing what the appropriate averaging interval should be for the arrival rate. Now we support the following:

- *Open* processes ignore the service time of requests they issue, and continue to generate requests at the same rate regardless of what the storage system response is. This means that there is no a priori upper bound on the number of outstanding I/O requests in flight at a time. Here, the `interArrivalTime` attribute dominates, and the `numOutstanding` attribute merely represents a measured value (e.g., it may not represent what is achieved in a new assignment).
- *Closed* processes have a fixed upper bound on the number of outstanding I/O requests in flight at a time. If they have a non-zero `interArrivalTime`, the number of outstanding requests may drop below this maximum. An "as fast as possible" arrival process can be specified by `{interArrivalTimeClosed 0}` together with some upper bound on the `numOutstanding`.

Not shown are new measures we are developing for use with the large data caches that are found in disk arrays. The basic idea is to include a measure of the LRU stack depth, or a richer (but much more expensive to measure), re-reference distance histogram, but we are still calibrating these measures against real disk arrays.

3.7   Substreams

A *substream* represents a portion of, or view onto, an enclosing stream specification. (We sometimes call them *rills*, from the Scottish word for a small stream.) We speak of the substreams as being "filtered from" the enclosing stream. This filtering can occur in a number of different ways:

- by target shard (a shard is a portion of a store, such as one of the back-end disks that the store layout maps to);
- by operation type (`opType`), such as read or write;
- by `phase`, which captures the idea that the stream accesses can be characterized by one pattern for a while, and then by another, and so on. This is expressed by use of a Markov-like phase transition model, with individual phases having their own properties, including a phase duration and a list of transition probabilities to other phases. Phases can be nested, and apply to multiple different time scales. They grew out of a simple on:off model, and are applied to handle the day-to-day change in activity levels as well as shorter-term burstiness effects.
- by a performability specification (see below).

There is no requirement for the set of filtered substreams to "cover" all the possible substreams that could be extracted from the enclosing stream. For example, a block stream might have just one substream, filtering for write operations in addition to data about the stream's overall requirements or behavior.

3.8   Performability

The top-level stream attributes describe the desired behavior in the absence of failures. We refer to it as the *baseline performability specification*. Failure to meet the baseline performance goal is termed an *outage*. Some streams can tolerate such outages – especially if the outages can be bounded in duration or frequency or both [15]. For example, an application may be able to tolerate a short downtime period once a month; or may be able to operate with about half its usual workload for a while until a broken disk can be repaired. To represent this, the duration and frequency of these outage periods can be described, together with the tolerable levels of performance during the outages. Each such *performability specification* is written as a set of attributes that are override the baseline performance for the specified outage periods. The use of approx values in these attributes naturally supports probabilistic specifications.

outageDuration	$<approx_{seconds}>$
The longest tolerable outage duration.	
outageFrequency	$<approx_{number\ per\ year}>$
outageFraction	$<approx_{fraction\ 0-1}>$
The first specifies the allowed number of separate outages that is permitted (measured, etc) per year. The second specifies the fraction of the total time that can be outages, averaged over 1 year.	

3.9   Goals, Observations, and Designs

Although their specifications may look nearly identical when written down – indeed, our early tools took in workload observations and used them as QoS goals with no editing – we have learned that it is helpful to explicitly label QoS specifications with their purpose.

- *Goals* are desired target state(s) of the system. A goal is a form of requirements specification, or service level objective, with an associated utility function: the better the goal is met, the higher the utility function value (our use of approx attribute values seldom gives us binary goals). Goals are used as inputs to design tools, and as part of the input to evaluation tools that assess whether a design meets a set of goals, compares observations against the goals, or compares two or more designs against a set of goals. They may include cost bounds, or other constraints on a design step, and (potentially) durations for when they will apply.
- *Predictions* (and their associated designs) are the anticipated outcomes of offering the target workload to a design for a storage system. They represent estimates of future observations, if such a design were to be implemented, and allow comparative evaluations of designs. A *design* is a proposed realization of a way to achieve a goal. It captures the notion of "what if ..."
- *Observations*, are descriptions of a system's behavior during some time interval. (An observation may or may not fit the original goal.)   There can be multiple

observations – the system might have been observed at different times, or with different sampling techniques.

These are obviously closely related. For example, our current storage design testbed takes measurements (observations) of a running system; feeds these as QoS specifications (goals) into a design step that attempts to optimize the resource usage in a running system while minimizing the amount of data movement required to do so. The result is a new design, whose likely performance we can predict.

What the Rome QoS specification does not include is the system objective function: essentially, what tradeoffs to make in the design process when faced with too few resources, or more than necessary. Determining – let alone specifying – the objective functions that system designers use is currently somewhat of a black art. The nearest that Rome gets to this is the notion of *utility functions*, which express the benefit to be received from achieving a particular value for a specification parameter. It also turns out to be necessary to introduce priorities, or ranks: for example, in order to describe the notion of "business critical" applications in the face of disasters such as site outages.

And we have discovered that the objective functions often vary during the design process: although people may start by asking "what is a minimum cost design?" they then often switch to "how well-balanced can I make the result?" or "how fast can I make it go, if we fix the budget?" This remains an area of active research for us.

4   Related Work

There is a great deal of work taking place on the use of QoS in designing systems (see, for example, the survey [1]). Most of the external academic work appears to be focused on network behavior; ours targets storage systems. Indeed, we deal with issues of network design only after the data placement decisions have been made. Part of this is because of the need to simplify the problem, but – perhaps more importantly – storage area networks (SANs) typically cost only a few percent of the total storage system cost, so simple over-provisioning works quite well.

As suggested in the introduction, even though storage workloads have many similarities to their network counterparts (burstiness, etc.), storage system workloads exhibit much greater disparity in their effective loads on the underlying system from behaviors like spatial locality, and the manner in which workloads interleave. We are not aware of other work addressing this issue in the same way as our approach.

We believe that the mapping between QoS goals and the design of the resulting system is itself a significant differentiator from most other work in this area. Although there have been a few examples of prior work in the storage system space, they have tended to assume very simple QoS models. Most work that we are aware of in the network space simply punts on the mapping issues – for example, the recent switch of emphasis to DiffServ in the IETF community merely pushes the design problem out to the people provisioning and using the network infrastructure. This probably makes sense for an environment where dynamic adaptation and congestion control with dropped packets is a very successful approach, but it doesn't seem to work for storage systems.

There is some work taking place in languages for specifying goals. For example, Bearden et al [2] discuss how goals might be represented in a CIM-like information model. Frølund and Koistinen [5] describe a rich language for expressing service level agreements (e.g., it has features to tackle the probabilistic comparisons that Rome's approx values support fairly simply). Both efforts focus a great deal of their expressiveness on describing things that it never occurred to us to write down. For example, both of these languages make a point of explicitly stating that if a response time goal is 100ms, then only response times that are less than 100ms are acceptable. Although this may make sense when very general contracts are being described, it seems to be less of a good idea in the rather narrower domain of storage systems – or networks, for that matter. Instead, we take such "goodness" comparisons as self-evident – or, if you prefer, intrinsic properties of the design tools that use them. (Try inverting the sense of a comparison to see why this seems reasonable!)

The Quo system [8] supports different operating regimes (perhaps similar to Rome's "outages"), but appears to describe them in terms of visible implementation decisions that applications can pick between, or ask to be notified about. This is at odds with our slogan: "tell us what you want to accomplish, not how to do it".

Some people find similarities between our goal-directed system design and policy-based system management. We beg to differ: most work on policy-based systems has been on policy *rules*, not policy *goals*. (A policy rule is a statement of the form *if* *<condition> then <action>*.) Languages such as Ponder [4] make such policy rule-based systems easier to describe, but they don't help a great deal in the mapping from higher-level goals down to selection of which mechanisms to exercise – such as which policy rules to enable.

5   Summary and Conclusions

Rome is an information model for capturing the important parts of the design and management problem for storage systems. Part of that model is a representation of QoS goals, predictions, and observations – together with the infrastructure to allow these to be turned into designs for storage systems to meet stated goals. QoS specifications for storage systems appear to need to be rather richer than their counterparts in the networking space, probably because of the much greater potential for non-linear performance interactions on mechanical storage devices, and caches. By combining the QoS specification system with the other portions of the storage system design problem, the Rome tool set can manipulate a common information model, which increases the ease with which a large set of functionality can be put together and developed incrementally.

The Rome 2 object model is quite simple – yet surprisingly powerful. Making an `objectType` a first class object enables a powerful, convenient attribute inheritance model. The freedom to add and override attributes has proven crucial to allow our tools to evolve gracefully, and has helped us avoid domino effects that often result from the traditional approach of hard-wiring an object's programmatic interface on a change. The result is a great deal of expressive power with relatively little overhead.

Rome is a living design: for example, it is actively evolving to encompass our improving understanding of the most important QoS attributes required to capture the nuances of new behavior patterns. (For example, we have only recently added file-level specifications to Rome.) It is also being actively extended in storage device modeling area: a topic for which space limitations prevented a discussion in this document. We believe that its inherent flexibility will allow these changes to be accommodated with relatively little difficulty.

Finally, we hope to make Rome publicly available for feedback and collaboration.

References

[1] Cristina Aurrecoechea, Andrew T. Campbell, and Linda Hauw. A survey of QoS architectures. *Multimedia Systems* **6**:138-151 (1998).

[2] M. Bearden, S. Garg and W. Lee. Integrating goal specification in policy based management. In *Proc. Policies for distributed networks and systems (Policy 2001)*, Bristol (Jan. 2001). Springer-Verlag *Lecture Notes in Computer Science* **1995**, pp 153-170.

[3] E. Borowsky, R. Golding, A. Merchant, L. Schreier, E.Shriver, M.Spasojevic, and J. Wilkes. Using attribute-managed storage to achieve QoS. Presented at *5th Intl. Workshop on Quality of Service*, Columbia University, New York (June 1997). Available from http://www.hpl.hp.com/SSP/papers/

[4] N. Damianou, N. Dulay, E. Lupu and M. Sloman. The Ponder specification language. In *Proc. Policies for distributed networks and systems (Policy 2001)*, Bristol (Jan. 2001). Springer-Verlag *Lecture Notes in Computer Science* **1995**, pp 18-38.

[5] Svend Frølund and Jari Koistinen. Quality of service specifications in distributed object system design. In *Proc. 4th USENIX Conf. on object-oriented technologies and systems. (COOTS)*, April 1998.

[6] R. Golding, C. Staelin, T. Sullivan, J. Wilkes. "Tcl cures 98.3% of all known simulation configuration problems" claims astonished researcher! Presented at the *Tcl Workshop*, New Orleans, May 1994. Available from http://www.hpl.hp.com/SSP/papers/

[7] M. E. Gómez and V. Santonja. Self-similarity in I/O workload: analysis and modeling. In *Workshop on Workload Characterization* (held in conjunction with the 31st annual ACM/IEEE International Symposium on Microarchitecture). Dallas, 1998.

[8] Joseph Loyall, Richard E. Schantz, John A. Zinky and David E. Bakken. Specifying and measuring quality of service in distributed object systems. In *Proc. of ISORC'98*, Kyoto, Japan (April 1998).

[9] John K. Ousterhout. *Tcl and the Tk toolkit*. Addison-Wesley, Professional Computing series (April 1994).

[10] Chris Ruemmler and John Wilkes. UNIX disk access patterns. *Proceedings of the Winter'93 USENIX Conference*, pages 405-420 (January 1993).

[11] Chris Ruemmler and John Wilkes. An introduction to disk drive modeling. *IEEE Computer* **27**(3):17-28, March 1994.

[12] Elizabeth Shriver. A formalization of the attribute mapping problem. HP Laboratories technical report HPL–SSP–95–10 rev. D, (July 1996), available from http://www.hpl.hp.com/SSP/papers/

[13] Elizabeth Shriver, Arif Merchant and John Wilkes. An analytic behavior model for disk drives with readahead caches and request reordering. *Proceedings of SIGMETRICS'98* (Madison, WI, June 1998).

[14] Transaction Processing Performance Council. *TPC benchmarks: standard specifications*. Available from http://www.tpc.org

[15] John Wilkes and Raymie Stata. Specifying data availability in multi-device file systems. Position paper for the *4th ACM-SIGOPS European Workshop* (Bologna, Italy, 3-5 September 1990), published as *Operating Systems Review* **25**(1):56-59, January 1991.

[16] John Wilkes, Richard Golding, Carl Staelin, and Tim Sullivan. The HP AutoRAID hierarchical storage system. *ACM Transactions on Computer Systems* **14**(1):108-136, February 1996.

[17] John Wilkes. The Pantheon storage-system simulator. HP Laboratories technical report HPL–SSP–95–14 (rev. 1, May 1996), available from http://www.hpl.hp.com/SSP/papers/

Extending a Best-Effort Operating System to Provide QoS Processor Management

Hans Domjan and Thomas R. Gross

Departement Informatik
ETH Zürich, CH–8092 Zürich
hans.domjan@ethz.ch, thomas.gross@ethz.ch

Abstract. The benefits of QoS network features are easily lost when the end-nodes are managed by a conventional, best-effort operating system. Schedulers of such operating systems provide only rudimentary tools (like priority adjustment) for processor management. We present here a simple extension to a processor management system that allows an application to reserve a share of the processor for a specified interval. The system is targeted at applications with frequently changing resource demands or recurring, though non-periodic resource requests. An example of such an application is a network-aware image search and retrieval system, but other network-aware client-server applications also fall into the same category. The admission control component of the processor management system decides if a resource request can be satisfied. To limit the amount of time spent negotiating with the operating system, the application can present a ranked list of acceptable reservations. The admission controller then picks the best request that can still be satisfied (using the Simplex linear programming algorithm to find the best solution). If there are insufficient resources, the application must deal with the shortage. Any possible adaptation (if the accepted request was not the application's first choice) is left to the application. The processor management system has been implemented for NetBSD and been ported to Linux, and the paper includes an evaluation of its effectiveness. The overhead is low, and although reservations are not guaranteed, in practical settings the application almost always obtains the cycles requested.

1 Introduction

Many networks include either provisions or proposals to provide network services with defined QoS properties. Even if the QoS property cannot be guaranteed (in the sense that the network will ensure the properties even in the presence of catastrophic failures), nevertheless services with QoS properties aid tremendously in the construction of applications with defined timing behavior. E.g., a video conferencing system may use a bandwidth reservation to send voice and image data.

However, network QoS is just one aspect of providing true end-to-end QoS properties. The benefits of network QoS are easily lost if the end-node operating system does not cooperate. A conventional "best-effort" operating system (like many variants of Unix) provides only simple tools to assign appropriate processor cycles to applications with QoS network connections. Techniques such as boosting the priority or

L. Wolf, D. Hutchison, and R. Steinmetz (Eds.): IWQoS 2001, LNCS 2092, pp. 92–106, 2001.

over-provisioning of resources work if there is a single application, but cannot easily be extended to realistic scenarios [30]. Therefore, the operating system must allow an application to request some QoS. Such requests are usually for processor cycles, although other kinds of resources may be of importance as well.

This paper presents a simple extension to conventional best-effort operating systems which allows an application to request CPU resources for a time interval. Such requests are made at the time the demand is established, not in-advance [31]. This OS interface is geared towards applications that have recurring but non-periodic requests. An example of such an application is a client that presents the result of some query to the user. Queries are handled by a server and differ in the amount of data that have to be transferred to the client. If the client wants to present the result of the server-side query (e.g., a set of images or a video clip) within a fixed time (to offer predictable response time), then the amount of CPU resources needed by the server will be function of the volume and complexity of the data. Another type of applications that can benefit from QoS processor management are network-adaptive applications that are able to trade off one kind of resources (e.g., network bandwidth) for other resources (e.g., CPU cycles). E.g., network-aware adaptive applications can reduce their bandwidth requirements by transcoding (compressing) the data to be transmitted. However, such a transformation must be done within a specific time limit – the data must be transcoded when the communication subsystem is ready to transmit.

This paper is organized as follows: Section 2 presents the application model and the implications for a QoS CPU management system. Section 3 discusses the overall structure of the CPU management system. Section 4 contains the evaluation of this approach using three different real world applications. Related work is discussed in Section 5, and Section 6 contains the conclusions.

2   Application Model and Implications for CPU Management

2.1   Application Model

The proposed processor management system is targeted mainly at client-server type applications and provides short-notice, recurring and dynamic resource reservations at runtime. Applications *not* within this domain consist of multimedia applications (periodic resource requirements), real-time systems (static, hard resource requirements) and applications from the realm of video on demand or telecommunications, where resource reservation and allocation is typically made once at the beginning of a connection and remains in effect for the whole life time of the session.

In the model, the application must produce its result within a (user- or system-provided) time frame. The preparation of the result can be divided into one or several subtasks (which have their own time constraints inferred from the overall time frame), and there may be several algorithms (with differing resource requirements) at the application's disposal to carry out each subtask. Thus, reservation (and adaptation) decisions are not made only once at the beginning of the task, but can be recurring to take fluctuations of resource availability and demand into account.

If different algorithms impose different CPU requirements, then the application may choose the best algorithm that has a CPU requirement that can be satisfied by the processor management system. To cut down on the overhead of negotiating with the operating system, the application presents a ranked list of CPU requirements. This list contains the CPU demands for all possible algorithm options. The OS then decides which option is admissible, based on the option's resource requirements and overall system resource availability. The CPU requirements are expressed as a request for a number of cycles within a specific interval. The length of the interval is determined by an estimate of the corresponding subtask's length, and the interval may start either "now" or sometime in the future. The OS notifies the application which option can be admitted so that the application can take appropriate action. As long as the OS delivers the requested cycles in the interval, the application needs are satisfied. The OS thus is free to deliver all the requested resources at the beginning, evenly distributed, or at the end of the interval.

2.2   Implications for CPU Management

The precise resource requirements of the applications, as well as the number and timing constraints of their subtasks are unknown in advance; the subtask's resource reservations are recurring, though non-periodic in time. These details depend on many factors known only at run-time, like user input or the details of the available network QoS. Additionally, the application makes several reservation decisions while processing a task to take changes in resource availability into account, and may switch between best-effort mode for non time-critical administrative work and reserving mode for time-critical productive work. Furthermore the number of reserving applications competing for end system resources may vary, and even a single application may be multi-threaded. Finally, it is unrealistic to devote a whole host to support only one single application. The thread that produces a result should be given the CPU resources it needs while allowing other threads to proceed as far as possible. A CPU resource reservation in a best-effort OS is a good approach to address these characteristics because reservations provide a predictable QoS for applications that need it, yet accommodate conventional applications without modifications. Therefore a dynamic scheduler that accepts *reservation* requests at runtime, (re-)calculates the schedule on-the-fly and accommodates best-effort processes as well is a viable method for supporting this end-system QoS.

Reserving applications are driven by a model of their network and end system resource requirements, and estimates of future resource availability, both of which may not always be completely accurate. Thus, mismatches between predicted and actual resource consumption can be expected, and the processor management system should be able to handle such situations. *Over-reservation* is not a problem for a particular application, since it will receive the allocated resources anyway, but over-reservation by many applications degrades the overall end system throughput noticeably. *Under-reservation*, on the other hand, is more serious for an application and should be handled gracefully.

Since applications are allowed to use all end system operating system features, they may block on I/O or other events. The OS should be able to handle this case so that the reserved application that has blocked is later allowed to catch up the backlog without delaying other processes with reservations.

3 The R-Scheduler: A Pragmatic Approach to End System QoS

The abstraction provided by the R-Scheduler reflects the boundary conditions introduced in the previous section: To make a resource reservation, the application provides the OS with a vector of reservation requests. Each reservation request R_i consists of a pair (I_i, C_i) where I is an interval (defined by its *Start* and *End* time, where *Start* can be either "now" or in the future) and C indicates how many cycles this application wants to obtain in the specified interval. How and when the CPU is actually allocated within the interval I remains opaque to the application. The submitted reservation requests are sorted by the application in decreasing preference.

To determine which requests are feasible, the admission controller processes the vector by solving a system of linear equations that capture the resource requests for all time intervals, and then picks 0 or 1 of the individual reservation requests [10]. The application is informed about the admission controller's choice and may then take an appropriate action, like executing the algorithm with a resource consumption that corresponds to the granted request. A process that has been granted a reservation request $R(I, C)$ is referred to as an R-process.

The resource demands C_i are often only estimates, and under-reservations may pose a problem to the application. To keep the R-Scheduler simple (and low-overhead), the R-Scheduler gives preference to R-processes with under-reservation over best-effort processes instead of a notification of the application and a possible re-negotiation of a reservation. On the other hand, in case of over-reservation, the R-process can yield no longer needed resources by means of a system call and make them available for new reservation requests. A detailed resource accounting scheme ensures that blocking R-processes are allowed to catch up their delay at the expense of best-effort processes, but not other R-processes.

The ability to obtain reservations is offered as an additional service to the user. Therefore all conventional, best-effort-type applications can be run unmodified on the system; only applications that want to take advantage of the reservations must be programmed accordingly. To ensure a minimal progress of best-effort applications in the system, only a pre-set fraction of the CPU is made available to R-Processes.

The R-Scheduling system has been implemented for the NetBSD [16] operating system and comprises of about 4200 lines of C code. In the actual kernel, only a few lines needed to be modified, mainly for the introduction of miscellaneous call-backs. In addition, the uneventful port of the R-Scheduler to Linux confirmed our assessment that adding such a scheduler to any reasonable operating system is straightforward.

4 Practical Experience

This section presents a comprehensive comparison of the QoS-enabled operating system (NetBSD) with the standard best-effort system. We report data for usage scenarios of three example applications, namely a resource-aware Internet server, a distributed image search and retrieval system, and an adaptive image decoder. All experiments were carried out on a 200MHz Intel Pentium Pro PC with 128MByte of RAM running NetBSD version 1.3 in an out-of-the-box configuration. We chose what is today a low-end PC to demonstrate that it is possible also by modest platforms to provide application-beneficial QoS. In the experiments, at most 90% of the CPU are made available for R-processes, unless stated otherwise. The graphs show the mean values of five experiment repetitions, and error bars denote $\pm\sigma$.

Microbenchmarks have shown that the overall overhead of the admission controller and scheduler is modest even if the system is considerably loaded (below 0.7% with an average of four requests per second submitted and calculated).

4.1   Resource-Aware Internet Server

Description: The first example is a resource-aware Internet server. The clients submit requests to the server and expect a reply within a certain user time limit. Given that time limit, and the estimated RTT between server and client, the server has a request processing window at its disposal within which to generate the reply (Figure 1(a)). Assuming that the client prefers a prompt request rejection notification instead of a response later than the sum of the user time limit and a time limit tolerance, and that the server-side reply generation is a CPU-bound task, resource reservation in the server is a viable solution to achieve this desired quality of service. As an alternative to request rejection, the server may redirect the request to another server and thus use the resource reservation mechanism for predictable load balancing.

Evaluation: Upon arrival of a new request, the main server thread immediately forks off a new child process to handle that request. Under the best-effort OS, the child always tries to process the request. It either manages to finish before the end of the window (in this case a "success"-message is sent back to the client), or it exceeds the window, discards the result and sends back a "fail"-message. With the QoS-enabled OS, the child tries to allocate the CPU before processing; and in case of a rejection due to over-reservation it immediately sends back the "fail"-message to the client without any further attempt to process the request. Under both OSes, the main server thread can only be scheduled in best-effort mode due to the random and thus unpredictable request arrival.

Both servers are subjected to four different request streams of 100 requests each (with exponentially distributed interarrival time with a mean of 200 ms, 500 ms, 1000 ms and 1500 ms respectively [1]; a user time limit of 5 sec [5], a time limit tolerance of 10%, a gamma distribution of RTTs according to [8, 20] and an exponentially distributed request processing time with a mean of 3000 ms (similar to [9]). The quality metric is the number of requests processed successfully within the time limit and tolerance.

Figure 1(b) shows the number of requests processed successfully for both OSes as a function of request interarrival time (for the QoS-enabled OS, the maximum percentage of the CPU allotted to reservations is given in the graph; see next paragraph for explanation). Performance increases with larger interarrival time because it leaves a larger overall timeframe for the server to process the identical requests. The QoS-enabled OS performs consistently better, but its advantage diminishes with less contention for resources (i.e., increasing interarrival time).

Figure 1(c) shows the request distribution as a function of the percentage of the CPU dedicated to R-processes, i.e., reservations, in the QoS-enabled OS. On one hand,

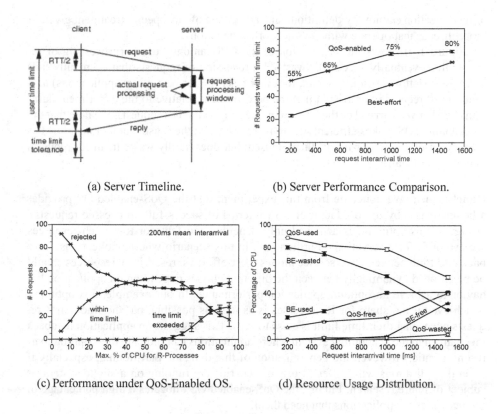

(a) Server Timeline.

(b) Server Performance Comparison.

(c) Performance under QoS-Enabled OS.

(d) Resource Usage Distribution.

Fig. 1. Resource-Aware Internet Server.

if only a small proportion of the overall resources (i.e., less than 20% of the CPU) is allocated to reservations, the QoS-enabled OS performs worse than the best-effort OS because the latter can use up to 100% of the CPU, whereas with such a configuration of the QoS-enabled OS a large percentage of the CPU must not be used by reservations, and thus too many requests are rejected. On the other hand, if too many resources (e.g., more than 70%) are allocated to reservations, the performance of the QoS-enabled OS starts to decrease and eventually becomes worse than that of the best-effort OS again. This behavior is due to the hybrid nature of the application where the main thread accepting connections is always running in best-effort mode, whereas the child thread actually processing the request is in reserved mode. Thus, if too many resources are allotted to (and used by) R-processes, the main thread has hardly any chance to run. The tasks of accepting the connection and forking off the child are delayed, and then there may not be enough time to provide the reply in time.

The percentage of resources that should be devoted to best-effort threads depends on the application scenario. These data, however, illustrate a subtle problem that may also be an issue for other processor management schemes that attempt to support network services with QoS properties. Administrative activities that establish the QoS properties of a connection cannot, by definition, take advantage of any special treatment given to those threads that operate with QoS network connections.

Figure 1(d) shows the distribution of the CPU among "used" (processing of requests that eventually succeed), "wasted" (processing of requests that eventually fail) and "free" (leftover resources, e.g., for other reserving or best-effort applications) as a function of request interarrival time; the percentage is relative to the experiment duration (i.e., between arrival of the first and processing of the last request). We note that the QoS-enabled OS makes efficient use of the resources in the sense that it either allocates them for useful work, or leaves them unused; but does hardly waste them for requests that eventually fail.

Conclusions: We conclude from this experiment that the QoS-enabled OS provides a better service by up to a factor of 2.3 (in terms of successfully processed requests). However, this optimum is achieved only if neither too few nor too much resources are available for reservations. Furthermore, in this scenario where applications move back and forth between two service modes (best-effort vs. reserving), resources should be partitioned dynamically between the different classes depending on application behavior to achieve an optimal application performance. In our example, this optimum is achieved if the share for R-processes is as high as possible, but the fraction of requests exceeding their time limit is close to zero. This suggests an application feedback mechanism to the scheduling system where applications can indicate their optimal partitioning ratio. However, the generalization of this dynamic partitioning, especially if several applications with different usage scenarios are running on a single system, is subject of future research. Finally, the QoS-enabled OS allocates a high percentage of the resources to applications that need them.

4.2   Distributed Image Search and Retrieval System

Description: The second example is a distributed image search and retrieval system that attempts to adapt its behavior in response to changes in network resource availability [7, 29]. A client formulates a query for images, the system's search engine identifies matching images, and the adaptive servers deliver the images in the best possible quality, considering network performance, system load, and a user-specified time limit.

The goal of the adaptation is to meet the user-specified limit on delivery time while maximizing the content quality of the images delivered. Content is correlated with size, so the system attempts to use its available bandwidth as well as possible. Therefore, while one thread transmits an object, concurrently a different thread prepares (transcodes) the next object(s) for transmission. To maximize the utilization of the available network bandwidth, the prepare thread should always have an object ready for transmission when the transmit thread can take another object. Therefore the application associates with each prepare step a deadline for completion that is derived from the model's estimate of the duration of the concurrent transmission step.

This scheme of adaptation is not limited to this particular application but can be applied successfully to many network-aware applications with request-response communication. The core mechanisms have in fact been factored into a framework for network-aware applications [7].

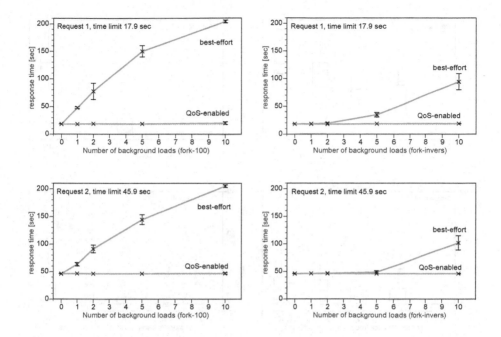

Fig. 2. Single Requests with Varying Background Loads.

Evaluation: For all experiments, the image server executes once under the best-effort OS, and once under the QoS-enabled OS. The same request for 25 images is used. Additionally, a varying number (0, 1, 2, 5, 10) of two different types of background loads is imposed. The aggressive load "fork-100" forks off a child every 100 ms and lets it run for 100 ms. The lighter load "fork-invers" forks off children with a more realistic lifetime according to [13]. The choice of forking background loads is justified since in a typical server scenario, new processes are created to handle client requests.

Experiment 1: One Server; Single Request. For this experiment, two different time limits are used: Request 1 specifies 17.90 sec (which yields overall image prepare costs corresponding to 26% average CPU usage), and Request 2 has a time limit of 45.9 sec (11% average CPU usage).[1]

Figure 2 plots the response time of the image retrieval system, i.e., its ability to meet the user-specified time limit (dotted line), as a function of the number of actually imposed background loads. *Without* background load, there is no significant difference between the two OSes, because on one hand there are ample resources in the system, and on the other hand the QoS-enabled OS does not incur any noticeable system overhead. *With* background load, the situation is different. Under the best-effort OS, the server's performance continually degrades as the number of background loads increases. The "fork-100" background load has a considerably more severe impact than "fork-invers".

---

[1] The time limits are chosen so that for Request 1 a high image reduction ratio and for Request 2 a low reduction ratio is obtained. For more details, see [6].

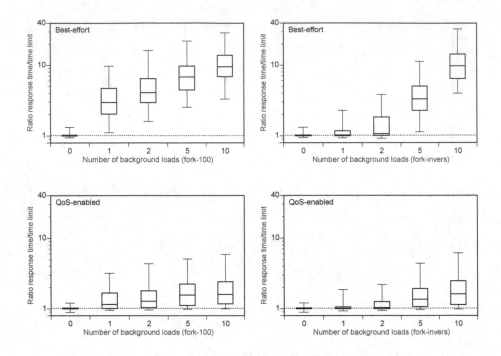

Fig. 3. Random Requests with Varying Background Loads; Distribution of Response Time.

Request 2 with the tighter time limit and thus higher CPU usage is more affected than Request 1. This is because the best-effort OS distributes the resources evenly among all competing applications. Under the QoS-enabled OS, however, the response time remains largely unaffected by both the kind and number of background loads, and additionally shows little variance. Thus the QoS-enabled OS is well able to provide a predictable service for reserving applications at the expense of best-effort applications.

An interesting observation is that using the given configuration, it would have taken 102 sec to transmit *all* the images *untransformed*, i.e., in their full size. The reason that it takes considerably more than 102 sec in the cases of the "fork-100" background load 5 and higher is twofold: On one hand, the simple cost model does not take into account that also the transmission of the images consumes little, albeit an under such background load not negligible amount of processing power. On the other hand, the simple CPU availability estimator fails in the case of this background load which consumes a disproportionate amount of CPU compared to pure compute-bound workloads. We conclude that under the QoS-enabled OS, the image server's working range with the simple cost and load model can be extended to work also with aggressive background loads.

Experiment 2: Servers with Multiple Random Requests. In this experiment, the server is subjected to 50 randomly generated requests with an exponentially distributed interarrival time (mean 10 sec). The time limit is also exponentially distributed between

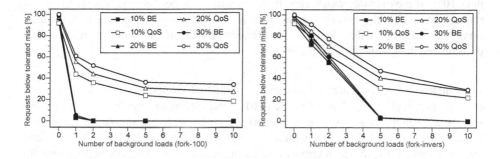

Fig. 4. Random Requests with Varying Background Loads; Percentage of Requests below Tolerated Time Limit Miss.

8.5 sec and 102 sec (mean 36.6 sec). For the given server configuration, these two values denote the two boundary cases where all images must compressed to minimal quality, or not reduced at all, respectively. This kind of requests produces an overall end system load with a dynamically varying number of concurrently running servers (from 0 to 6) and is designed to reflect a real server situation with bursts of requests.

Figure 3 shows the distribution of the aggregated ratio[2] $\frac{\text{Achieved Response Time}}{\text{User Imposed Time Limit}}$ (with a log-scaled y-axis). The top, bottom and line through the middle of the box correspond to the 75th, 25th and 50th percentile, respectively. The whiskers on the bottom extend from the 10th and top 90th percentile. Without background load, the response time under the QoS-enabled OS is slightly better than for the best-effort OS because the former has, due to the reservations, a more precise knowledge about the application's exact resource requirements and is thus better able to allocate resources at the right time to the appropriate application. As expected, performance degrades with increasing severity and number of the background loads. For the "fork-100" load, the QoS-enabled OS performs better by almost an order of magnitude.

In Figure 4 the "end-user" view is presented. This figure shows the percentage of requests that are handled on time, i.e., within the user-specified time limit plus a tolerances value of 10%, 20%, and 30%, respectively. Since the QoS-enabled OS does not provide hard guarantees, it may exceed a reservation's time limit it by a few %; just reporting "success" or "failure" might present an inaccurate picture of the system's capabilities. Note that the success to meet the time limit does not only depend on the end system OS but also on the application and its ability to adapt to changing network conditions.

Figure 4 shows that the best-effort OS is rarely able to allow the application to finish within the limit. With the "fork-100" background load, the failure is almost complete; for "fork-invers" there is a continuous decline as the load increases up to 5 background processes. A higher load implies that only a few percent of the requests succeed. With the QoS-enabled OS, however, between 43% and almost 18% (for a number of background loads between 1 and 10) of all requests do not exceed their time limit by more

---

[2] A value $X < 1$ thus means that the actual delivery was finished before the user-imposed time limit.

than 10%. If the user tolerates a time limit miss of 30%, between 61% and 35% of all requests are within this bound. For the "fork-invers" background load, the R-Scheduled server performs better than the best-effort scheduled one, although the difference is less pronounced than with "fork-100".

Conclusions: We conclude from the experiments that the QoS-enabled OS is able to effectively shield applications from the detrimental effects of any kind of background load. Furthermore, the image server's working range with the simple cost and load model can be extended to work with a larger variety of adversary loads. Finally, end user satisfaction (expressed in terms of time limit miss) is considerably better than with the best-effort OS despite the lack of hard guarantees.

4.3   Image Decoder

Description: The third example is an adaptive image decoder based on the Berkeley software MPEG–1 player [22]. An MPEG movie consists of three different frame types, namely self-contained I-frames, P-frames that depend on the pervious and/or next I-frame(s), and B-frames that depend on the previous and/or next I- and P-frames. A simple way of adaptation to available end system resources is to decode the movie at one of three quality levels corresponding to the decoding of all frames (IPB), IP-frames, and I-frames only. The decoder determines the decoding resource requirements of a future movie sequence (whose length should be on the order of seconds) in the three different quality levels based on linear regression of the frame sizes [4] and submits this vector to the scheduler that decides, based on the available CPU resources, which level can be decoded. Subsequently, the frames are decoded into a buffer, from which they are displayed using any periodic scheduler with the specified movie frame rate.

Due to the large variability in decoding time characteristics as well as fluctuating end system resource availability, it makes little sense to reserve a certain bandwidth of CPU for the whole movie in advance (this might even be impossible in case of a live broadcast with undetermined end time), but resource reservations must continually be reconsidered.

Evaluation: The goal of this experiment is to show that the QoS-enabled OS provides effective support for dynamically adapting applications in a resource contention situation. The movie chosen for the experiments has a play time of 16.35 sec, and uses an average CPU bandwidth of 38% for decoding all frames (IPB), 17% for I and P frames, and 4% for the I frames alone. Under the QoS-enabled OS, the decoder allocates resources for an interval of 1 sec with the four levels IPB, IP, I and one I. Under the best-effort OS, the decoders run in different static adaptation configurations.

Table 1 shows the quality metrics "decoding time" and "percentage of decoded frames" for the different configurations. For two parallel decoders, both OSes perform equally well since there are enough resources to decode both movies. With increasing number of decoders, we note that to decode all frames, more time than the desired decoding time of 16.35 sec is needed. On the other hand, for a particular statically configured quality level, the desired decoding time is met, but at the cost of dropped frames.

Table 1. Image Decoder Results.

#Decoders OS types (static configuration)	Overall CPU (%)	time (16.35 sec)	frames decoded (%) I	P	B
2 Best-effort OS (IPB/IPB)	77	16.30	100	100	100
QoS-enabled OS	75	16.09	100	100	100
3 Best-effort OS (IPB/IPB/IPB)	100	18.45	100	100	100
QoS-enabled OS	95	16.19	99	98	76
Best-effort OS (IPB/IPB/IP)	94	16.33	100	100	66
4 Best-effort OS (IPB/IPB/IPB/IPB)	100	24.37	100	100	100
QoS-enabled OS	95	16.32	97	94	40
Best-effort OS (IPB/IPB/IP/I)	98	16.75	100	75	50
Best-effort OS (IPB/IP/IP/IP)	91	16.30	100	100	25
5 Best-effort OS (IPB/IPB/IPB/IPB/IPB)	100	30.66	100	100	100
QoS-enabled OS	88	16.05	100	100	0
Best-effort OS (IPB/IPB/I/I/I)	89	16.30	100	40	40
Best-effort OS (IPB/IP/IP/IP/I)	95	16.44	100	80	20
Best-effort OS (IP/IP/IP/IP/IP)	87	16.30	100	100	0
6 **QoS-enabled OS**	95	16.26	98	92	0
Best-effort OS (IPB/IP/IP/IP/I/I)	95	16.50	100	67	17
Best-effort OS (IP/IP/IP/IP/IP/I)	91	16.30	100	83	0
7 **QoS-enabled OS**	86	16.05	99	64	0
Best-effort OS (IPB/IP/IP/IP/I/I/I)	100	16.62	100	57	14
Best-effort OS (IPB/IP/IP/I/I/I/I)	90	16.34	100	43	14
Best-effort OS (IP/IP/IP/IP/IP/I/I)	96	16.54	100	71	0

Conclusions: The main conclusions from this experiment are that the QoS-enabled OS supports applications in a graceful dynamic degradation of the quality if there is resource contention, without prior knowledge about end system utilization. Furthermore, the QoS-enabled OS permits applications to dynamically achieve a quality level comparable to static adaptation policies. The modifications necessary to turn the static image decoder into an adaptive one were modest, adding evidence that it is both feasible and worthwhile to change applications to use the features of the QoS-enabled OS.

4.4   Experiment Conclusions

The experiments in this section have shown that the QoS-enabled OS

- provides a superior service for applications (in terms of end user metrics) which is consistently visible in a number of usage scenarios with one or more QoS-aware applications, as well as a variation of background loads.
- allocates resources effectively to applications that need and can make use of them.
- extends the working range of adaptive applications having simple resource models.
- supports applications in dynamic adaptation in case of resource contention.
- can be integrated with minimal effort, and adds negligible runtime overhead to a best-effort OS.
- provides an easy-to-use interface for a variety of applications.

Furthermore, we have identified the issue of dynamic application-adaptive resource partitioning among best-effort and reserving resource usage classes as an issue of future research.

5 Related Work

CPU reservations are central to real-time systems that schedule a fixed set of periodic, independent, non-blocking tasks with known, constant execution times [15, 24, 25]. In contrast to real-time systems, the resource requirements of the target applications are dynamic, aperiodic and known in detail only shortly before the resources are needed, and not necessarily contiguous. Additionally, there are a dynamically varying number of best-effort processes and processes with reservations that have to be scheduled. Furthermore, the tasks may interact or block on I/O.

Schedulers for multimedia systems (like Processor Capacity Reserves [17, 18, 19], SMART [21], the Rialto Scheduler [14], ETI [3]) pose less stringent restrictions on the scheduled process set than real-time systems and can accommodate a dynamically changing number of processes with varying resource requirements, but they too are mainly targeted at supporting tasks with periodic resource demands. Furthermore, some schemes enforce adaptivity by dropping single periods — a behavior that is application-specific and does not fit the application model considered here.

Proportional-share schedulers like *lottery* and *stride scheduling* are resource allocation mechanisms providing efficient, responsive control over the relative execution rates of computations [12, 26, 27, 28]. In contrast to the proportional share model, the system presented here offers absolute, time-bounded resource reservations which can be contracted with the system in advance.

Conventional operating systems like UNIX [23] use dynamic priorities with decay-usage scheduling. This scheme gives high responsiveness for I/O-intensive applications and prefers them over long-running CPU-bound processes making it a good choice for interactive systems [2, 11, 16]. They do not, however, provide resource reservations or other means of predictable resource availability.

6 Conclusions

This paper presents the evaluation of a low-cost, pragmatic approach to processor management based on resource reservations. The R-Scheduler consist of an interface for applications to negotiate their future needs and a reservation-based scheduling scheme that prevents system overload. This R-Scheduler, which extends the notion of QoS from the network into the OS, co-exists with a best-effort scheduler and has been implemented with modest effort for NetBSD and ported to Linux.

Experiments with three different applications identify CPU scheduling as a key success factor for the effectiveness of a QoS-enabled OS. They show that such a QoS-enabled OS provides adequate support for resource-aware applications. Applications that attempt to limit their response time perform considerably better when using reservations than when execution is controlled by a traditional best-effort OS. Finally, the effort required to make applications take advantage of the features of the QoS-enabled OS as well as to add those features to any reasonable operating system is modest.

As the importance of network-aware adaptive applications and differentiated services increases, operating systems are challenged to provide low-cost support for the CPU resource reservations. The scheduler presented here provides an approach that can be easily integrated into existing systems and can co-exist with current best-effort scheduling disciplines.

References

[1] M. F. Arlitt and C. L. Williamson. Internet web servers: Workload characterization and performance implications. *IEEE/ACM Transactions on Networking*, 5(5):631–645, Oct. 1997.

[2] M. J. Bach. *The Design of the UNIX Operating System*. Prentice Hall, 1986.

[3] M. Baker-Harvey. ETI resource distributor: Guaranteed resource allocation and scheduling in multimedia systems. In *Proceedings of the Third Symposium on Operating Systems Design and Implementation (OSDI '99)*, pages 131–144, Feb. 1999.

[4] A. C. Bavier, A. B. Montz, and L. L. Peterson. Predicting MPEG execution times. In *Proceedings of the SIGMETRICS '98/PERFORMANCE '98 Joint International Conference on Measurement and modeling of Computers Systems, June 22-26, 1998; Madison WI*, pages 131–140, May 1998.

[5] N. Bhatti, A. Bouch, and A. Kuchinsky. Integrating user-perceived quality into web server design. Technical Report HPL-2000-3, Internet Systems and Applications Laboratory, HP Laboratories Palo Alto, Jan. 2000.

[6] J. Bolliger. *A framework for network-aware applications*. PhD thesis, ETH Zürich, Apr. 2000. No. 13636.

[7] J. Bolliger and T. Gross. A framework-based approach to the development of network-aware applications. *IEEE Transactions on Software Engineering (Special Issue on Mobility and Network-Aware Computing*, 24(5):376–390, May 1998.

[8] R. L. Carter and M. E. Crovella. Dynamic server selection using bandwidth probing in wide-area networks. Technical Report BU-CS-96-007, Computer Science Department, Boston University; 111 Cummington St., Boston MA 02215, Mar. 1996.

[9] J. Dilley. Web server workload characterization. Technical Report HPL-96-160, Hewlett-Packard Laboratories, Dec. 1996.

[10] H. Domjan and T. R. Gross. Providing resource reservations for adaptive applications in a best-effort operating system. Technical report, ETH Zürich, Feb. 2001.

[11] B. Goodheart and J. Cox. *The Magic Garden explained: the Internals of UNIX System V Release 4, an Open-Systems design*. Prentice Hall, 1993.

[12] P. Goyal, X. Guo, and H. M. Vin. A hierarchical CPU scheduler for multimedia operating systems. In *Proceedings of the Second USENIX Symposium on Operating Systems Design and Implementation (OSDI). 28-31 Oct. 1996; Seattle, WA, USA*, pages 107–121, Oct. 1996.

[13] M. Harchol-Balter and A. B. Downey. Exploiting process lifetime distributions for dynamic load balancing. In *Proceedings of the 1996 ACM SIGMETRICS Conference*, pages 13–23, 1996.

[14] M. B. Jones, D. Roşu, and M.-C. Roşu. Cpu reservations and time constraints: Efficient, predictable scheduling of independent activities. In *Proceedings of the Sixteenth ACM Symposium on Operating System Principles (SOSP-16), Saint-Malo, France, October 5–8*, pages 198–211, Oct. 1997.

[15] C. L. Liu and J. W. Layland. Scheduling algorithms for multiprogramming in a hard-real-time environment. *Journal of the Association of Computing Machinery*, 20(1):46–61, Jan. 1973.

[16] M. K. McKusick, K. Bostic, M. J. Karels, and J. S. Quarterman. *The Design and Implementation of the 4.4BSD Operating System*. Addison-Wesley Publishing Company, 1996.

[17] C. W. Mercer. *Operating System Resource Reservation for Real-Time and Multimedia Applications*. PhD thesis, School of Computer Science, Carnegie Mellon University Pittsburgh, PA 15213-3890, June 1997.

[18] C. W. Mercer, S. Savage, and H. Tokuda. Processor capacity reserves: An abstraction for managing processor usage. In *Proceedings of the Fourth Workshop on Workstation Operating Systems*, Oct. 1993.

[19] C. W. Mercer, S. Savage, and H. Tokuda. Processor capacity reserves: Operating system support for multimedia applications. In *Proceedings of the IEEE International Conference on Multimedia Computing and Systems*, pages 90–99, May 1994.

[20] A. Mukherjee. On the dynamics and significance of low frequency components of internet load. *Internetworking: Practice and Experience*, 5(4):163–205, Dec. 1994.

[21] J. Nieh and M. Lam. The design, implementation and evaluation of SMART: A scheduler for multimedia applications. In *Proceedings of the 16th ACM Symposium on Operating System Principles, St. Malo, October 1997*, pages 184–197, Oct. 1997.

[22] K. Patel, B. C. Smith, and L. A. Rowe. Performance of a software MPEG video decoder. In *Proceedings of the First International Conference on Multimedia; 1–6 August 1993; Anaheim, CA, USA*, pages 75–82. ACM, Aug. 1993.

[23] D. M. Ritchie and K. Thompson. The UNIX time-sharing system. *Communications of the ACM*, 17(7):365–375, July 1974.

[24] L. Sha, R. Rajkumar, and J. P. Lehoczky. Priority inheritance protocols: An approach to real-time synchronization. *IEEE Transactions on Computers*, 39(9):1175–1185, Sept. 1990.

[25] B. Sprunt, L. Sha, and J. Lehoczky. Aperiodic task scheduling for hard-real-time systems. *The Journal of Real-Time Systems*, 1(1):27–60, June 1989.

[26] I. Stoica, H. Abdel-Wahab, K. Jeffay, S. K. Baruah, J. E. Gehrke, and C. G. Plaxton. A proportional share resource allocation algorithm for real-time, time-shared systems. In *Proceedings of the 17th IEEE Real-Time Systems Symposium*, pages 288–299, Dec. 1996.

[27] C. A. Waldspurger and W. E. Weihl. Lottery scheduling: Flexible proportional-share resource management. In *Proceedings of the First Symposium on Operating Systems Design and Implementation*, pages 1–11, 1994.

[28] C. A. Waldspurger and W. E. Weihl. Stride scheduling: Deterministic proportional-share resource management. Technical Memorandum MIT/LCS/TM-528, MIT Laboratory for Computer Science, Cambridge, MA 02139, June 1995.

[29] R. Weber, J. Bolliger, T. Gross, and H.-J. Schek. Architecture of a networked image search and retrieval system. In *Proceedings of the ACM Conference on Information and Knowledge Management (CIKM'99)*, pages 430–441, Nov. 1999.

[30] L. C. Wolf. *Handbook of Multimedia Computing*, chapter Resource Management in Multimedia Systems, pages 891–912. CRC Press, Boca Raton, FL, USA, 1998.

[31] L. C. Wolf, L. Delgrossi, R. Steinmetz, S. Schaller, and H. Wittig. Issues of reserving resources in advance. In *Proceedings of the Fifth International Workshop on Network and Operating System Support for Digital Audio and Video; April 19–21, 1995; Durham, New Hampshire, USA*, Apr. 1995.

User Focus in Consumer Terminals
and Conditionally Guaranteed Budgets

Reinder J. Bril and E. (Liesbeth) F.M. Steffens

Philips Research Laboratories Eindhoven (PRLE), Prof. Holstlaan 4
5656 AA Eindhoven, The Netherlands
{Reinder.Bril,Liesbeth.Steffens}@philips.com

Abstract. Media processing in software enables consumer terminals to become open and flexible. Because consumer products are heavily resource constrained, this processing is required to be cost-effective. Our QoS approach aims at cost-effective media processing in software. QoS resource management is based on multilevel control, corresponding to different time-horizons, and resource allocation below worst-case using periodic budgets provided by a budget scheduler.

Multilevel control combined with budgets below worst-case gives rise to a problem related to user focus. Upon a sudden increase in load of an application with user focus, its output will have a quality dip. To resolve this user focus problem, we present the novel concept of a *conditionally* guaranteed budget (CGB). A feasible extension of our budget scheduler with CGBs is briefly described.

1 Introduction

From a business perspective, high volume electronics (HVE) consumer terminals, such as TV sets and set-top boxes (STBs), are required to become open and flexible. This openness and flexibility can be achieved by replacing dedicated single-function hardware components, which are typically contained in present-day hardware architectures, by powerful programmable components. In this way many media functions, such as audio and video decoding, or picture improvement, can be performed in software, and can be more easily adapted to changing standards or modifications of functionality.

Media processing in software opens the way to use dynamically scalable functions, trading resources for quality, to provide more functionality on a given platform and to support the construction of product families. Hence, functionality normally only found in high-end consumer terminals can also be provided in mid-range and low-end consumer terminals, albeit at a lower quality. We use the term Quality of Service (QoS) for an enhanced form of scalability, in which the overall perceptual quality is optimised at run-time, and in which seamless switching between different modes of operation (with different functionality) is supported; e.g. ATSC (digital TV input), or NTSC (analogue TV input) with picture improvement [3].

L. Wolf, D. Hutchison, and R. Steinmetz (Eds.) : IWQoS 2001, LNCS 2092, pp. 107-120, 2001.
© Springer-Verlag Berlin Heidelberg 2001

1.1   Media Processing in Consumer Terminals

The focus of our work is on consumer terminals such as digital TV sets, digitally improved analogue TV sets and STBs, i.e. receivers in a broadcast environment providing high-quality digital audio and video. A brief overview of scalable video algorithms and QoS resource management (QoS RM) for consumer terminals is given in [7], and examples of scalable video algorithms may be found in [18] and [22]. This domain has a number of distinctive characteristics when compared to mainstream multimedia processing in, for example, a workstation environment.

Consumer products are heavily resource constrained, with a high pressure on silicon cost and power consumption. In order to be able to compete with dedicated hardware solutions, the available resources will have to be used very cost-effectively, while preserving typical qualities of HVE consumer terminals, such as robustness, and meeting stringent timing requirements imposed by high-quality digital audio and video processing.

In HVE consumer terminals, software media processing is done using dedicated media processors, such as TriMedia™ Technologies Inc.'s family of very long instruction word (VLIW) processors; see [6]. Compared to dedicated hardware solutions, these media processors are expensive, both in cost and power consumption. Therefore, *cost-effectiveness* is a major issue in HVE consumer terminals. Cost-effectiveness requires a high average resource utilisation.

Current HVE consumer terminals provide robust behaviour, and users expect the same *robustness* when media processing is performed in software. For the time being, users do not have similar expectations of multimedia applications on desktops and internet appliances (and it is also not uncommon that these applications exhibit non-robust behaviour).

High-quality video has a field/frame-rate of 50 – 120 Hz, no tolerance for jitter (i.e. frame-rate fluctuations), and low tolerance for frame skips, properties that are characteristic of the *hard real-time* domain. In contrast, mainstream multimedia applications are characterised by low frame rates (with a maximum of 30 Hz) and high jitter tolerance, and in addition accept frequent frame skips, properties that are characteristic of the soft real-time domain. It is conceivable, however, that future users will expect guaranteed timing behaviour from multimedia applications on desktops and internet appliances as well (see, for example, [19]).

1.2   General Approach

Our approach has much in common with QoS for mainstream multimedia systems [2]. Media processing applications can be scaled dynamically (trading resources for quality), and these applications provide (estimated) resource requirements for each quality level. The QoS resource manager (QoS RM) adapts the quality levels at which the applications are executed, so as to maximise the perceived quality of the combined outputs, given the available resources. Maximisation of perceived quality is based on a model similar to the one described in [11]. Because rapidly changing quality levels are perceived as non-quality, quality levels must be adjusted sparingly. Currently, we focus on a single resource, the media processor.

QoS RM is conceived as a multilevel structure, in order to address dynamic behaviour at different time scales. For more details, see [14]. A central concept in QoS RM is the notion of *resource budget* (which is similar to reservations in [15]). The higher levels determine and adjust quality levels and resource budgets to maximise perceived output quality. The lowest level, the *budget scheduler*, provides, guarantees and enforces the allocated budgets. Thus, the higher levels build their policies on a mechanism provided by a lower level.

In our approach, QoS RM and media applications have separate, but complementary responsibilities in meeting system requirements, which makes our approach inherently co-operative. The distinctive characteristics of media processing in HVE consumer terminals give rise to conflicting requirements with respect to cost-effectiveness and robustness in the time domain. Cost-effectiveness requires average-case rather than worst-case resource budgets, but average-case resource allocation leads to robustness problems in the time domain. Robustness *between* applications is addressed by QoS RM, by guaranteeing and enforcing budgets. The remaining robustness problems *within* applications are to be resolved by the applications themselves [3].

1.3 User Focus and Relative Importance

A TV set may support a variable number of windows, such as a main window (showing, for example, a movie) and one (or more) secondary windows, e.g. PiP (Picture-in-Picture), videophone, or a web-browser. The user's focus is (typically) on one thing at a time, but changes dynamically from one window to another. Windows having user focus are evaluated differently by a user than other windows, and thus the applications with user focus should be treated differently with respect to quality. Hence, user focus induces a *relative importance* of the applications of consumer terminals. This relative importance is taken into account during the overall system optimisation, and a change of user focus requires a re-optimisation (see also [17]).

Stable output quality is a primary quality requirement for the application with user focus. If an application with user focus is confronted with a structural load increase, QoS RM will sacrifice the quality of an application without user focus, in order to keep the output of the user focus application at the same level. However, this is not sufficient to keep the output quality of the user focus application stable in case of a sudden load increase. Since it takes time to detect the structural nature of the load increase, and subsequently perform the necessary re-optimisation, the output quality of the application with user focus will have a temporary dip. In the remainder of this paper, the problem of the temporary quality dip in the output of the user focus application will be referred to as the "*user focus problem*". It will be shown that the notion of *conditionally guaranteed budget*, a supplement to the budget mechanism, can be used to solve the user focus problem.

1.4 Overview of the Paper

In Section 2, we briefly describe those aspects of QoS RM for HVE consumer terminals that are relevant for the user focus problem. The user focus problem is presented in Section 3. An analysis of the problem and an outline of the solution using conditionally guaranteed budgets is presented in Section 4. The extension of our budget mechanism with conditionally guaranteed budgets is the topic of Section 5. In Section 6, our work on user focus and conditionally guaranteed budgets is compared with other work. Finally, Section 7 provides conclusions and discusses future work.

2 QoS Resource Management

In this section, we discuss the dynamic load in relation to resource allocation below worst-case, present the structure of QoS RM, and give a brief introduction to the budget scheduler, including a comparison with related approaches. More details can be found in [14].

2.1 Dynamic Load

In the high quality video domain, the load of a system varies dynamically on multiple time scales. In this paper, our main focus is on changes induced by the incoming stream. Changes initiated by the service provider, such as the interruption of a movie by a commercial, take place at a time scale of minutes. Data dependent changes in the average load of applications take place at a time scale of seconds, e.g. scene changes in a movie. Finally, many media processing functions, such as MPEG encoding and decoding, and natural motion [12], have a load that show large data dependent variations over time. These data dependent load variations take place at a time scale of tens of milliseconds. In summary, there are variations around a quasi-fixed average load, and variations that involve a change of the average load; see Figure 1.

Average-case resource allocation or close-to-average-case resource allocation will result in occasional transient overload situations when peak loads exceed the budget, and in structural overload situations when the average load increases beyond the allocated budget. In our co-operative approach, transient overloads are to be dealt with by the applications. We say that applications have to *get by* on their budgets, which necessarily implies that the output quality is slightly lower than expected given the set quality level. In case of structural overloads, the budget must be adapted to maintain optimal overall behaviour of the system.

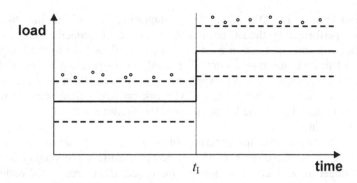

Fig. 1. The average load (*bold line*) suddenly increases at t_I. On either side of the load increase, there are high-frequency variations (*shaded areas*), with occasional high peaks (*small dots*). A typical budget would accommodate most of the variations, but might fail for occasional peak loads.

2.2   Main Components of QoS RM

In our approach, QoS RM is partitioned into two main components, a *Quality Manager* (QM) and a *Resource Manager* (RM). The notion of resource budget is common to both. The Quality Manager uses an additional notion of quality level, which it shares with the applications. An application may consist of one or more so-called entities. An *entity* is a set of tasks; it may constitute a continuous thread of control (as described in [10]), but that is not a prerequisite. Quality levels and resource budgets are associated to entities.

The QM maintains a mapping from quality levels to resource budgets, based on the estimated resource requirements from the applications and on information on the dynamic resource needs provided by the RM. This mapping is used in finding a set of quality levels that maximises the overall perceived quality given the available resources, taking into account the relative importance of the different applications. After such a (re-) optimisation, the QM assigns the selected quality levels to the entities, and allocates the corresponding resource budgets. Changes in the number of applications, relative importance of applications, and requests for quality level adaptations from the RM require re-optimisations. Since frequent changes in quality are perceived as non-quality, especially for applications with user focus, such re-optimisations should not be frequent.

The RM has two distinct parts, a budget scheduler (BS) that provides a run-time environment to the application, and a budget controller (BC). The BS provides guaranteed periodic budgets to the entities. This guarantee is based on an admission test that checks the feasibility of scheduling a set of budgets, and an enforcement mechanism that prevents entities from interfering with the budgets of other entities. The functionality of the BS closely resembles the functionality of the resource kernel described in [19]. The BS only enforces the budgets, without judging their appropriateness. Based on measurements, the BC adapts the budgets to their optimal values, and informs the QM about these adaptations, or its inability to adapt. The adaptations performed by the BC are based on classical feedback control.

Together with the application entities that adjust their load to the budget, the BC and the QM provide multilevel control, roughly corresponding to the different time horizons presented in Section 2.1. Application control works at a time-scale of tenths of milliseconds, whereas the BC and QM work on larger time scales, ranging from seconds to minutes. Given the larger time-scales, the interventions of the BC and QM may take more time.

QoS RM is responsible for ensuring *smooth transitions*, again together with the application entities. Assume that, after a re-optimisation, the quality level and the resource budget for a given entity must be increased. BC increases the budget, and BS effectuates the budget increase. When the new budget is in effect, QM assigns the new quality level to the entity. Finally, the entity makes sure that the transition is performed smoothly.

2.3 Budget Scheduler

We use a middleware approach for our QoS RM for HVE consumer terminals, based on a commercial off-the-shelf (COTS) real-time operating system (RTOS) providing pre-emptive priority-based scheduling. The budget scheduler is implemented on top of this COTS RTOS, providing periodic budgets that support applications from the high-quality video domain. Our scheduling mechanism is currently based on rate monotonic priority assignment [13].

An example of the usage of a middleware approach for QoS resource management for media processing in a (networked) workstation environment, based on Unix, may be found in [16]. The scheduling mechanism is based on time-slicing, and supports applications from the soft real-time domain. In [19], resource kernels are described in the context of QoS resource management for (hard) real-time and multimedia systems. The description abstracts from (and hence allows different types of) scheduling mechanisms.

3 User Focus Problem

Consider two applications that are running in the system. The outputs of both applications are visible to the user. The output of one application (identified by UF) has user focus, whereas the output of the other application (identified by ¬UF) does not have user focus. Upon a load increase of UF, the output quality of UF is restored at the cost of the output quality of ¬UF. In addition to these two applications, there may be other, so-called neutral, applications, which are unaffected by the load increase of UF. For ease of presentation, we restrict ourselves to UF and ¬UF in this section. We assume that both UF and ¬UF consist of a single entity, which in turn consists of a single task.

3.1   Problematic Behaviour

Figure 2 visualises the user focus problem by showing the load induced by the input data and the perceived output quality of both applications as a function of time. When a sudden increase of the induced load of UF occurs (at time t_I), UF faces a structural overload situation, as described in Section 2.1. The BC detects the structural overload. If BC cannot accommodate the structural overload by adapting budgets, it signals the problem to the QM. Subsequently, the QM determines the new optimal quality levels at which UF and ¬UF will run. We assume that the quality level for UF remains the same. Thus, after a certain reaction time (from t_I to t_R), the quality and the budget of ¬UF are reduced, in this order, and, subsequently, the budget of UF is increased. At time t_S, a new equilibrium is reached. In the mean time (from t_I to t_S), the perceived quality of UF's output is degraded, because UF's resource budget is temporarily insufficient to cope with the increased load, and UF has to get by on its budget, which necessarily results in some form of quality reduction at the output. Thus, the perceived quality at the output of UF is temporarily reduced, even though the quality level for UF remains the same.

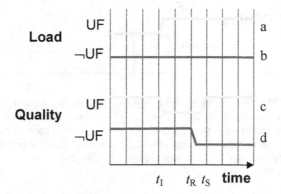

Fig. 2. The load of the user-focus application UF (*line a*) increases at time t_I. The load of the non-user-focus application ¬UF (*line b*) does not change. The system reacts at time t_R by decreasing the quality level, and thus the perceived output quality, of ¬UF (*line d*), and reallocating the required resources to UF. In the mean time, the perceived output quality of UF (*line c*) shows a dip.

3.2   Desired Behaviour

In Figure 3, the desired behaviour is shown. UF's perceived output quality is not affected by the sudden load increase, whereas the quality of ¬UF is degraded smoothly. There are two major difficulties in obtaining this desired behaviour.
- *Very quick reaction required* $\Rightarrow$ *cuts through layers.*
  The desired stable output quality for UF can only be achieved if the additional resources become available instantaneously. However, for stability reasons, it is undesirable to react quickly without being sufficiently certain that the overload was

indeed the result of a structural load increase. Therefore, a sufficiently quick reaction to support the desired behaviour for UF can only be achieved by relying on built-in behaviour of the BS.

- *Smooth degradation of ¬UF may not be feasible.*
  To achieve a smooth transition for ¬UF, the quality of ¬UF must be reduced first, followed by a reduction of the budget of ¬UF. However, in order to address the needs of UF, the resources allocated to ¬UF will be taken away instantaneously, and the quality level of ¬UF will be adjusted after a certain reaction time. In the mean time, ¬UF will have to get by on its remaining resources, and the perceived quality very much depends on ¬UF's ability to do so.

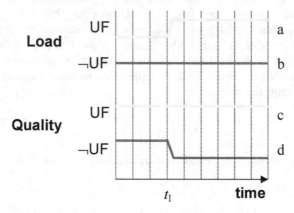

Fig. 3. The load of the user-focus application UF (*line a*) increases at time t_1. The load of the non-user-focus application ¬UF (*line b*) does not change. The perceived output quality of UF (*line c*) remains stable, in spite of the load increase. The load increase is met by decreasing the quality level, and thus the perceived output quality, of ¬UF (*line d*).

4   Towards a Solution: Conditionally Guaranteed Budgets

The user focus problem can be traced down to two complementary causes.

- *UF is confronted with a structural resource shortage.*
  After the (structural) load increase, the UF has insufficient resource capacity to meet the requirements of the set quality level. This results in an overload situation in which UF has to get by on its budget, which in turn results in a degradation of UF's perceived output quality.
- *The reaction time is too long.*
  The system needs a particular reaction time to detect the structural overload and take appropriate measures (i.e. degrade the quality level of ¬UF, decrease its budget correspondingly, and increase the budget of UF). The reaction time is too long in the sense that it gives rise to the dip in the quality of the output of UF.

By eliminating one of these causes, the problem can be solved. In this section we show that elimination of only one of the causes is feasible in our situation and present an outline of a solution using conditionally guaranteed budgets.

4.1 Load Increase Anticipation

If either the reaction time of the system can be reduced or the structural resource shortage for UF can be prevented, the problem will not occur. Unfortunately, it is uncommon that media data in broadcast environments provide the necessary information to detect upcoming structural load changes timely. In the absence of such information, the structural overload must be detected by the BC. As we have seen in Section 3.2, detecting the structural overload necessarily takes time. So we must seek the solution in eliminating the structural resource shortage.

Hence, rather than assigning a budget well below worst-case, we have to revert to a higher budget for UF, which is large enough to accommodate a potential structural load increase. This higher budget includes a *budget margin* for an anticipated structural load increase. When this anticipated load increase occurs, it can be accommodated without delay. The quality of UF will remain stable without a dip. The higher budget of UF can be accommodated by giving ¬UF a lower budget. Thus, the budgets allocated to UF and ¬UF, and the corresponding quality settings, anticipate the load increase for UF. However, due to the budget margin, UF will generate slack before the load increase, reducing the cost-effectiveness of the system.

4.2 Unused Resource Capacity Allocation

To gain back on the cost-effectiveness loss and to compensate ¬UF for its budget loss, ¬UF receives an additional budget with a *conditional* guarantee. This so-called *conditionally guaranteed budget* (CGB) is available to ¬UF when UF does not use its budget margin. To distinguish a normal budget, which has an absolute guarantee, from a CGB, we sometimes use the term *absolutely guaranteed budget* (AGB).

In order to provide the CGB, the BS must be extended with a mechanism to schedule the CGB at run time without interfering with the guarantees for the AGBs, and a corresponding additional admission test. This admission test must ensure that the CGB is available when the condition is met, i.e. when UF does not use its budget margin.

When the QM allocates both an AGB and a CGB to ¬UF, it informs ¬UF that it *can* run on a quality level matching the AGB, and that it *may* run on a quality level matching the combination of budgets. In this way, the CGB provides an option for *controlled* quality improvement.

4.3 Budget Withdrawal Anticipation

Whereas ¬UF is informed about the existence of the situations in advance, it is confronted with a structural change from one situation to the other, because the system needs a particular reaction time to detect such a change. When confronted with a change in the availability of the CGB, ¬UF is assumed to adjust to the new situation by selecting the appropriate quality level. Note that even when ¬UF consistently receives the CGB, peak loads of UF may cause occasional unavailability of the CGB. Smooth transitions may therefore not always be feasible, and we have to revert to a best-effort approach for ¬UF.

Hence, the long reaction time causing the user focus problem still takes its toll, but we shifted the resulting problem to a place where it is less severe. Moreover, the application facing the problem may be selected for its ability to handle it.

5   Extension to Budget Scheduler

The implementation of the conditionally guaranteed budget strongly depends on the implementation of the budget scheduler. Therefore, a concise description of our current implementation of the budget scheduler is presented, followed by a brief discussion of one possible extension.

5.1   Description of Current Implementation

Budgets are implemented by means of priority manipulations. In-budget execution is performed at high priority, and out-of-budget execution is done at low priority. This gives rise to two main priority bands, a high-priority band (HP) for in-budget executions and a low-priority band (LP) for out-of-budget executions. An entity that consists of multiple tasks gives rise to a sub-priority band, so that tasks within the entity can be prioritised. Priority bands of entities are disjoint (i.e. they do not overlap). Budgets are periodic, and the budget periods may be different for each entity. The budget for entity E_i is denoted by $<B_i, T_i>$, where T_i is the budget period, and B_i the budget per period for E_i.

In HP, the entities are scheduled in rate-monotonic priority order, i.e. entities with smaller budget periods get higher priorities. At the start of each new period, the priority of an entity is raised to its rate-monotonic priority within HP. When the budget is exhausted, or when the entity releases the processor, the entity's priority is lowered to LP. In case of a multi-task entity, the complete sub-priority band is raised or lowered, leaving the internal priority ordering intact. We found inspiration for our implementation of budgets in [1] and [21].

The admission test of our resource scheduler is based on rate monotonic analysis (RMA), which may be considered state-of-the-art given the existence of an RMA handbook [9]. Since we strive for a processor utilisation that exceeds the utilisation bound derived in [13], the implementation of our admission test is based on [8]. Assume a set of entities $(E_1, E_2, ..., E_n)$, with budgets $<B_1, T_1>$, $<B_2, T_2>$, ..., $<B_n, T_n>$. E_i's priority in HP is denoted by HP_i. The priorities are rate monotonic, i.e. if $T_i < T_j$, then $HP_i > HP_j$. The admission test is passed, if for all entities E_i a worst-case response time R_i can be found that satisfies equations (1) and (2). Note that when the admission test is passed, all entities can consume their budget within their period.

$$R_i = B_i + \sum_{HPj>HPi} \lceil R_i/T_j \rceil \times B_j \qquad (1)$$

$$R_i \leq T_i \qquad (2)$$

5.2 Extension

In order to support load increase anticipation for applications with user focus, we experimented with several alternative implementations of CGBs. The extension of our budget scheduler with an additional, middle-priority band (MP) for CGBs is briefly described below. For the description we assume a set of three entities, UF, ¬UF, and a neutral entity N. The AGBs for these entities are $<B_{UF}, T_{UF}>$, $<B_{\neg UF}, T_{\neg UF}>$, and $<B_N, T_N>$, respectively, where B_{UF} includes a budget margin BM_{UF}. In addition, ¬UF has a conditionally guaranteed budget $<CGB_{\neg UF}, T_{\neg UF}>$, which it will receive when the load of UF is consistently lower than $<B_{UF} - BM_{UF}, T_{UF}>$.

At run time, the normal procedure is followed, except in the following situations. When ¬UF has exhausted its AGB, its priority is not lowered to LP, but to MP. When ¬UF has exhausted its conditionally guaranteed budget, its priority is lowered to LP. If the next budget period starts before ¬UF has exhausted its conditionally guaranteed budget, the priority is raised to HP.

The additional admission test for the $CGB_{\neg UF}$ is passed, if a worst-case response time $CR_{\neg UF}$ can be found that satisfies equations (3) and (4).

$$CR_{\neg UF} = CGB_{\neg UF} + \sum_{j \neq UF} \lceil CR_{\neg UF}/T_j \rceil \times B_j + \lceil CR_{\neg UF}/T_{UF} \rceil \times (B_{UF} - BM_{UF}) \qquad (3)$$

$$CR_{\neg UF} \leq T_{\neg UF} \qquad (4)$$

The introduction of MP potentially leads to a non-rate monotonic priority assignment, which yields a non-optimal solution [13]. For instance, in the example presented above, T_N may be much larger than $T_{\neg UF}$, and even B_N may be larger than $T_{\neg UF}$. In the latter case, it is immediately obvious that no feasible value for $CGB_{\neg UF}$ can be found. With the utilisation we are aiming for, it is very probable that non-rate monotonic priority assignments will not be feasible. The QM has to deal with this problem in some way or other, possibly using a technique described in [20]. Exploitation of the CGB by the QM is subject to further investigation.

The extension of the budget scheduler as presented in this section naturally generalises to entity sets with one UF entity, n ¬UF entities, and m neutral entities. With this extension, the risk of a non-rate monotonic priority order is even more pronounced.

When all budget periods are equal, the problem of the non-rate monotonic priority ordering does not arise. This is not a completely hypothetical case. In present-day TV sets, for instance, there is generally one dominant frequency. In Europe, high quality TV sets are advertised as being 100 Hz.

6   Related Work

The notion of user focus and relative importance can also be found in other work (e.g. [15], [17], and [11]). The extension of the interpretation of relative importance with load increase anticipation has not been addressed in the literature before.

Resource budgets (also termed reservations) are a well-known and proven concept (see, for example, [15], [19], or [4]). In the reservation model described in [19], three different types of reservations (hard, firm, and soft) are distinguished based on the way they compete for the slack. Budget sharing for overrun control, as described in [4], is another way to allocate slack. In that article, the primary goal of the slack allocation algorithm is the improvement of the robustness of a system.

CGBs are (very) different from AGBs and slack allocation algorithms. They differ from AGBs by being inherently conditional as expressed in the (conditional) admission test. They differ from slack allocation algorithms by their very nature of being budgets, having an admission test and being enforced. CGBs may be viewed as a refinement of the models currently supported by existing resource kernels. The notion of a CGB has not been addressed in the literature before.

7   Conclusions and Future Work

In this paper, we briefly described media processing in software in HVE consumer terminals, and its distinctive characteristics when compared to mainstream multimedia processing in workstations. These distinctive characteristics require our QoS approach to be different from similar approaches in the mainstream multimedia domain in a number of respects. In particular, applications receive a resource budget below worst-case for cost-effectiveness reasons in our approach. We showed that such a budget allocation combined with multilevel control and dynamic load gives rise to a so-called user focus problem. Upon a sudden increase of load of an application having user focus, its output will have a quality dip. We introduced the notion of conditionally guaranteed budgets and showed how this notion can be used to resolve the user focus problem at the cost of the quality of an application that does not have user focus (i.e. we shifted the problem to a place where it is less severe). A feasible implementation of CGBs has been briefly described.

In this paper, we only considered situations with a single application having user focus and one or more applications without user focus. The user focus problem is not restricted to a window having user focus. Other applications, such as delayed viewing, encounter the user focus problem as well. To solve the user focus problem for multiple user focus applications, another implementation of CGBs is needed, which allows the provision of the unused budget margin of one specific entity to one or more other specific entities. This implementation is the topic of a forthcoming paper.

Although we performed a number of initial experiments, the actual validation of CGBs by media processing applications is a subject of further research, to be performed in close co-operation with our colleagues from the high-quality video domain.

Finally, the focus of this paper has been on CGBs as a mechanism. The exploitation of this mechanism by policies residing at higher levels in QoS RM requires future work.

Acknowledgements

We would like to thank our colleagues from the high-quality video domain for stimulating discussions, in particular Kees van Zon (who planted the original seed for CGBs), Christian Hentschel, Maria Gabrani, John Zhong, John Lan, and Yingwei Chen. We further thank Alejandro Alonso and Angel Groba from the dit/UPM (Technical University of Madrid) for acting as a sounding-board from a real-time system's perspective, Hannie C.P. van Iersel for her contributions in illuminating the concept, Clara M. Otero Pérez for implementing and comparing various alternatives of CGBs, and Jos T.J. van Eijndhoven and Wim F.J. Verhaegh for valuable comments on the draft version of this paper. The work reported upon in this paper is funded by ITEA/Europa [5].

References

1. Audsley, N.C., Burns, A., Richardson, M.F., Wellings, A.J.: Incorporating unbounded algorithms into predictable real-time systems, Computer Systems Science & Engineering 8(2): 80 – 89, April 1993.
2. Aurrecoecha, C., Campbell, A., Hauw, L.: A Survey of QoS Architectures, ACM/Springer-Verlag Multimedia Systems Journal, 6(3): 138 – 151, May 1998.
3. Bril, R.J., Gabrani, M., Hentschel, C., van Loo, G.C., Steffens, E.F.M.: QoS for consumer terminals and its support for product families, To appear in: Proc. International Conference on Media Futures (ICMF), Florence, Italy, May, 2001.
4. Caccamo, M., Buttazzo, G., Sha, L.: Capacity Sharing for Overrun Control, In: Proc. of the 21st Real-Time Systems Symposium (RTSS), Orlando, Florida, pp. 295 – 304, November 27-30, 2000.
5. Gelissen, J.H.A.: The ITEA project EUROPA, To appear in: Proc. International Conference on Media Futures (ICMF), Florence, Italy, May 2001.
6. Glaskowsky, P.N.: Philips Advances TriMedia Architecture – New CPU64 Core Aimed at Digital Video Market, Microdesign Resources, 12(14): 33 – 35, October 1998.
7. Hentschel, C., Gabrani, M., Bril, R.J., Steffens, L.: Scalable video algorithms and Quality-of-Service resource management for consumer terminals, To appear in: Digest of Technical Papers IEEE International Conference on Consumer Electronics (ICCE), Los Angeles, CA, USA, June 2001.
8. Joseph, M., Pandya, P.: Finding response times in a real-time system, The Computer Journal, 29(5):390-395, 1986.

9. Klein, M.H., Ralya, T., Pollak, B., Obenza, R., González Harbour, M.: A Practitioner's Handbook for Real-Time Analysis: Guide to Rate Monotonic Analysis for Real-Time Systems, Kluwer Academic Publishers, 1993.
10. Lee, C., Rajkumar, R., Mercer, C.: Experiences with Processor Reservation and Dynamic QoS in Real-Time Mach, In: Proc. of Multimedia Japan, March 1996.
11. Lee, C., Lehoczky, J., Rajkumar, R., Siewiorek, D.: On Quality of Service Optimization with Discrete QoS Options, In: Proc. of the 5th IEEE Real-time Technology and Applications Symposium (RTAS), Vancouver, Canada, pp. 276 – 286, June 1999.
12. Lippens, P., De Loore, B., de Haan, G., Eeckhout, P., Huijgen, H., Löning, A., McSweeney, B., Verstraelen, M., Pham, B., Kettenis, J.: A video signal processor for motion-compensated field-rate upconversion in consumer television, IEEE Journal of Solid-State Circuits, 31(11): 1762 – 1769, November 1996.
13. Liu, C.L., Layland, J.W.: Scheduling algorithms for multiprogramming in a hard real-time environment, JACM 20(1): 46 – 61, January 1973.
14. van Loo, S., Steffens, L., Derwig, R.: Quality of Service Resource Management in consumer Terminals, Philips Research Laboratories Eindhoven (PRLE), Doc. id. NL-MS 21166, available from   , May 2001.
15. Mercer, C.W., Savage, S. Tokuda, H.: Processor Capability Reserves: Operating System Support for Multimedia Applications, In: Proc. of the International Conference on Multimedia Computing and Systems (ICMCS), pp. 90 – 99, May 1994.
16. Nahrstedt, K., Chu, H. Narayan, S.: QoS-aware Resource Management for Distributed Multimedia Applications, Journal on High-Speed Networking, Special Issue on Multimedia Networking, IOS Press, Vol. 8, No.3-4, pp. 227 – 255, 1998.
17. Ott, M., Michelitsch, G., Reininger, D. Welling, G.: An architecture for adaptive QoS and its application to multimedia systems design, Computer Communications, 21(4): 334-349, 1998.
18. Peng, S.: Complexity scalable video decoding via IDCT data pruning, To appear in: Digest of Technical Papers IEEE International Conference on Consumer Electronics (ICCE), Los Angeles, CA, USA, June 2001.
19. Rajkumar, R., Juvva, K., Molano, A., Oikawa, S.: Resource Kernels: A Resource-Centric Approach to Real-Time and Multimedia Systems, In: Proc. of the SPIE/ACM Conference on Multimedia Computing and Networking, January 1998.
20. Sha, L., Lehoczky, J., Rajkumar, R.: Solutions for some practical problems in prioritized preemptive scheduling, In: Proc. of the 7th IEEE Real-Time Systems Symposium (RTSS), pp. 181-191, December 1986.
21. Sprunt, B., Sha, L. Lehoczky, J.P.: Aperiodic Task Scheduling for Hard Real-Time Systems, The Journal of Real-Time Systems 1: 27-60, 1989.
22. Zhong, Z., Chen, Y.: Scaling in MPEG-2 decoding loop with mixed processing, To appear in: Digest of Technical Papers IEEE International Conference on Consumer Electronics (ICCE), Los Angeles, CA, USA, June 2001.

Extending BGMP for Shared-Tree Inter-Domain QoS Multicast*

Aiguo Fei and Mario Gerla

Department of Computer Science
University of California, Los Angeles, CA 90095, USA
{afei,gerla}@cs.ucla.edu

Abstract. QoS support poses new challenges to multicast routing especially for inter-domain multicast where network QoS characteristics will not be readily available as in intra-domain multicast. Several existing proposals attempt to build QoS-sensitive multicast trees by providing multiple joining paths for a new member using a flooding-based search strategy which has the draw-back of excessive overhead and may not be able to determine which join path is QoS feasible sometimes. In this paper, first we propose a method to propagate QoS information in bidirectional multicast trees to enable better QoS-aware path selection decisions. We then propose an alternative "join point" search strategy that would introduce much less control overhead utilizing the root-based feature of the MASC/BGMP inter-domain multicast architecture. Simulation results show that this strategy is as effective as flooding-based search strategy in finding alternative join points for a new member but with much less overhead. We also discuss extensions to BGMP to incorporate our strategies to enable QoS support.

1 Introduction

Over the years, many multicast routing algorithms and protocols have been proposed and developed for IP networks. Several routing protocols[20,17,3,12] have been standardized by the IETF (Internet Engineering Task Force), with some of them having been deployed in the experimental MBone and some Internet Service Providers'(ISP) networks[1]. Some of these protocols[20,17] are for intra-domain multicast only while the others[3,12] have scalability limitation and/or other difficulties when applied to inter-domain multicast though they were designed for that as well.

To provide scalable hierarchical Internet-wide multicast, several protocols are being developed and considered by IETF. The first step towards scalable hierarchical multicast routing is Multiprotocol Extensions to BGP4(MBGP)[4] which extends BGP to carry multiprotocol routes. In the MBGP/PIM-SM/MSDP architecture[1], MBGP is used to exchange multicast routes and PIM-SM(Protocol Independent Multicast - Sparse Mode) [12] is used to connect group members across domains, while another protocol, MSDP(Multicast Source Discovery Protocol)[15], is developed to exchange information of active multicast sources among RP (Rendezvous Point) routers across domains.

---

* This work was supported in part by grants from Cisco Systems and NSF.

L. Wolf, D. Hutchison, and R. Steinmetz (Eds.): IWQoS 2001, LNCS 2092, pp. 123–139, 2001.
© Springer-Verlag Berlin Heidelberg 2001

The MBGP/PIM-SM/MSDP architecture has scalability problems and other limitations, and is recognized as a near-term solution[1]. To develop a better long-term solution, a more recent effort is the MASC/BGMP architecture[16,21]. In this architecture, MASC(Multicast Address Set Claim)[13] allocates multicast addresses to domains in a hierarchical manner and BGMP(Border Gateway Multicast Protocol)[21] constructs a bidirectional group-shared inter-domain multicast tree rooted at a root domain. The bidirectional shared tree approach is adopted because of its many advantages[16].

In recent years, great effort has been undertaken to introduce and incorporate quality of service(QoS) into IP networks. Many multicast applications are QoS-sensitive in nature, thus they will all benefit from the QoS support of the underlying networks if available. The new challenge is how to build multicast trees subject to multiple QoS requirements. However, all the existing multicast protocols mentioned above are *QoS-oblivious*: they build multicast trees based only on connectivity and *shortest-path*. QoS provision is thus "opportunistic": if the default multicast tree can not provide the QoS required by an application then it is out of luck.

Several QoS sensitive multicast routing proposals [8,14,22,10] attempt to build QoS-sensitive multicast trees by providing multiple paths for a new member to choose in connecting to the existing tree through a "local search" strategy which finds multiple connecting paths by flooding TTL(Time-To-Live)-controlled search messages. In these alternate path search strategies, connecting path selection is solely based on QoS characteristics of the connecting paths. This may not necessarily yield the best connecting path selection and sometimes may not provide a QoS feasible path at all. Moreover, the flooding nature of local search strategy can cause excessive overhead if the TTL gets large.

In this paper, first we argue that the limitations of path selection with current proposals originates from the lack of QoS information concerning the existing multicast tree (for example, maximum inter-member delay and residual bandwidth of the existing tree). We then propose a method to propagate QoS information in group-shared bi-directional multicast trees so that better routing decisions can be made. In the mean time, we propose a new alternative "join point" search strategy that can significantly reduce the number of search messages utilizing the root-based feature of the recent MASC/BGMP inter-domain multicast architecture, which feature is also shared by other multicast routing protocols including CBT (Center Based Trees)[3] and PIM-SM[12].

In our new "scoped on-tree search" strategy, the on-tree node receiving a *join* message from a new member starts a search process to notify nearby on-tree nodes to provide alternative connecting points for the new member if the default one is not QoS feasible. This strategy starts a search only if the default route is not feasible and the on-tree search nature will only introduce limited overhead. Simulation results show that this strategy is as effective as flooding-based search strategy in finding alternative connecting "points" for a new joining member but with much less overhead.

Our work was motivated by the realization of the limitations of current alternate path selection method and excessive overhead of flooding based search strategies. The purpose of this paper is not to propose a new multicast routing protocol, but rather to show how our QoS information propagation method and alternative join point search strategy can be incorporated into BGMP to extend it for QoS support. Our methods are general in principle and can be applied to other protocols such as PIM-SM or other future routing protocols for QoS support as well.

The rest of this paper is organized as follows. Section 2 reviews the current Internet multicast architecture especially the MASC/BGMP architecture, QoS-aware multicast routing and several existing proposals. Section 3 discusses the limitations with current local-search-based QoS multicast routing proposals and Section 4 proposes a QoS information propagation mechanism for shared-tree multicast to address those limitations. Section 5 proposes our improved alternative join point search strategy with simulation results. Section 6 discusses modifications and extensions to BGMP to support QoS and Section 7 concludes the paper.

2   Background and Related Work

2.1   Current IP Multicast Architecture

In the current IP multicast architecture, a host joins a multicast group by communicating with the designated router using Internet Group Membership Protocol (IGMP[9]) (by sending a membership report or answering a query from the router). To deliver multicast packets for a group, IP multicast utilizes a tree structure which is constructed by multicast routing protocols. In MOSPF[17], routers within a domain exchange group membership information. Each router computes a source-based multicast tree and corresponding multicast tree state is installed. In CBT[3] or PIM-SM[11,12], a group member sends an explicit join request towards a core router or an RP router. The request is forwarded hop-by-hop until it reaches a node which is already in tree or the core/RP, and multicast states are installed at intermediate routers along the way.

The Internet consists of numerous Autonomous Systems (AS) or domains, which may be connected as service provider/customers in a hierarchical manner or connected as peering neighbors, or both. Normally a domain is controlled by a single entity and can run an intra-domain multicast routing protocol of its choice. An inter-domain multicast routing protocol is deployed at border routers of a domain to construct multicast trees connecting to other domains. A border router capable of inter-domain multicast communicates with its peer(s) in other domain(s) via inter-domain multicast protocols and routers in its own domain via intra-domain protocols, and forwards multicast packets across the domain boundary. Currently there are two prominent inter-domain multicast protocol suits: MBGP/PIM-SM/MSDP and MASC/BGMP[1,16,21].

In this paper, we are concerned with multicast routing at the inter-domain level. By **receiver** or **multicast group member**, we refer to a router that has multicast receiver(s) or multicast group member(s) in its domain. We are not concerned with how a receiver host joins or leaves a group and how an intra-domain multicast tree is constructed. Similarly, by **source**, we refer to a router that has a multicast traffic source host in its domain.

2.2   The MASC/BGMP Architecture

In the MASC/BGMP[16,21] architecture, border routers run BGMP to construct a bidirectional "shared" tree for a multicast group. The shared tree is rooted at a root domain that is mainly responsible for the group (e.g., the domain where the group communication initiator resides). BGMP relies on a hierarchical multicast group address allocation protocol (MASC) to map a group address to a root domain and an inter-domain routing protocol (BGP/MBGP) to carry "group route" information (i.e., how to reach the root domain of a multicast group).

MASC is used by one or more nodes of a MASC domain to acquire address ranges to use in its domain. Within the domain, multicast addresses are uniquely assigned to clients using intra-domain mechanism. MASC domains form a hierarchical structure in which a child domain (customer) chooses one or more parent (provider) domains to acquire address ranges using MASC. Address ranges used by top-level domains (domains that don't have parents) can be pre-assigned and can then be obtained by child domains. This is illustrated in Fig. 1, in which A, D, and E are backbone domains, B and C are customers of A while B and C have their own customers F and G, respectively. A has already acquired address range 224.0.0.0/16 from which B and C obtain address ranges 224.0.128.0/124 and 224.0.1.1/25, respectively.

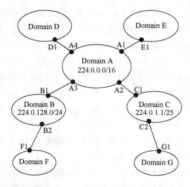

Fig. 1. Address allocation Using MASC, Adopted from [16].

Using this hierarchical address allocation, multicast "group routes" can be advertised and aggregated much like unicast routes. For example, border router B1 of domain B advertises *reachability* of root domains for groups in the range of 224.0.128.0/124 to A3 of domain A, and A1(A4) advertises the aggregated 224.0.0.0/ 16 to E1(D1) in domain E(D). Group routes are carried through MBGP and are injected into BGP routing tables of border routers. BGMP then uses such "group routing information" to construct shared multicast tree and distribute multicast packets.

BGMP constructs a bidirectional shared tree for a group rooted at its root domain through explicit join/prune as in CBT and PIM-SM. A BGMP router in the tree maintains a *target list* that includes a *parent target* and a list of *child targets*. A parent target is the next-hop BGMP peer towards the root domain of the group. A child target is either a BGMP peer or an MIGP (Multicast Interior Gateway Protocol) component of this router from which a join request was received for this group. Data packets received for the group will be forwarded to all targets in the list except the one from which data packet came. BGMP router peers maintain persistent TCP connections with each to exchange BGMP control messages (join/prune, etc.).

In BGMP architecture, a source doesn't need to join the group in order to send data. When a BGMP router receives data packets for a group for which it doesn't have forwarding entry, it will simply forward packets to the next hop BGMP peer towards the root domain of the group. Eventually they will hit a BGMP router that has forwarding state for that group or a BGMP router in the root domain. BGMP can also build source-specific branches, but only when needed (i.e., to be compatible with source-specific trees used by some MIGPs), or to construct trees for source-specific groups. A source-specific branch stops where it reaches either a BGMP router on the shared tree or the source domain. This is different from the source trees by some MIGPs in which source-specific state is setup all the way up to the source.

2.3 QoS-Aware Multicast Routing

Currently the Internet only provides shortest-path routing based on administrative costs and policies. Providing QoS in this routing architecture is "opportunistic": if the default route can not provide the QoS required by an application then it is out of luck. To effectively support QoS, QoS routing may be a necessity in future QoS-enabled IP networks. Two main issues involved in QoS routing are how to compute or find QoS specific paths and how to enforce their use. Because of the network-stateless nature of conventional IP networks, the second issue (i.e., enforcing specific paths, which are different from the default unicast path, for different connections) might be a more difficult one to tackle within the current Internet architecture. In this regard, multicast is more suited for QoS-aware routing than unicast connections because the network maintains multicast states which provide a natural way to "pin-down" a QoS-feasible path or tree.

A multicast tree grows and shrinks through joining and pruning of members. Several QoS sensitive multicast routing proposals [8,14,22,10] attempt to provide multiple paths for a new member to choose in connecting to the existing tree, hoping that one of them is QoS feasible. We call a join path QoS feasible if it satisfies the specific QoS requirements (for example, has enough bandwidth). In these proposals, some kind of search mechanism is used to explore multiple paths for the new member. YAM[8] proposes an inter-domain join mechanism called "one-to-many join". In QoSMIC[14], there are two ways for a new router to select an on-tree router to connect to the tree. One is called "local search" which is very similar to "one-to-many join" in YAM[8] and is further studied in [22]. In "local search", the new joining router floods a BID-REQUEST message with scope controlled by the TTL(Time-To-Live) field, an on-tree routers receiving search messages unicasts BID messages back to the new member. The other one is called "multicast tree search" in which the new joining router contacts a manager which then starts a "bidding" session by sending BID-ORDER messages to the tree, a subset of routers which receive the message become candidates. A candidate router unicasts a BID message to the new member. In both cases, a BID message "picks up" QoS characteristics of the path and the new member collects BID messages and selects the "best" candidate according to the QoS metrics collected, then it sends JOIN message to setup the connection. One drawback of both YAM and QoSMIC is the large amount of messages flooded, and consequently the large number of nodes which have to participate in the route selection process. In another proposal, QMRP[10], a more elaborate search strategy is proposed to restrict the number of nodes involved and the amount of search messages flooded. In YAM and QoSMIC, messages are only controlled by TTL. While in QMRP, they are also controlled by restricting branching degree – the number of REQUESTs that can be sent to neighbors by a node; more importantly, a REQUEST message is only forwarded to nodes that have the *required resources*.

The "multicast tree search" in QoSMIC[14] was also proposed to limit the overhead of flooding: a joining node starts a local search with a small TTL first, it contacts a tree manager to start a tree search if the local search fails. However this tree search strategy has its own drawbacks: (1)centralized manager may have to handle many join requests and be a potential hot spot; (2)there is no easy scalable way to distribute the information of mapping of multicast groups to tree managers for inter-domain multicast – this is one motivation for the development of inter-domain multicast protocols instead of simply using PIM-SM; (3)a manager has to multicast a search message to the group which is considerably expensive since each on-tree router has to decide whether it should be a candidate[14] (unless the manager has global network topology and group membership information so it can select a few connecting candidates for the new member itself – the problem is, though group membership information can be obtained and maintained, it is not possible with global network information in inter-domain multicast), and it is not easy for an existing tree node to do so because it needs some kind of distance information [14] (to the new member) which the inter-domain routing protocol (BGP) is unable to provide or such information is very coarse.

Among these proposals, YAM's one-to-many join is specifically proposed for inter-domain multicast while the others are targeted for both intra-domain and inter-domain. For routing within an AS domain, a link-state routing protocol is often used. It is possible that intra-domain QoS-sensitive routing in future QoS-enabled networks is still link-state-based[2]. A link-state routing protocol gives all network nodes the complete knowledge of network connectivity, and QoS characteristics in case of QoS-sensitive routing. Thus a local search strategy might be neither necessary, nor technically attractive, in intra-domain QoS multicast. One might argue that, QoS information propagated by a link-state protocol wouldn't be as up-to-date and accurate as what discovered by local search. The counter argument is that, if QoS multicast gains widespread use, then the overhead caused by local search of the many new joins would be more than that of flooding QoS link-state information to achieve the same timeliness and accuracy. However, for inter-domain multicast, link-state-based routing is not an option today and a local search or other search strategies must be pursued in constructing QoS multicast trees.

3   Limitations of Current QoS Multicast Routing Proposals

Besides the drawback of control overhead, there are other limitations with the several current QoS multicast routing proposals discussed in Section 2.3. Consider a multicast tree and a new joining member as illustrated in Fig. 2. Discussions here apply to both intra-domain and inter-domain multicast. In inter-domain multicast this paper is focused on, a link between two on-tree neighbors in the figure can be an actual physical link if these two are just one hop away, but can also be a logical link which may have several other nodes in between (for example, two intra-domain peers as to be discussed later). Existing QoS-sensitive multicast routing proposals[8,14,22,10] provide multiple connecting paths for a new joining member to choose. For example, new member N can connect to the tree by connecting to node A or B or C. Also note that there are two connecting paths to node B. This is possible if a search strategy as in [10] is used, or if BID-REQUEST messages in "local search"[14] are allowed to "pick up" multiple paths to an on-tree node and BID messages are sent back hop-by-hop according to the paths picked up.

In these routing schemes, a connecting candidate is selected according to the QoS properties of the connecting paths. Now consider the four available paths in Fig. 2. Assume that the QoS metric we care about here is delay and the connecting path from A to N has the shortest delay. Thus A would be chosen as the candidate for N to connect to. However, A might have the shortest delay to N, but might not be the best candidate after all; for example, the link between A and D is a slow one with long delay. Now assume the QoS metric in question is bandwidth. It is reasonable to assume that an on-tree node can easily learn how much traffic it is receiving from the group (say, 256kb/s) and intermediate nodes can detect bandwidth availability. So this will not a be a problem for the local search scheme if the new joining member is a receiver only, though it is arguable how a local *resource availability* check would work in [10] before a REQUEST reaches an on-tree node and learns the bandwidth requirement. The problem arises if the new member is a source as we assume bidirectional shared-tree multicast here. It is easy to detect if the connecting path can accommodate the amount of traffic that is going to be injected by N, but how does N or A(or any other on-tree node) know if the existing tree can accept at node A(or any other on-tree node) the amount of traffic that is going to be injected by N?

In the above discussion, we illustrate two problems with candidate selection based solely on QoS properties of connecting paths: (1) the "best" connecting path might not necessarily be the best choice; (2) QoS information collected by local search itself is not enough to determine how and whether a new member can be connected to the existing tree

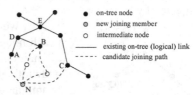

Fig. 2. Multiple Connecting Paths.

if the new member is a source. We need some mechanism to convey the QoS properties and resource (e.g., bandwidth) availability of the existing multicast tree to member nodes to address the above two problems. Without such information readily available, a local search strategy would have to resort to a brutal-force flooding of connection request to the whole tree to find out.

One may think of using RSVP[7]. But because of its reliance on the routing protocol to construct a tree first, it is not a viable solution here. In RSVP, a multicast source periodically sends out PATH messages downstream along the tree, carrying traffic specifications. Receivers can then send RESV messages upstream to reserve required resources. As a signalling protocol, RSVP requires a tree already in place, while the problem here is how to construct a tree that is capable of supporting the QoS required.

4   QoS Information Propagation

In this section we discuss how QoS information can be propagated along a multicast tree to address the problems we have with local-search-based or other alternative path routing strategies. Here first we use bandwidth as an example, which is likely to be the most important metric to consider if QoS support is ever going to be realized. Other metrics, such as end-to-end delay and buffer spaces, can be easily accommodated by our scheme with little or no modification. The procedures discussed here apply to both BGMP-based inter-domain multicast and other bidirectional shared-tree multicast.

For an on-tree node to determine if it is a "feasible" connecting point for a new member, it maintain the following information: total bandwidth of traffic injected by existing members (B_{recv}^t) and the bandwidth that can be injected to the tree "at this point" (B_{feasb}^t, "feasible bandwidth"). Let's look at a very simple tree with four nodes as illustrated in Fig. 3. Assume A and D are source nodes and each has 0.2Mb/s data to transmit, and every (logical) link has available bandwidth 1Mb/s on both directions before the multicast group is using any bandwidth. So A and D are receiving 0.2Mb/s from the group and B and D are receiving 0.4Mb/s from the group. It is also easy to calculate that B_{feasb}^t at node B is 0.6Mb/s: the residual bandwidth from B to A or D is 0.8Mb/s and it is 0.6Mb/s from B to C. B_{feasb}^t at node A and D is also 0.6Mb/s while B_{feasb}^t at node C is 0.8Mb/s. With such information, when receiving a connection request from a new joining receiver member (say, N), a node (say, C) can inform the new member the required available bandwidth and determine whether the connecting path (from C to N) has enough bandwidth. If the new member is also going to be a source, node C also has to check if it can accept the amount of traffic that is going to be injected by N.

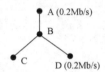

Fig. 3. A Tree with Four Nodes.

Now we describe how such information can be maintained at each node and propagated along the tree. An on-tree router will maintain B_{recv} and B_{feasb} for each neighbor. B_{recv} is the bandwidth of traffic received from that neighbor, and B_{feasb} is the amount of traffic that neighbor can accept (into the tree towards downstream). In BGMP, a neighbor is a *target* in the *target list* if it is a peer in other domain, or another BGMP router within the same domain as to be discussed in more details later. A node periodically sends messages to a neighbor with bandwidth information summarized from B_{recv} and B_{feasb} information of all other neighbors: B_{recv} sent will be the sum of B_{recv} for all other nodes plus bandwidth of local sources, and B_{feasb} sent will be the minimum of all other nodes,

$$\begin{cases} B_{recv} = B(local) + \sum_j B_{recv}(j), \forall neighbor\ j\ except\ x \\ B_{feasb} = min\{B_{feasb}(j), \forall neighbor\ j\ except\ x\}. \end{cases} \quad (1)$$

When node j receives a B_{feasb} from neighbor i, it should replace it with the minimum of B_{feasb} and $B_{available}(link\ j \rightarrow i)$:

$$B_{feasb}(i) = min\{received\ B_{feasb}(i),\ B_{available}(link\ j \rightarrow i)\} \quad (2)$$

In the example discussed above, node A sends B:

$$\begin{cases} B_{recv} = 0.2Mb/s \\ B_{feasb} = \infty, \end{cases} \quad (3)$$

since A has no other neighbor nodes. Information B stores for A will be:

$$\begin{cases} B_{recv}(A) = 0.2Mb/s \\ B_{feasb}(A) = min\{\infty,\ 1Mb/s - 0.2Mb/s\} = 0.8Mb/s. \end{cases} \quad (4)$$

Node B sends C:

$$\begin{cases} B_{recv} = B_{recv}(A) + B_{recv}(D) = 0.4Mb/s \\ B_{feasb} = min\{B_{feasb}(A), B_{feasb}(D)\} = 0.8Mb/s. \end{cases} \quad (5)$$

In the above example, we illustrated how *group shared* bandwidth information is maintained at tree nodes and propagated along the tree. The same can be easily done with delay information. Now assume what we care is the maximum inter-member delay. A node i maintains the following information for each neighbor j: $D^{fr}(j)$=maximum delay from the farthest member to i through j, and $D^{to}(j)$=maximum delay to reach the farthest member through j in the branch down from j. Summary information is maintained as:

$$\begin{cases} D^{fr}_{max} = max\{D^{fr}(j), \forall neighbor\ j\} \\ D^{to}_{max} = max\{D^{to}(j), \forall neighbor\ j\}. \end{cases} \quad (6)$$

The information for i to propagate to neighbor x is:

$$\begin{cases} D^{fr} = max\{D^{fr}(j) + delay(i \rightarrow x), \forall neighbor\ j\ except\ x\} \\ D^{to} = max\{D^{to}(j), \forall neighbor\ j\ except\ x\}. \end{cases} \quad (7)$$

When neighbor x receives the D^{to} from i, it gets

$$D^{to}(i) = received\ D^{to}(i) + delay(x \rightarrow i). \quad (8)$$

4.1   Discussions

Source-Specific QoS Information. BGMP supports source-specific branch which is recursively built from a member sending such a request and stops where it reaches either a BGMP router on the shared tree or a BGMP router in the source domain. To facilitate QoS management on the source-specific branches, source-specific QoS information can be propagated to and maintained at BGMP routers where source-specific group states reside. To do so, a source can send a source-specific QoS update down the tree; all members forward this update with QoS information related to this source in addition to group shared summary to downstream nodes. Any node with source-specific state for this source will store such source-specific QoS information locally while all other nodes don't need to do so and only need to pass it down.

Intra-domain BGMP Peers. In BGMP, internal BGMP peers (BGMP border routers within the same multicast AS) participate in the intra-domain tree construction through an intra-domain multicast protocol. They don't exchange *join* or *prune* messages with each other as with external peers. To support QoS, they would have to exchange QoS update information as above. They connect to each other through an intra-domain multicast tree and can be considered as tree "neighbors" connected through virtual links. When exchanging QoS update information with an internal peer, a tree node should only send "summary" for updates received from external neighbors. While it should send summary for all updates received including that via virtual links when exchanging QoS updates with external peers. Fig. 4 illustrates portion of a multicast tree with an intra-domain tree shown. When node A2 sends QoS information to node A1 and A3, it should only send summary for what received from node C1 and F1; but when it sends QoS information to node C1 and F1, it should include what received from A1 and A3.

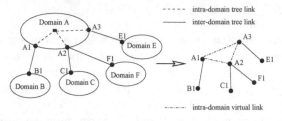

Fig. 4. Portion of an Inter-Domain Multicast Tree with an Intra-Domain Tree Shown.

In the above discussion we assumed that QoS information is propagated by periodical update message exchange between neighbors. The longest propagation delay for a change to reach all routers in the tree is about $D \times T$ in average and $2 \times D \times T$ for worst case, where T is the update timer and D is "diameter" of the tree (i.e., hop-count between two farthest nodes). This delay can be significant if the group gets large or spreads out far. When a member newly joins, especially if it is a source, such change should be reflected at all nodes as soon as possible. To do so, the new member could "order" an immediate flush of update down the whole tree. Any other significant change in the tree such as leaving of a source node or loss of connectivity should also prompt such immediate update flush.

So far we are only concerned with how QoS information is propagated and maintained. Apparently we need some mechanism to provide such information. For example, available bandwidth information on a "virtual link" between two internal peers may be obtained through a *bandwidth broker* as in DiffServ[5]. Availability of bandwidth and other resources may also subject to policy constraints. For example, even physically the available bandwidth between two external BGMP peers is 10Mb/s, a policy or service level agreement may specify that at most 2Mb/s can be used for a multicast or at most 1Mb/s can be used by a single multicast group. Some other metric (for example, delay) can be measurement-based.

5   Alternative Join Point Search

In the existing BGMP, to join a group, the new member sends a join message towards the root domain of the group. It is forwarded hop-by-hop and corresponding group states are installed at each hop, until it reaches a router which already has state for that group or a BGMP router of the root domain. This way, the join path to the existing tree is the unique default route, which may not be able to provide the QoS required. To effectively support QoS, here we propose a search strategy that can provide alternative "join points" in case the default one doesn't work out. This strategy utilizes the "root-based" feature of the MASC/BGMP architecture: mapping of a multicast group to a root domain and the ability to provide a default route to the root domain.

5.1   Scoped On-Tree Search

Our proposed strategy is called *scoped on-tree search*. We discuss this strategy within the BGMP routing architecture and the discussion actually applies to CBT or PIM-SM architecture with little modifications. In this strategy, a new joining member sends join request for the group towards the root domain as usual and it is forwarded by other BGMP routers hop-by-hop. The join request collects QoS information along the way and will eventually hit a BGMP router on the tree or in the root domain. The on-tree (or root domain) router checks QoS feasibility based on the group-shared QoS information concerning the current tree and the QoS information collected by the join request. It sends an acknowledgment message to the new member if QoS requirements are met; otherwise it sends a search request to all neighbors in the tree, with a TTL (time-to-live) field to control the search depth. For example, if the TTL is 2, then a search request is forwarded at most 2 hops away from its sender. An on-tree nodes receiving a search request forwards the request to all neighbors except the one from which the request came, unless TTL reaches 0. It also sends back (by unicast) a reply message to the new member whose address is carried in the search request. Before sending a reply, a BGMP router does a QoS feasibility check if possible and sends reply only if the check is passed.

The way that a reply message is forwarded depends on the functionalities of the routing protocol. To effectively use this search technique, the routing protocol must be able to provide a *multicast path*, which can be different from unicast forwarding path especially in inter-domain cases, back to the joining new member. BGMP provides such functionalities. The protocol specification[21] specifies that: "For a given source-specific group and source, BGMP must be able to look up the next-hop towards the source in the Multicast RIB (routing information base)". This means that a reply message can be effectively sent hop-by-hop by BGMP routers to the joining member since the new member's address can essentially be treated as a source address and the next (multicast) hop can be looked up from multicast RIB. Along the way, BGMP routers can insert the corresponding QoS information. If the joining member doesn't get an acknowledgment, it collects reply messages and selects one satisfying the QoS requirement (or the best one) to send a renewed join. This join message will not be forwarded along the default route; instead it will be forwarded along the route picked up by the reply message.

In Fig. 5, a new joining member (N) sends a join message towards root domain until it reaches on-tree node B. B discovers that the default route cannot provide the required QoS (e.g., delay on that path is too much) or this connecting point cannot provide the required QoS (e.g., B cannot accept the amount of traffic that

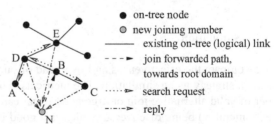

Fig. 5. On-Tree Search for Alternative Join Point for a New Joining Member.

is going to be injected by N), it then sends a *tree-search* request to its neighbors with TTL=2. This message reaches nodes A, C, D and E. They then send reply messages back to N for N to select one to connect to.

5.2   Discussions

Our strategy is proposed to work with the existing routing protocol BGMP (may also work with others that provide similar functionalities), thus it relies on a hierarchical multicast addressing and forwarding (MASC/MBGP) architecture as BGMP. This hierarchical routing architecture eliminates the need and drawbacks of requiring a tree manager in the *multicast tree search* strategy proposed in [14], and provides a nature way to start an on-tree search. In our strategy, a search process is started by an on-tree node that a join request first reaches on its way towards the root domain, and is started only if the default route is not QoS feasible. The search process also only involves a small set of nodes in its neighborhood. Both these features help improve the scalability – or, say, "hurt" the scalability less as every QoS proposal imposes this and that additional requirements and ours is no exception.

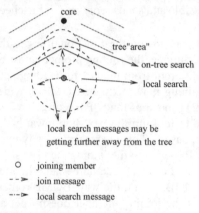

Fig. 6.   Rationale   behind   On-Tree Search.

This strategy is based on the rationale that, the first on-tree node "hit" by a join request is the one close to the joining member, and other on-tree nodes close to this new member must be around the neighborhood. Flooding-based local search is "direction less": it floods search messages to all directions. It is the most effective strategy in the sense it can discover all on-tree nodes within a certain range. But it is also very expensive because it forwards many messages "blindly". This rationale is illustrated in Fig. 6. Simulation results in the next subsection shows that this rationale is indeed very reasonable and the search strategy based on it can find most of the on-tree nodes discovered by flooding-based local search.

5.3   Simulation

We compare the effectiveness and overhead of our strategy with flooding-based local search strategy using simulation. The metric used to compare *effectiveness* is the number of valid alternative join on-tree nodes (that can serve as a connecting point for the new member) being discovered, which is a good indication of the effectiveness since more alternative join nodes being discovered means a better chance for one of them to provide a QoS feasible join path. We only count nodes that are *valid* join points; for example, in Fig. 5, if the join path from N to D has to go through node B, then D is not a valid connecting point (while B is) and will not be counted. This number is called *number of hits* in the following presentation.

One may attempt to compare the *quality* of the corresponding join paths. However, this would require to make assumptions about different QoS characteristics(available bandwidth, delay, etc.) of the networks at inter-domain level. Given the current status of QoS support in real networks, it is extremely difficult to make any realistic or representative assumptions to conduct such simulation.

We use the number of search messages being forwarded counted on a per-hop basis as the overhead. For example, in Fig. 5, a message is forwarded by node B to D and then to A and E, the overhead is counted 3. We do not count the number of reply messages which is proportional to the number of on-tree nodes discovered in our on-tree search and is far less a concern compared with the number of search messages in flooding-based search.

In unrestricted local search, forwarding of a message is terminated only if its TTL reaches 0 or it hits an on-tree node. Thus a node may receive multiple search messages for the same search and forward them multiple times: in Fig. 7, joining node A sends a message to B and C, B and C would again forward the message to each other; message looping may also happen: node A forwards message to B then B to C, which then forwards it back to A, if initial TTL>3. This is necessary if we want to discover not only all nearby on-tree routers but also all the possible paths from the new member to them. This can generate significant overhead and may be hardly necessary. So we also simulate a restricted local search in which a node will forward a search message (for a given joining member) only once (to all neighbors except the one from which it came) and do so only if that neighbor is not already in the path the message travelled so far (loop prevention).

Fig. 7.   A Loop.

The topology used in our simulation consists of 3325 nodes obtained from a BGP routing table dump available from [18], which was collected by the Oregon Route Views Project[19]. A similar set of data was used for the simulation topology in [16]. For each simulation instance, we first construct a shortest-path tree (from the root domain) of a random multicast group of a given size. Then a random non-tree node is chosen as the new member and search strategies are simulated to find alternative join points. Each data point in the figures shown is the average of 1000 instances.

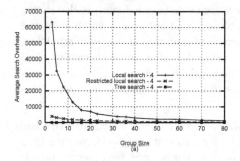
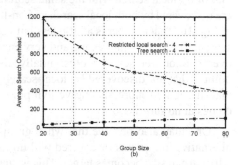

Fig. 8. Average overhead of local search and scoped on-tree search vs. group size. The number following a search method name is the TTL or search depth.

Fig. 8 shows the comparison of average search message overhead with varied group size. Group size varies from 3 to 80; 80 members at inter-domain level represents a fairly large group. The overhead of unrestricted local search is significantly larger than restricted local search and scoped on-tree search with the same TTL, especially when the multicast group is small in which case search messages forwarded may hardly "hit" any on-tree routers before the TTL reaches 0. When the group size increases, there is a better chance for a search message to reach an on-tree node and terminates earlier thus the overhead is significantly reduced. Restricted local search does help reduce the overhead significantly while requiring much more sophisticated message handling (i.e., a node needs to "remember" if it has already forwarded search message for a new joining node). But its overhead is still much larger than scoped on-tree search with the same search depth. Another fine point is that, on-tree search is only conducted when the default *shortest-path* route is not QoS feasible. So if the default route is always "good" enough (say, has enough bandwidth), then the search will not be necessary at all.

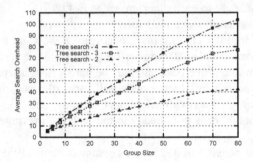

Fig. 9 shows the overhead vs. group size of on-tree search with different search depths. One can see that, when group size increases, so does the overhead. The reason is apparent: when the group becomes larger, so does the *branch degree* of on-tree nodes and thus the number of search messages being forwarded increases.

Fig. 9. Average overhead of scoped on-tree search with different search depths vs. group size.

Fig. 10 presents the average number of hits of local search and on-tree search vs. group size. Both restricted and unrestricted search with the same TTL will discover the same number of join points since both will reach all nodes within the specified TTL range. Fig. 10 shows that the number of hits by on-tree search is a little less than, but very close to, that of by local search with the same search depth (4). The average number of hits that are discovered by both local search and on-tree search with search depth 4 is also shown in Fig. 10. That number is a little less but very close to that of by local search or on-tree search. This is very interesting and very encouraging for on-tree search: with much less overhead, on-tree search are essentially discovering the same set of alternative "join points" that are discovered by local-search. The results for on-tree search also shows that, it can find more alternative join points when the search depth is increased. This comes at no surprise.

One may also observe that when group size is large enough, the number of valid alternative join points discovered will no longer increase but actually decrease a little when group size becomes larger. This is surprising at first thought, but actually there is a reason behind. When multicast group becomes larger, it tends to "spread" closer to the new joining member. A search strategy may be able to find more on-tree nodes, but many of them go through some other on-tree nodes to reach the new member, and the number of on-tree nodes that provide connecting paths without going through any other on-tree node still remains a small number. What that number will be for a specific router depends a lot on its connectivity and location on the network topology. For example, a router of a stub domain may always have only one connecting point to connect to while a router at the backbone may be able to find a fairly large number of alternative connecting points in its neighborhood. When a multicast group becomes sufficiently large (say, almost all nodes are members), then the average number of alternative connecting points will become close to the average node out-degree (when all neighbors are already in the tree). In topology used in our simulation, the average node out-degree is about 3.5.

6   Extending BGMP for QoS Support

Here we discuss the modifications and extensions that need to be made to BGMP to support QoS. First of all, we need to use the method described in Section 4 to propagate QoS information along the tree. In BGMP, peers periodically exchange *join* messages to keep multicast group state (soft-state protocol). Thus periodical QoS update information can be piggybacked with join messages. Of course, message processing at each node now becomes more complicated and more state in-

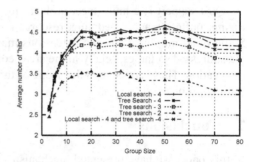

Fig. 10. Average Number of "Hits" vs. Group Size.

formation has to be stored. While it is still an open question how often QoS update should be exchanged so that QoS information is accurate within a certain error margin or "useful" to a meaningful extend. To quickly convey some important QoS update (for example, a new joining source is transmitting at 100Kb/s), a new type of QoS update message should be introduced. This update message originates from a node where such update occurs (e.g., the new source node) and is multicasted along the tree. Any on-tree node receiving this message should process and forward it promptly. In order to scale to support large group, use of such update messages should be limited. In practice, dramatic changes that will trigger this message may not be very frequent after all.

In BGMP, a source can send data to a group without joining that group. For multicast with QoS support, there are reasons to require a source to join the group first before it transmits data, especially if some kind of guaranteed service is required. First of all, without joining a group, a source may not know if the rate of data it is going to transmit can be supported by the network or the existing tree (or delay requirement can be met). At the same time, it is not desirable for other members to see a new source disrupt the service quality in the existing tree. Access control, authentication requirement are among the other reasons: e.g., a group may not want to receive data from any others or only only want to communicate with someone with an authentic identity (for example, a company's internal conference).

A couple of modifications to the join process are also necessary. First of all, to support alternate path routing, a BGMP must be able to support the on-tree search strategy and be able to forward a join message along a join path specified by the new joining member – this is necessary when the default shortest-path route is not feasible and a search procedure finds a feasible one. Moreover, distinctions between a join update and a new join should also be introduced: a new join involves searching for alternative join points if necessary and possibly resource reservation, admission control and an authentication process while a join update is to keep the state and renew the reservation. To determine whether the specified QoS requirements can be met, a new join also involves join acknowledgement and thus be a multiple-phase process. The leaving of a group member can be handled as the same (by sending prune messages). However, if the leaving node is a source node, it should order an immediate QoS information update before it leaves so the the the corresponding resource used can be released and state information (about available bandwidth etc.) can be updated as soon as possible.

QoS support is a multi-facet problem involving traffic classification and/or prioritizing, buffering and scheduling, resource reservation, signaling, routing, and others. Some other modifications and extensions may also be necessary in addition to what we addressed so far, especially in the areas concerning admission control and resource reservation. These are beyond the scope of this paper in which we are focusing on the routing problem.

7   Conclusions

In this paper we have presented mechanisms to support QoS-aware multicast routing in group-shared bidirectional multicast trees within the BGMP inter-domain routing architecture. First we reviewed several existing QoS multicast routing proposals and discussed their drawbacks. We then analyzed some important limitations of these proposals imposed by making join path selection decisions based solely on the QoS characteristics of the alternative join paths, and proposed a QoS information propagation mechanism to address such limitations. After that, we proposed a new scoped on-tree search strategy to discover alternative "join points" for a new joining member if the default join path is not QoS feasible. Simulation results demonstrates that this strategy is as effective as the flooding-based local search strategy but with much less overhead. Finally we briefly discussed some modifications and extensions to BGMP that are necessary to effectively support QoS.

Contributions of our work lie in the followings: (a)a QoS information propagation method in bidirectional shared multicast tree to enable better routing decision making, (b)a low-overhead on-tree search strategy for alternative "join points", (c)applying these to BGMP-based multicast. Though they were presented within the BGMP architecture, both the QoS information propagation method and the alternative join point search strategy can be applied to other similar routing protocol for QoS support as well.

References

1. K. Almeroth, "The Evolution of Multicast: From the MBone to Inter-Domain Multicast to Internet2 Deployment", IEEE Network, January/February 2000.
2. G. Apostolopoulos, S. Kama, D. Williams, R. Guerin, A. Orda, and T. Przygienda, "QoS routing mechanisms and OSPF extensions", IETF RFC 2676, August 1999.

3. A. Ballardie, "Core based trees (CBT version 2) multicast routing: protocol specification", RFC2189, Septermber 1997.
4. T. Bates, R. Chandra, D. Katz, and Y. Rekhter, "Multiprotocol extensions for BGP-4", IETF RFC 2283, Feburary 1998.
5. S. Blake, D. Black, et al., "An architecture for differentiated services", RFC 2475, December 1998.
6. R. Braden, D. Clark, and S. Shenker, "Integrated services in the Internet architecture: an overview", IETF RFC 1633, 1994.
7. R. Braden, L. Zhang, S. Berson, et al., "Resource reservation protocol (RSVP) – version 1 functional specification", IETF RFC 2205, September 1997.
8. K. Carlberg and J. Crowcroft, "Quality of multicast service (QoMS) by yet another multicast (YAM) routing protocol ", in *Proceedings of HIPARCH'98*, June 1998.
9. B. Cain, S. Deering, and A. Thyagarajan, "Internet group management protocol, Version 3", Internet draft: draft-ietf-idmr-igmp-v3-01.txt, Feburary 1999.
10. S. Chen, K. Nahrstedt, and Y. Shavitt, "A QoS-aware multicast routing protocol", in *Proceedings of IEEE INFOCOM'00*, March 2000.
11. S. Deering, D. Estrin, D. Farinacci, et al., "The PIM architecture for wide-area multicast routing", *IEEE/ACM Transaction on Networking*, Vol.4(2), pp.153-162, April 1996.
12. S. Deering, D. Estrin, D. Farinacci, et al., "Protocol independent multicast-sparse mode (PIM-SM): motivation and architecture", Internet draft: draft-ietf-idmr-pim-arch-05.txt{ps}, August 1998.
13. D. Estrin, M. Handley, and D. Thaler, "Multicast-Address-Set advertisement and Claim mechanism", IETF Internet draft: draft-ietf-malloc-masc-05.txt, July 2000.
14. M. Faloutsos, A. Banerjea, and R. Pankaj, "QoSMIC: quality of service sensitive multicast internet protocol", in *Proceedings of ACM SIGCOMM'98*, pp.144-153, September 1998.
15. D. Farinacci, Y. Rekhter, D. Meyer, P. Lothberg, H. Kilmer, and J. Hall, "Multicast source discovery protocol (MSDP)", IETF Internet draft: draft-ietf-msdp-spec-06.txt.
16. S. Kumar, P. Radoslavov, D. Thaler, C. Alaettinoğlu, D. Estrin, and M. Handley, "The MASC/BGMP architecture for inter-domain multicast routing", in *Proceedings of ACM SIGCOMM'98*, pp.93-104, September 1998.
17. J. Moy, "Multicast routing extensions to OSPF", RFC 1584, March 1994.
18. Measurement and Operations Analysis Team, National Laboratory for Applied Network Research: http://moat.nlanr.net/AS/.
19. Oregon Route Views Project: http://www.antc.uoregon.edu/route-views/.
20. C. Partridge, D. Waitzman, and S. Deering, "Distance vector multicast routing protocol", RFC 1075, 1988.
21. D. Thaler, D. Estrin, and D. Meyer, "Border gateway multicast protocol (BGMP): protocol specification", Internet Draft, draft-ietf-bgmp-spec-02.txt, November 2000.
22. D. Zappala, "Alternate path routing in multicast", in *Proceedings of IEEE INFOCOM'00*, March 2000.

Granularity of QoS Routing in MPLS Networks

Ying-Dar Lin[1], Nai-Bin Hsu[1], and Ren-Hung Hwang[2]

[1] Dept of Computer and Info. Science
National Chiao Tung University, Hsinchu, Taiwan
{ydlin,gis84811}@cis.nctu.edu.tw
[2] Dept of Computer Science and Info. Eng.
National Chung Cheng University, Chiayi, Taiwan
rhhwang@cs.ccu.edu.tw

Abstract. This study investigates how the Constraint-based routing decision granularity significantly affects the scalability and blocking performance of QoS routing in MPLS network. The coarse-grained granularity, such as per-destination, has lower storage and computational overheads but is only suitable for best-effort traffic. On the other hand, the fine-grained granularity, such as per-flow, provides lower blocking probability for bandwidth requests, but requires a huge number of states and high computational cost.

To achieve cost-effective scalability, this study proposes using hybrid granularity schemes. The *Overflowed cache* of the per-pair/flow scheme adds a per-pair cache and a per-flow cache as the routing cache, and performs well in blocking probability with a reasonable overflow ratio of 10% as offered load=0.7. *Per-pair/class* scheme groups the flows into several paths using routing marks, thus allowing packets to be label-forwarded with a bounded cache.

1   Introduction

The Internet is providing users diverse and essential Quality of Services (QoS), particularly given the increasing demand for a wide spectrum of network services. Many services, previously only provided by traditional circuit-switched networks, can now be provided on the Internet. These services, depending on their inherent characteristics, require certain degrees of QoS guarantees. Many technologies are therefore being developed to enhance the QoS capability of IP networks. Among these technologies, the *Differentiated Services* (DiffServ) [1,2,3,4] and *Multi-Protocol Label Switching* (MPLS) [5,6,7,8] are the enabling technologies that are paving the way for tomorrow's QoS services portfolio.

The DiffServ is based on a simple model where traffic entering a network is classified, policed and possibly conditioned at the edges of the network, and assigned to different behavior aggregates. Each behavior aggregate is identified by a single *DS codepoint* (DSCP). At the core of the network, packets are fast-forwarded according to the *per-hop behavior* (PHB) associated with the DSCP. By assigning traffic of different classes to different DSCPs, the DiffServ network provides different forwarding treatments and thus different levels of QoS.

L. Wolf, D. Hutchison, and R. Steinmetz (Eds.): IWQoS 2001, LNCS 2092, pp. 140–154, 2001.

MPLS integrates the label swapping forwarding paradigm with network layer routing. First, an explicit path, called the *label switched path* (LSP), is determined, and established using a signaling protocol. A label in the packet header, rather than the IP destination address, is then used for making forwarding decisions in the network. Routers that support MPLS are called *label switched routers* (LSRs). The labels can be assigned to represent routes of various *granularities*, ranging from as coarse as the destination network down to the level of each single flow. Moreover, numerous traffic engineering functions have been effectively achieved by MPLS. When MPLS is combined with DiffServ and Constraint-based routing, they become powerful and complementary abstractions for QoS provisioning in IP backbone networks.

Constraint-based routing is used to compute routes that are subject to multiple constraints, namely explicit route constraints and QoS constraints. Explicit routes can be selected statically or dynamically. However, network congestion and route flapping are two factors contributing to QoS degradation of flows. To reduce blocking probability and maintain stable QoS provision, dynamic routing that considers resource availability, namely QoS routing, is desired.

Once the explicit route is computed, a signaling protocol, either Label Distribution Protocol (CR-LDP) or RSVP extension (RSVP-TE), is responsible for establishing forwarding state and reserve resources along the route. In addition, LSR use these protocols to inform their peers of the label/FEC bindings they have made. Forwarding Equivalence Class (FEC) is a set of packets which will be forwarded in the same manner. Typically packets belonging to the same FEC will follow the same path in the MPLS domain.

It is expected that both DiffServ and MPLS will be deployed in ISP's network. To interoperate these domains, EXP-LSP and Label-LSP models are proposed [7]. EXP-LSP provides no more than eight Behavior Aggregates (BA) classes but scale better. On the other hand, Label-LSP provides finer service granularity but results in more state information.

Path cache [9] memorizes the Constraint-based routing decision and behaves differently with different granularities. The coarse-grained granularity, such as per-destination, all flows moving from different sources to a destination are routed to the same outgoing link, has lower storage and computational overheads but is only suitable for best-effort traffic. On the other hand, the fine-grained granularity, such as per-flow, each individual flow is computed and routed independently, provides lower blocking probability for bandwidth requests, but requires a huge number of states and high computational cost. In per-pair granularity, all traffic between a given source and destination, regardless of the number of flows, travels the same route. Note that in cases of explicit routing, per-destination and per-pair routing decisions are identical.

This study investigates how the granularity of the routing decision affects the scalability of computation or storage, and the blocking probability of a QoS flow request. To reduce the blocking probability without sacrificing per-flow QoS requirement, two routing mechanisms are proposed from the perspective of granularity. The *Per_Pair_Flow* scheme adds a per-pair cache (P-cache) and an over-

flowed per-flow cache (O-cache) as routing cache. The flows that the paths of
P-cache cannot satisfy with the bandwidth requirement are routed individually and their routing decisions overflowed into the O-cache. The *Per_Pair_Class*
scheme aggregates flows into a number of forwarding classes. This scheme reduces the routing cache size and is suitable for MPLS networks, where packets
are labeled at edge routers and fast-forwarded in the core network.

The rest of this paper is organized as follows. Section 2 describes two on-
demand path computation heuristics which our study based on. Section 3 describes the overflowed cache *Per_Pair_Flow* scheme. Section 4 then describes
the *Per_Pair_Class* scheme, which uses a mark scheme to reduce cache size. Subsequently, Sect. 5 presents a simulation study that compares several performance
metrics of various cache granularities, Finally, Sect. 6 presents conclusions.

2   Path Computation

WSP_Routing(F, s, d, b, D)
topology $G(V, E)$; /* width b_{ij} associate with $e_{ij} \in E$ */
flow F;   /* from s to d with req. b and D */
routing entry S_d;
/* set of tuple$(length, width, path)$ from s to d */
shortest path σ;
Begin
    initialize $S_d \leftarrow \phi$, prune e_{ij} if $b_{ij} < b, \forall e_{ij} \in E$
    for hop-count $h = 1$ to H
        $B_s \leftarrow \infty$, $D_s \leftarrow 0$
        find all paths $(s, ..., x, d)$ with h hops, and
        Begin
        update $D_d \leftarrow h$
        $B_d \leftarrow Max\{Min[B_x, b_{xd}]\}, \forall x$
        $\sigma_d \leftarrow \sigma_x \cup d$
        $S_d \leftarrow S_d \cup (D_d, B_d, \sigma_d)$
        End
        if $(S_d \neq \phi)$ pick path σ_d with widest B_d, **stop**
    endfor
End

Fig. 1. Widest-Shortest Path (WSP) Heuristic.

This paper assumes that link state based and explicit routing architecture
are used. Link state QoS routing protocols use reliable flooding to exchange
link state information, enabling all routers to construct the same Link State
Database (LSDB). Given complete topological information and the state of resource availability, each QOS-capable router finds the least costly path that still
satisfies the resource requirements of a flow. Two on-demand shortest path computation heuristics are described as the basis in this study. QOSPF [10] uses the
Widest-Shortest Path (WSP) selection criterion. Figure 1 shows the algorithm

to respond to the route query. Each routing entry of $S_d^h = (D_d^h, B_d^h, \sigma_d^h)$ consists of the shortest length D_d^h, width B_d^h and path σ_d^h to node d, with minimum hops h.

This algorithm iteratively identifies the optimal (widest) paths from itself (i.e. s) to any node d, in increasing order of hop-count h, with a maximum of H hops, and where H can be either the value of diameter of G or can be set explicitly. Afterwards, WSP picks the widest σ of all possible shortest paths to node d as the routing path with minimum hops.

> **CSP_Routing**(F, s, d, b, D)
> topology $G(V, E)$;/* width b_{ij} associate with $e_{ij} \in E$ */
> flow F;   /* from s to d with req. b and D */
> label L;   /* set of labeled nodes */
> shortest path σ;
>   /* obtained by backtracking the inspected nodes */
> Begin
>   1) prune e_{ij} if $b_{ij} < b, \forall e_{ij} \in E$
>   2) initialize $L \leftarrow \{s\}, D_i \leftarrow d_{si}, \forall i \neq s$
>   3) find $x \notin L$ such that $D_x = Min_{i \notin L}[D_i]$
>   /* examine tentative nodes */
>   4) if $D_x > D$, "path not found", **stop**
>   5) $L \leftarrow L \cup \{x\}$
>   6) if $L = V$, return(σ) with $length(\sigma) = D_d$, **stop**
>   7) update $D_i \leftarrow Min[D_i, D_x + d_{xi}], \forall i$ adjacent to x
>   8) go to 3)
> End

Fig. 2. Constrained Shortest Path (CSP) Heuristic.

Another heuristic, *Constrained Shortest Path* (CSP), shown in Fig. 2, uses "minimum delay with abundant bandwidth" as the selection criterion to find a shortest path σ for flow F. Step 1 eliminates all links that do not satisfy the bandwidth requirement b. Next, the CSP simply finds a shortest path σ from itself (i.e. s) to destination d, as in steps 2–8. Step 3 chooses a non-labeled node x with minimum length and x is labeled in step 5. Step 7 updates the length metric for each adjacent node i. Meanwhile, CSP is terminated either in step 4, as the length exceeds the threshold D before reaching destination d, or in step 6, as all nodes are labeled. Consequently, CSP finds a QoS path, $\sigma = s \ldots d$, such that $width(\sigma) \geq b$ and $length(\sigma) \leq D$, to satisfy the bandwidth requirement h and length requirement D.

3   Cache with Per-Pair/Flow Granularity

This section introduces a routing scheme with per-pair/flow hybrid cache granularity. The architecture presented herein uses source routing and hop-by-hop signaling procedure such as CR-LDP or RSVP-TE. Loop-free can be guaranteed in source routing and the signaling procedure prevents each packet of the flow from carrying complete route information. Sets of labels distinguish destination address, service class, forwarding path, and probably also privacy. In MPLS, edge devices perform most of the processor-intensive work, performing application recognition to identify flows and classify packets according to the network policies.

Upon a flow request during the signaling phase, the path-query can be through with by computing path on-demand, or extracting path from the cache. When the query is successful, the source node initiates the hop-by-hop signaling to setup forwarding state and destination node initiates bandwidth reservation backward on each link in the path.

The routing path extracted from the cache could be *misleading*, i.e., flows following a per-destination cache entry might not find sufficient resources along the path, although there exist alternative paths with abundant resources. This lack of resources is attributed to flows of the same source-destination (S-D) pair are routed on the same path led by the cache entry, which is computed merely for the first flow. Therefore, this path might not satisfy the bandwidth requirements of subsequent flows. Notably, the blocking probability increases rapidly when a link of the path becomes a bottleneck.

On the other hand, although no such misleading (assume no *staleness* of link state) occur in the per-flow routing, flow state and routing cache size could be enormous, ultimately resulting in poor scalability. Furthermore, due to the overheads of per-flow path computation, on-demand path finding is hardly feasible in real networks (with high rate requests.) Therefore, path pre-computation is implemented in [10], which asynchronously compute feasible paths to destinations.

The routing cache of this scheme is functionally divided into three parts, a per-pair cache (*P-cache*), an overflowed per-flow cache (*O-cache*), and a per-destination cache (*D-cache*). Shortest paths on the P-cache and the D-cache are pre-computed at the system start-up or can be flushed and computed on-demand under the network administration policy. Entry of the O-cache is created when a request arrives and cannot find sufficient bandwidth on the path in P-cache. By looking up the next-hop in the D-cache, best-effort traffic is forwarded as in a non-QoS support OSPF router. QoS paths are extracted from the P-cache in this scheme.

Figure 3 shows the *Per_Pair_Flow* scheme, is detailed as follows. When a path query with multiple constraints is executed at the ingress LSR, lookup the P-cache for routing information. If the lookup is a miss, it implies that no routing path is stored for the particular request. Therefore, in this situation the *Per_Pair_Flow* invokes the *FindRouteLeastCost* function to find a QoS path. If the path σ is found, this path is stored in the P-cache and the flow request F is sent through σ explicitly. Otherwise, if no path can be found, the request is blocked.

However, if the lookup of the P-cache is a hit, a resource availability check must be made according to the latest link states to ensure the QoS of the flow.

If the check is successful, the signaling message of F is sent according to the P-cache. Meanwhile, if the check fails, function *FindRouteLeastCost* is invoked to find an alternative path based on the information in LSDB and on the Residual Bandwidth Database (RBDB). If a QoS path σ is found, the path is stored in the O-cache, i.e. *overflowed* to the O-cache and signaling of F is sent through σ. Finally, if no path can be found, the flow is blocked.

```
Per_Pair_Flow(F, s, d, b, D)
flow F; /* from s to d with req. b and D */
path σ;
Begin
    case miss(P-cache):
        σ ← FindRouteLeastCost(s, d, b, D)
        if (σ found)
            insert(P-cache), label(F) & route(F) through σ
        else "path not found"
    case σ ← hit(P-cache):
        if (width(σ) ≥ b) and (delay(σ) ≤ D)
            label(F) & route(F) through σ
        else /* overflow */
        Begin
        σ ← FindRouteLeastCost(s, d, b, D)
        if (σ found)
            insert(O-cache), label(F) & route(F) through σ
        else "path not found"
        End
End
```

Fig. 3. *Per_Pair_Flow* Routing.

Function *FindRouteLeastCost* in Fig. 3 on-demand finds a QoS path using WSP or CSP heuristics in Sect. 2. The link cost function in this computation can be defined according to the needs of network administrators. For example, hop counts, exponential cost [11], or distance [12] can be used as the link cost metric in the computing function.

Assuming a flow arrival F from s to d with requirement b, the probability of overflowing the P-cache, namely θ, can be defined as $\theta = Prob(width(\sigma) < b)$, where σ is the path between s and d held in the P-cache. Simulation results show that θ is between zero to 0.3 in a 100 nodes network, depending on the offered load in the *Per_Pair_Flow* scheme. For example, θ is 10% as offered load ρ=0.7, if the forwarding capacity of the router is $100K$ flows, the *Per_Pair_Flow* scheme can reduce the number of routing cache entries from $100K$ to $20K$, including P-cache and O-cache. Additionally, number of cache entries could be further bounded by the scheme, *Per_Pair_Class*, presented in Sect. 4.

4   Cache with Per-Pair/Class Granularity

This section presents another hybrid granularity scheme using a routing *mark* as part of the label in MPLS. Herein, when a flow request arrives at an edge router, it is routed to the nearly best path given the current network state, where the "best" path is defined as the least costly feasible path. Flows between an S-D pair are routed on several different paths and marked accordingly at the source and edge routers. Notably, flows of the same routing path may require different qualities of service. The core router in a MPLS domain uses the label to determine which output port (interface) a packet should be forwarded to, and to determine service class. Core devices expedite forwarding while enforcing QoS levels assigned at the edge.

By limiting the number of routing marks, say to m, the routing algorithm can route flows between each S-D pair along a limited number of paths. The route *pinning* is enforced by stamping packets of the same flow with the same *mark*. Rather than identifying every single flow, the forwarding process at intermediate or core routers is simplified by merely checking the label. The size of the routing cache is bounded to $O(n^2 m)$, where n is the number of network nodes. Note that if the Constraint-based routing is distributed at the edge nodes, this bound reduce to $O(nm)$.

Figure 4 illustrates the structure of the routing cache, which provides a maximum of m feasible routes per node pair. The first path entry LSP_1 can be pre-computed, or the path information can be flushed and computed on-demand under the network administration policy. Besides the path list of LSP, each path entry includes the residual bandwidth ($width$), maximum delay ($length$), and utilization (ρ). Information on the entry can be flushed by the management policy, for instance, refresh timeout or reference counts. Regarding labeling and forwarding, the approach is scalable and suitable for the *Differentiated Services* and MPLS networks.

src, dst	LSP_1	.....	LSP_m
s, d_1	π_{11}, $width_{11}$, $delay_{11}$, ρ_{11}	.....	π_{1m}, ...
s, d_i	π_{i1}, $width_{i1}$, $delay_{i1}$, ρ_{i1}	.....	.....

Fig. 4. Routing Cache in the *Per_Pair_Class* Routing.

Figure 5 describes that upon a flow request F, the *Per_Pair_Class* algorithm first attempts to extract the least costly feasible path π from the routing cache. If the extraction is negative, the scheme attempts to compute the least costly feasible path, termed σ. If σ is found, *Per_Pair_Class* assigns a new mark to σ, inserts this new mark into the routing cache, and then labels/routes the flow request F explicitly through σ. Meanwhile, if π is found and the path is only

```
Per_Pair_Class(F, s, d, b, D)
flow F;                 /* from s to d with req. b and D */
cache entry Π(s, d);/* set of routing paths from s to d */
extracted path π;
computed path σ;
Begin
    initiate cost(NULL) ← ∞
    extract π ∈ Π(s, d)
        that cost(π) is the least & satisfy constraint
            { width(π) ≥ b, length(π) ≤ D, ... }
    case (π not found):
        σ ← FindRouteLeastCost(s, d, b, D)
        if (σ not found) then "path not found"
        insert/replace(σ, Π(s, d)),
        label(F) & route(F) through σ
    case (π is found):
        if (ρ(π) lightly utilized) then
            label(F) & route(F) to π
        endif
        σ ← FindRouteLeastCost(s, d, b, D)
        if (σ not found) then "path not found"
        if (cost(σ) < cost(π)) then /* σ better */
            insert/replace(σ, Π(s, d)),
            label(F) & route(F) through σ
        else /* π better */
            label(F) & route(F) to π
        endif
End
```

Fig. 5. *Per_Pair_Class* Routing with Marks.

lightly utilized, the *Per_Pair_Class* marks the flow F and routes it to path π. Otherwise the flow is blocked. If the utilization of path $\rho(\pi)$ exceeds a pre-defined threshold, the *Per_Pair_Class* can either route F to a π held in the cache, or route F through a newly computed path σ, whichever is least costly. Therefore, traffic flows can be aggregated into the same forwarding class (FEC) and labeled accordingly at the edge routers. Notably, flows of the same FEC may require different service class. Consequently, flows between an S-D pair may be routed on a maximum of m different paths, where m is the maximum number of routing classes. In the *Per_Pair_Class* algorithm, function *FindRouteLeastCost* compute the least cost path using the *constrained shortest path* or *widest-shortest path* heuristics in Sect. 2.

5   Performance Evaluation

This section evaluates the performance of unicast QoS routing, and particularly its sensitivity to various routing cache granularities. The performance of the proposed *Per_Pair_Flow* and *Per_Pair_Class* schemes are evaluated.

5.1   Network and Traffic Model

Simulations were run on 100-node random graphs based on the Waxman's model [13]. In this model, n nodes are randomly distributed over a rectangular coordinate grid, and the distance between each pair of nodes is calculated with the *Euclidean* metric. Then, edges are introduced between pairs of nodes, u, v, with a probability depending on the distance between u and v. The average degree of nodes in these graphs is in the range [3.5, 5]. Each link is assumed to be STM-1 or OC-3 with 155Mbps.

The simulations herein assume that the token rate is used as the bandwidth requirement which is the primary metric. Furthermore, this study assumes that there are two types of QoS traffic, GS_1 has a mean rate of 3 Mbps, while GS_2 has a mean rate of 1.5 Mbps. The flow arrival process is assumed to be independent at each node, following a Poisson model. Flows are randomly destined to the else nodes. The holding time of a flow is assumed to be exponentially distributed with mean μ. The mean holding time can be adapted to keep the offered load at a constant. The link-states of adjacent links of a source router are updated immediately while the states of other links are updated by periodically receiving link state advertisements (LSAs).

5.2   Performance Metrics

From the perspective of cache granularity, this study expects to find QoS routing techniques with a small blocking probability while maintaining scalable computational costs and storage overheads. Thus, several performance metrics are interesting here: (1) Request bandwidth blocking probability, P_{req}, is defined as

$$P_{req} = \frac{\sum rejected\_bandwidth}{\sum rquest\_bandwidth} = P_{rout} + P_{sig} \ . \tag{1}$$

P_{rout} is the routing blocking probability, defined as the probability that a request is blocked due to no existing path with sufficient resources, regardless of cache hit or miss. P_{sig} denotes the signaling blocking possibility, namely the probability that a successfully routed flow gets rejected during the actual backward reservation process, during the *receiver-initiated* reservation process of RSVP. (2) Cache misleading probability, P_{misl}, is the probability of a query hit on the routing cache but the reservation signaling being rejected due to insufficient bandwidth. (3) Normalized routing cache size, or N_{cache}, is the storage overhead per flow for a caching scheme. (4) Normalized number of path computations, or $\widetilde{N}_{comp}$, is the number of path computations per flow in the simulated network.

5.3   Simulation Results

The simulation results are mainly to examine the behavior of the flows under moderate traffic loading (e.g., $\rho=0.7$) where most of the blocking probabilities would not go beyond 20%. Moreover, the 95% *confidence interval* lies within 5% of the simulation average for all the statistics reported here.

Blocking Probability. This experiment focuses on the effects of inaccurate link-state information, due to their update periods, on the performance and overheads of QoS routing. Figure 6 and 7 show the blocking probabilities, P_{req}, on the 100-node random graph with an offered load ρ=0.7; the flow arrival rate λ=1; the mean holding time is adjusted to fix the offered load; the refresh timeout of cache entry (*flush*)=100 units. As described in Sect. 2, CSP and WSP heuristics are used. As expected, larger update periods basically increase flow blocking. The larger update period results in the higher degree of inaccuracy in the link-state, and more changes in network could be unnoticed. As links approach saturated under the inaccuracy, link states and residual bandwidth databases viewed from the source router are likely unchanged, might mistake infeasible path as feasible. In that case, flows are blocked in signaling phase and only admitted if other flows leave.

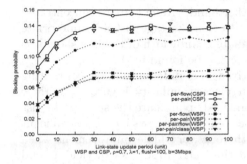
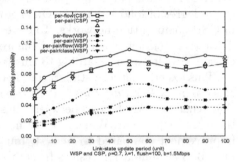

Fig. 6. Blocking Probability with Large Requirement.

Fig. 7. Blocking Probability with small Requirement.

However, blocking does not appear to grow even higher as the update period goes beyond a critical value. Figure 7 shows results of the experiment as in Fig. 6 but with the less bandwidth requirement and longer mean duration of the flow. The climbing of the curves grow slower than those in Fig. 6. This phenomenon suggests that to get more accurate network state and better QoS routing performance, update period (namely the value *MaxLSInterval* in OSPF) should not go beyond the mean holding time of the admitted flows.

Per-pair routing gets higher blocking probability than other granularities. Traffic in the pure per-pair network tend to form bottleneck links and is more imbalanced than in other networks. Conversely, in the per-flow and per-pair/flow networks, the traffic obtains a QoS path more flexibly and has more chances to get alternative paths in large networks.

Intuitively, the finest granularity, per-flow scheme should result in the lowest blocking probability. However, it is not always true in our experiments. In Fig. 6, indeed, the per-flow scheme with CSP has the strongest path computation ability; it could find a feasible route for a flow under heavy load but with a longer length. A flow with longer path utilizes more network resources than a flow with shorter path. Though we limit the number of hops, namely H, of the selected path to the network diameter, the per-flow scheme still admits as many flows as it can. Eventually, network resources are exhausted, new incoming flow is only admitted if other flows are terminated. That is why the per-flow scheme performs similarly to or somewhat poor than the per-pair/flow and per-pair/class schemes that we proposed.

In addition, with the large update periods, stale link-state information reduces the effectiveness of path computation of the per-flow scheme. It is possibly to mistake infeasible path as feasible (*optimistic-leading*), or mistake feasible path as infeasible (*pessimistic-leading*). Thus, it will get more *signaling blocks* in former case and *routing blocks* in latter case; both are negative to the performance of per-flow routing.

Obviously, by comparing the statistics of CSP with WSP, WSP heuristic performs better than CSP in this experiment. WSP uses breadth-first search to find multiple shortest paths and pick one with the widest bandwidth, and achieves some degree of load balancing. On the other hand, traffic is more concentrated in CSP-computation networks. To cope with this shortage in CSP, appropriate link cost functions which consider the available bandwidth of the link should be chosen. Studies with regarding to this issue can be found in [14].

This experiment also studies the effectiveness of different numbers of routing classes, m, of *Per_Pair_Class* with per-pair/class granularity. Figure 8 illustrates that when $m = 1$, all flows between the same S-D pair share the same path, just the same as per-pair. When $m = 2$, the per-pair/class shows its most significant improvement compared to $m = 1$, but there is very little improvement when $m \geq 3$. The simulation results reveal that the *Per_Pair_Class* can yield a good performance with only a very small number of routing alternatives.

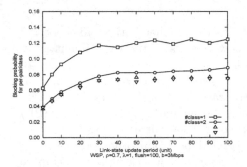
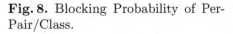
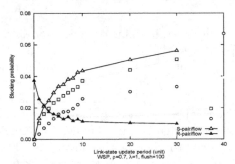

Fig. 8. Blocking Probability of Per-Pair/Class.

Fig. 9. Routing Blocks vs. Signaling Blocks.

Misleading and Staleness. In our architecture, the path cache is used under the link-state update protocol. Both cache-leading and staleness of network state may cause misbehavior of routing schemes. Routing paths extracted from the cache could be optimistic-leading or pessimistic-leading that we mentioned previously. The extracted path also could be misleading, that is, flows following a misleading path might not find sufficient resources along the path, although there exist alternative paths with abundant resources.

In Fig. 9, we have insight into the blocking probability, blocked either in routing phase (prefix "R-") or in signaling phase (prefix "S-"). Look at the performance under accurate network state (i.e. period=0), the *routing blocks* account for all blocked flow requests. As the update period getting larger, more and more flows mistake an infeasible path as feasible path. Therefore, those flows cannot reserve enough bandwidth in the signaling phase and will be blocked. Situation of blocking shifting from routing to signaling phase is caused by the staleness of network state. Rising (P_{sig}) and falling (P_{rout}) curves of each scheme cross over. The cross point is postponed in the per-pair cache scheme. As caching mechanism usually does not reflect accurate network state immediately and thus sensitivity of staleness is reduced.

Figure 10 shows the cache misleading probabilities due to caching, i.e.,

$$P_{misl}(\texttt{scheme}) = P_{req}(\texttt{scheme}) - P_{req}(\texttt{per\_flow}), \qquad (2)$$

of various routing schemes. In case of no staleness of the network state, since the per-flow routing requires path computation for every single flow, there is no misleading, i.e., $P_{misl}(\text{per-flow})=0$, regardless of network load and size. Obviously, the per-pair (or per-destination) scheme obtains the highest misleading from the cache. On the other hand, the per-pair/flow and the per-pair/class have little misleading probability. This is because when looking up the P-cache, if the feasibility check fails due to insufficient residual bandwidth along the path held in the P-cache, per-pair/flow and per-pair/class schemes will find another feasible path for the flow. However, feasible path finding could fail due to some paths are nearly exhausted and trunk reservation in [6] can slow down links from being exhausted.

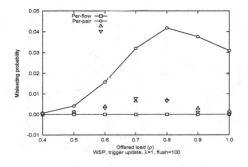

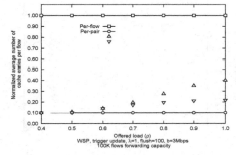

Fig. 10. Cache Misleading Probability.

Fig. 11. Average Number of Cache Entries per Flow.

Cache Size. Figure 11 gives the average number of cache entries for each single flow, i.e. normalized $\widetilde{N}_{cache}$. It indicates that the $\widetilde{N}_{cache}$ of per-flow, per-pair, and per-pair/class schemes remain nearly constant regardless of traffic loading. On the other hand, $\widetilde{N}_{cache}$ of per-pair/flow increases as the traffic load increases. Statistics in Fig. 11 can be verified by the storage complexities as follows.

The cache size of per-pair is bounded by $(n-1)^2$ with $O(n^2)$ complexity, where n is the number of nodes. Metric $\widetilde{N}_{cache}$(per-pair) is relative to the network size and forwarding capacity. Assume the wire-speed router has the forwarding capacity of $100K$ flows, $\widetilde{N}_{cache}$(per-pair) is near to 0.1. Similarly, cache of per-pair/class is bounded by $(n-1)^2m$ and has a complexity of $O(n^2m)$, where m is the number of classes. $\widetilde{N}_{cache}$(per-pair/class) is $(n-1)^2m$ divided by the number of forwarding flows.

Finally, $\widetilde{N}_{cache}$(per-flow)=1 in Fig. 11, and thus, the cache size increases dramatically as the number of flows increases in per-flow scheme, which disallows it to scale well for large backbone networks. Compared to the per-flow, hybrid granularity is used in the per-pair/flow and the per-pair/class schemes, both significantly reducing the cache size to about 10% (in light load) to 20–40% (in heavy load) without increasing the blocking probability, compared to the per-flow in Fig. 7.

Number of Path Computations. Figure 12 compares the average number of path computations per flow, i.e. normalized $\widetilde{N}_{comp}$, of various schemes. This metric primarily evaluates the computational cost. Note that in order to evaluate the effect of granularity in QoS routing, the simulation only uses on-demand WSP and CSP path computation heuristics. However, only plotting curves of WSP are shown, statistics of CSP are almost the same as WSP. Obviously Fig. 12 and 13 have an upper bound, i.e. $\widetilde{N}_{comp}$(per-flow)=1, and a lower bound, i.e. $\widetilde{N}_{comp}$(per-pair) which increases as the number of blocked flows increases. Note that the $\widetilde{N}_{comp}$(per-pair/flow) and $\widetilde{N}_{comp}$(per-pair/class) are quite influenced by the refresh timeout of entry (i.e. *flush*).

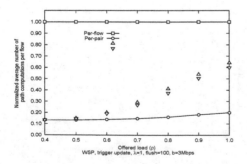

Fig. 12. Average Number of Path Computations per Flow.

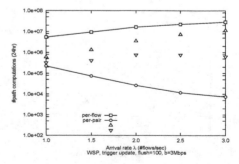

Fig. 13. A Snapshot of Number of Path Computations.

Figure 13 shows the number of path computations regarding to different flow request rate within a specific twenty-four hours period. Statistics of the per-flow, form the upper bound curve and the per-pair form the lower bound. The per-pair/flow and the per-pair/class schemes are in between, and as the loading increases they either increase and approach the upper bound, or decrease and approach the lower bound. The upward-approaching routing schemes, whose number of path computations is dominated by the number of flow request, including per-flow and per-pair/flow schemes. On the other hand, the number of path computations is dominated by its network size and connectivity, including per-pair and per-pair/class schemes.

Figure 13 reveals that "caching" in the per-pair scheme, reduces number of path computations as the loading increases. This is because there are sufficient flows to relay and survive the life time of cache entry, and a greater percentage of succeeding flow requests do not invoke path-computation. These requests simply look up the cache to make routing decisions. On the other hand, in cases of per-flow granularity, since every flow request requires path computation, the number of path computations increases with the offered load. Notably, there is an *inflection* point in the curve of per-pair/class. Multiple entries in the per-pair/class cache lead to this phenomenon. The property of the basic per-pair cache produces the *concave* plotting while the multiple entries produce the *convex* plotting.

6   Conclusions

This study has investigated how granularity affects the Constraint-based routing in MPLS networks and has proposed hybrid granularity schemes to achieve cost effective scalability. The *Per_Pair_Flow* scheme with *per-pair/flow* granularity adds a *P-cache* (per-pair) and an *O-cache* (per-flow) as the routing cache, and performs low blocking probability. The *Per_Pair_Class* scheme with *per-pair/class* granularity groups the flows into several routing paths, thus allowing packets to be label-forwarded with a bounded cache size.

Table 1. Summary of the Simulation Results.

Cache granularity	Compu. overhead	Storage overhead	Blocking	Misleading
Per-pair	★★★	★★★	★	★
Per-flow	★'	★	★★★	★★★
Per-pair/flow	★★	★★	★★★	★★★
Per-pair/class	★★★	★★★	★★★	★★★

★★★: good, ★★:medium, ★:poor
★' can be improved by using path pre-computation.

Extensive simulations are run with various routing granularities and the results are summarized in Table 1. Per-pair cache routing has the worst blocking probability because the coarser granularity limits the accuracy of the network state. The *per-pair/flow* granularity strengthens the path-finding ability just as the per-flow granularity does. Additionally, the *per-pair/class* granularity has small blocking probability with a bounded routing cache. Therefore, this scheme is suitable for the Constraint-based routing in the MPLS networks.

References

1. D. Black, S. Blake, M. Carlson, E. Davies, Z. Wang, and W. Weiss. An architecture for differentiated services. *RFC 2475*, December 1998.
2. K. Nichols, S. Blake, F. Baker, and D. Black. Definition of the differentiated services field (DS Field) in the ipv4 and ipv6 headers. *RFC 2474*, December 1998.
3. T. Li and Y. Rekhter. A provider architecture for differentiated services and traffic engineering(PASTE). *RFC 2430*, October 1998.
4. Y. Bernet, J. Binder, S. Blake, M. Carlson, S. Keshav, E. Davies, B. Ohlman, D. Verma, Z. Wang, and W. Weiss. A framework for differentiated services. *draft-ietf-diffserv-framework-02.txt*, February 1999.
5. R. Callon, P. Doolan, N. Feldman, A. Fredette, G. Swallow, and A. Viswanathan. A framework for multiprotocol label switching. *draft-ietf-mpls-framework-05.txt*, September 1999.
6. Eric C. Rosen, Arun Viswanathan, and Ross Callon. Multiprotocol label switching architecture. *draft-ietf-mpls-arch-06.txt*, August 1999.
7. Francois Le Faucheur, Liwen Wu, Bruce Davie, Shahram Davari, Pasi Vaananen, Ram Krishnan, and Pierrick Cheval. MPLS support of differentiated services. *draft-ietf-mpls-diff-ext-02.txt*, October 1999.
8. Eric C. Rosen, Yakov Rekhter, Daniel Tappan, Dino Farinacci, Guy Fedorkow, Tony Li, and Alex Conta. MPLS label stack encoding. *draft-ietf-mpls-label-encaps-07.txt*, September 1999.
9. G. Apostolopoulos, R. Guerin, S. Kamat, and S. K. Tripathi. On Reducing the Processing Cost of On-Demand QoS Path Computation. In *Proc. of International Conference on Networking Protocols (ICNP)*, Austin, October 1998.
10. G. Apostolopoulos, D. Williams, S. Kamat, R. Guerin, A. Orda, and T. Przygienda. QoS Routing Mechanisms and OSPF extensions. *RFC 2676*, August 1999.
11. J. A. Shaikh. *Efficient Dynamic Routing in Wide-Area Networks*. PhD thesis, University of Michigan, May 1999.
12. Q. Ma, P. Steenkiste, and H. Zhang. Routing High-Bandwidth Traffic in Max-Min Fair Share Networks. In *SIGCOMM'96*. ACM, August 1996.
13. B. M. Waxman. Routing of Multipoint Connections. *IEEE JSAC*, 6(9):1617–1622, December 1988.
14. D. Zappala, D. Estrin, and S. Shenker. Alternate Path Routing and Pinning for Interdomain Multicast Routing. Technical Report 97-655, USC, 1997.

Fault Tolerance and Load Balancing in QoS Provisioning with Multiple MPLS Paths

Scott Seongwook Lee and Mario Gerla

Computer Science Department
University of California, Los Angeles, CA 90095-1596, U.S.A.
{sslee, gerla}@cs.ucla.edu

Abstract. The paper presents approaches for fault tolerance and load balancing in QoS provisioning using multiple alternate paths. The proposed multiple QoS path computation algorithm searches for *maximally disjoint* (i.e., *minimally overlapped*) multiple paths such that the impact of link/node failures becomes significantly reduced, and the use of multiple paths renders QoS services more robust in unreliable network conditions. The algorithm is not limited to finding fully disjoint paths. It also exploits partially disjoint paths by carefully selecting and retaining common links in order to produce more options. Moreover, it offers the benefits of load balancing in normal operating conditions by deploying appropriate call allocation methods according to traffic characteristics. In all cases, all the computed paths must satisfy given multiple QoS constraints. Simulation experiments with IP Telephony service illustrate the fault tolerance and load balancing features of the proposed scheme.

1 Introduction

The problem we address in this paper is how to provide robust QoS services in link/node failure prone IP networks by provisioning multiple paths. So far, successful results have been reported for the QoS support with multiple constraints [1,2]. In certain applications, however, it is important also to guarantee *fault tolerant* QoS paths in the face of network failures. Examples include: video/audio conference interconnection of space launch control stations around the world; conduction of time critical experiments in a virtual collaboration; commander video-conferencing in the battlefield, etc. To achieve this goal, we propose to extend the single path computation algorithm with multiple QoS constraints to a multiple path computation algorithm which still looks for QoS satisfying paths with multiple constraints. The proposed multiple path computation algorithm searches for *maximally disjoint*[1] multiple paths such that the paths are *minimally overlapped* with each other and link failures[2] have the least impact on established connections. The algorithm is not limited to finding fully disjoint paths. Multiple paths are intelligently derived by preserving common links which should be kept in the computation to produce more feasible multiple paths. Our fault tolerant solution exploits those computed multiple

---

[1] The term "maximally disjoint" is interchangeably used with "minimally overlapped."
[2] Node failures are also implied.

L. Wolf, D. Hutchison, and R. Steinmetz (Eds.): IWQoS 2001, LNCS 2092, pp. 155–169, 2001.
© Springer-Verlag Berlin Heidelberg 2001

paths with planned redundancy. Using multiple paths also brings the benefit of load balancing in the network. Load balancing over multiple paths has not been very popular in IP networks. Although the standard IP routing protocols such as OSPF [3] support multipath routing through the computation of equal minimum-cost paths, the actual packet forwarding is usually achieved on a single path. This choice is dictated by several concerns such as resequencing arriving packets, complexity in keeping per-flow states across multiple paths, end-to-end jitter, impact on TCP protocol, etc. In our proposed solution, we will outline and investigate various path management schemes such that the most appropriate scheme can be chosen to fit the application's specific QoS service requirements and traffic characteristics.

While the routing algorithms can find QoS feasible paths, when they exist, between arbitrary node pairs, the next-hop forwarding paradigm in the network is not supportive in making packets follow the computed paths. Thus, pinning down the computed paths is the complementary requirement addressed in this paper. This requirement can be fulfilled by using the explicit routing features of Multiprotocol Label Switching Protocol (MPLS) [4]. An advantage offered by MPLS is aggregated traffic support (a la DiffServ), relaxing the need to keep per-flow state at each router. A related advantage is the ability to dynamically and transparently reconfigure the allocated calls to another path without requiring explicit per flow state change. As implied, using multiple paths with the path pinning capability of MPLS provides not only low latency in response to link failures with high robustness and load balancing, but also fast QoS provisioning by allocating additional calls (in the same aggregate group) to one of the precomputed multiple paths which satisfy the same QoS constraints.

The paper is organized as follows; Section 2 reviews the single QoS path computation algorithm and proposes the multiple QoS path computation for better efficacy in fault tolerance and load balancing. Section 3 defines various path management schemes which make use of the single path and the multiple path algorithms, determine how many paths are to be computed, and provide different call allocation methods to meet the requirements of desirable QoS services according to traffic characteristics. Section 4 presents simulation experiments to show the benefits of provisioning multiple paths for fault tolerance and load balancing.

2   QoS Path Computation Algorithms

Finding QoS paths with multiple QoS constraints consists of two fundamental tasks [2]; distributing network state information and computing paths with the collected network state information. Distributing network state information allows every node in the network to capture the global picture of the current network status. This is usually carried out by the link state flooding mechanism of the conventional OSPF with two modifications; more frequent flooding such that every node in the network can have the latest states and compute QoS paths, and link state packet extensions to accommodate multiple pieces of network state information since each link in the network is associated with multiple

QoS conditions. Q-OSPF [2] is assumed to be ready for these extensions and we mainly focus on the second task; computing paths with multiple QoS constraints and the collected network state information in this section. We summarize the single QoS path computation algorithm and introduce the extensions with which multiple QoS paths can be computed for the specific conditions; fault tolerance and load balancing.

2.1   Single Path Computation

The single QoS path computation algorithm with multiple QoS constraints derives from the conventional Bellman-Ford algorithm as a breadth-first search algorithm minimizing the hop count and yet satisfying multiple QoS constraints [2,5]. This single path computation algorithm will be extended to perform the multiple QoS path computation. Each node in the network builds the link state database which contains all the recent link state advertisements from other nodes. In a Q-OSPF environment, the topological database captures dynamically changing QoS information such as available bandwidth of links, queueing delays, etc. The link state database accommodates all the QoS conditions, and we define each condition as a *QoS metric* and each link in the network is assumed to be associated with multiple QoS metrics which are properly measured and flooded by each node. Each of these QoS metrics has its own properties when operated upon in the path computation. The properties are thoroughly investigated in [1]. However, unlike the issues discussed in [1] (i.e., the optimization of an additional cost metric while minimizing the hop count), we deal with more than one additional cost metric with still the hop count minimized for the QoS path computation. The principal purpose of our single path and multiple path computations is to find the shortest (i.e., min hop) path among those which have enough resources to satisfy given multiple QoS constraints, rather than the shortest path with respect to another cost metric (e.g., maximizing available bandwidth or minimizing end-to-end delay). Each of the QoS metrics is manipulated in the same way as in [1] by increasing the hop count.

Definition 1. *QoS metrics.*

Consider a network represented by a graph $G = (V, E)$ where V is the set of nodes and E is the set of links. Each link $(i, j) \in E$ is assumed to be associated with R multiple QoS metrics. These QoS metrics are categorized mainly into *additive, transitive,* and *multiplicative* ones [6].

In Fig. 1, the QoS metric $q(s, j)$ from node s to node j can be computed by the concatenation of $q(s, i)$ and the QoS metric $q(i, j)$. The concatenation function depends on the QoS metric property. The following lists typical examples of the QoS metrics and concatenation functions.

$$
\begin{aligned}
q(s, j) &= q(s, i) + q(i, j) & \text{delay} \\
q(s, j) &= \min[q(s, i), q(i, j)] & \text{bandwidth} \\
q(s, j) &= q(s, i) \times q(i, j) & \text{delivery probability}
\end{aligned}
\tag{1}
$$

where delay, bandwidth, and delivery probability are called additive, transitive, and multiplicative respectively. Each QoS metric is manipulated by their corresponding concatenation functions, and regardless of the specific properties of

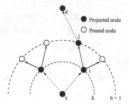

Fig. 1. Pruned and Projected Nodes through the Path Search.

the functions, we generalize and compound the functions such that we have $F = \{F_1, \ldots, F_R\}$ since R multiple QoS metrics are assumed to be associated with each link.

Definition 2. *QoS descriptor.*

Along with these individual QoS metrics, QoS descriptor $D(i,j)$ is defined as a set of multiple QoS metrics associated with link (i,j);

$$D(i,j) = \{q_1(i,j), \ldots, q_R(i,j)\} \tag{2}$$

With $D_{q_l}(i,j)$ defined as l^{th} QoS metric in $D(i,j)$, $D(s,j)$ becomes:

$$\{F_l(D_{q_l}(s,i), D_{q_l}(i,j))|1 \le l \le R\} \tag{3}$$

For simplicity, we assume that the nodes of G are numbered from 1 to n, so $N = \{1, 2, \ldots, n\}$. We suppose without loss of generality that node 1 is the source. $D(i)$ implies the QoS descriptor from the source to node i. Neighbors of node i are expanded with the QoS metrics associated with link (i,j) and $D(j)$ becomes $F(D(i), D(i,j))$ where $j \in N(i)$ and $N(i)$ is the set of neighbors of i. The algorithm iteratively searches for the QoS metrics of all reachable nodes from the source as the hop count increases. In this procedure, the QoS descriptors of the nodes must be checked if they satisfy given QoS constraints. If not satisfying, the nodes of the non-satisfying descriptors are *pruned* to search for only the constraint-satisfying paths.

Definition 3. *Constraint verification.*

A set of multiple QoS constraints is defined as Q and the set is in the same format of D; $Q = \{c_1, \ldots, c_R\}$ such that each QoS metric in D is verified with corresponding constraints. A Boolean function $f_Q(D)$ is also defined to verify if D *satisfies* Q; $f_Q(D) = 1$ if $c_l \in Q$, $q_l \in D$, c_l is satisfied by q_l for all $l \le l \le R$, otherwise $f_Q(D) = 0$. This reflects our initial premise; finding paths of sufficient resources satisfying all the multiple constraints without considering how sufficiently the individual metrics satisfy the constraints.

Being satisfied of c_l by q_l may be abstract and depend on their properties. For instance, if the QoS metric q_l is available bandwidth and the given constraint for that is c_l, c_l can be satisfied by q_l when $c_l \le q_l$, while if c_l is delay constraint, it should be $c_l \geq q_l$ for the constraint to be satisfied. We leave the specific characteristics of individual metrics abstract and assume that they are properly verified by corresponding operators. Now we focus only on verifying if the QoS descriptors of each node satisfy the QoS constraints, rather than verifying whether or not the individual QoS metrics are qualified since the QoS metrics all together must lie in the feasible region of the given QoS constraints. When $f_Q(D(i)) = 0$, node i gets *pruned* by the algorithm. Otherwise, it is *projected* and its neighbors are further expanded until the destination d is reached. Fig. 1 illustrates this operation.

We now define more concretely the single path computation algorithm which iteratively projects reachable nodes with qualified QoS metrics. To determine not only the validity of the QoS metrics, but also where the projected nodes pass, we add an extra field in the QoS descriptor to keep track of the preceding node from which the node of the descriptor is expanded, and $D(i)$ becomes $\{q_1(i), \ldots, q_R(i), p\}$. To find the complete path, we can follow the pointers p in the descriptors backwards from the destination to the source after the algorithm successfully finds the desired destination.

Definition 4. *The single path computation algorithm.*

The algorithm starts with the initial state; $T = \{D(1)\}$ and $P = \emptyset$ where T is a temporary set of QoS descriptors and initialized with the QoS descriptor of source node 1, and P is the set which collects all the qualified QoS descriptors of projected nodes. After this initialization, the algorithm runs a loop of steps like the following searching for the destination d.

```
         h = 0
         while D(d) ∉ P and T ≠ ∅ do
Step 0:     T' = ∅
            for each D(i) ∈ T do
                if f_Q(D(i)) = 1 then T' ← D(i)
Step 1:     for each D(i) ∈ T' do
                if D(i) ∉ P then P ← D(i)
                else discard D(i) from T'
Step 2:     T = ∅
            for each D(i) ⊂ T' do
                for each j ∈ N(i) do
                    if j ≠ D_p(i) then T ← F'(D(i), D(i, j))
                    else skip j
            h = h + 1
```

where h is the hop count. F' is a new generic concatenation function which performs the same operations as F and an additional operation on D_p. When the qualified nodes in T' are projected (i.e., $T \leftarrow F'(D(i), D(i, j))$), $D_p(j)$ becomes i and this is to record the preceding node of each projected node so that the final path can be tracked in reverse from destination node d to source node 1.

The algorithm consists of three substantial parts; pruning non-qualified nodes with improper QoS metrics (i.e., Step 0), projecting qualified nodes satisfying the QoS constraints (i.e., Step 1), and expanding the QoS descriptors by increasing hop count (i.e., Step 2). Note that Step 1 allows the qualified nodes to be projected only if the nodes are not yet projected. Therefore, when a node is projected into P, its hop count h at the moment is minimal since the nodes are projected as the algorithm runs by increasing the hop count. Other paths to the same node may be found later after the hop count becomes larger, but they will not be accepted since P already contains the node with a shorter hop count than any other possible paths. This fact leads the algorithm to finding nodes with the shortest paths to the nodes, and if the algorithm ends with the destination found, the path to the destination also becomes the shortest path from the source.

Eventually, when the loop ends, P becomes $\{D(i)|f_Q(D(i)) = 1\}$, which means that P contains all the qualified reachable nodes from the source. If the desired destination d has not been included in P, then it means that there is no such a path satisfying the multiple QoS constraints. Otherwise, the final path to the destination can be generated by selecting the precedences $D_p(i)$ of the nodes in P from the destination d, and with the previous discussion about the shortest path, it can be easily proved that the algorithm eventually finds the shortest path to the destination with the multiple QoS constraints all satisfied as long as the destination has been successfully projected.

2.2   Multiple Path Computation

In this section, we present the multiple path computation algorithm. There have been many studies under the *Finding K Shortest Paths* [7,8] issues since 1950 for various applications including fault tolerance and load balancing. Most algorithms are limited to defining *alternate* paths without consideration of QoS constraints.

Definition 5. *Path computation conditions.*

In order to search for multiple alternate paths to provide fault tolerance and load balancing yet satisfying QoS constraints, we define alternate paths with the following conditions:

- Satisfying given QoS constraints
- *Maximally disjoint* from already computed paths
- Minimizing hop count

The proposed QoS constrained and path disjoint problem is obviously NP-complete and very hard to solve optimally. We propose here a heuristic solution. Moreover, we do not limit ourselves to strictly "path disjoint" solutions. Rather, we search for multiple, maximally disjoint paths (i.e., with the least overlap among each other) such that the failure of a link in any of the paths will still leave (with high probability) one or more of the other paths operational. Having relaxed the "disjoint path" requirement, the path computation still consists in finding *QoS-satisfying* paths, with a modified objective function that includes the degree of path non-overlap. The multiple path computation algorithm can then be derived from the single path computation algorithm with simple modifications as shown in Section 2.1. This multiple path computation algorithm produces incrementally a single path at each iteration rather than multiple paths at once. All the previously generated paths are kept into account in the next path computation.

Definition 6. *New or old paths.*

We augment the QoS descriptor with two new variables, n for "new" and o for "old," to keep track of the degree of being disjoint. These two variables are updated by checking if the node of the QoS descriptor has been already included in any previously computed paths. n increases when the node is not included in any previously computed paths, and o increases when it is detected in those paths. These two variables play the most important role in the multiple path computation algorithm such that a path *maximally disjoint* from previously computed paths can be found. $D(i)$ becomes $\{q_1(i), \ldots, q_R(i), p, n, o\}$.

Definition 7. *The multiple path computation algorithm.*

We build the multiple path computation algorithm by extending the single path algorithm, and the algorithm becomes the following with the modifications underlined.

```
          h = 0
          while T ≠ ∅ do
Step 0:      T' = ∅
             for each D(i) ∈ T do
                if f_Q(D(i)) = 1 then T' ← D(i)
Step 1:      for each D(i) ∈ T' do
                if D(i) ∉ P then P ← D(i)
                else if D'_o(i) > D_o(i)
                         or (D'_o(i) = D_o(i) and D'_n(i) > D_n(i)) then P ← D(i)
                else discard D(i) from T
Step 2:      T = ∅
             for each D(i) ∈ T' do
                if i = d then skip i
                for each j ∈ N(i) do
                   if j ≠ D_p(i) then T ← F''(D(i), D(i,j))
                   else skip j
             h = h + 1
```

where $D'(i)$ in Step 1 means the QoS descriptor of node i which has been already projected and included in P. Compared to the single path computation, the termination condition becomes $T \neq 0$ since the algorithm always looks for a newer path by investigating all qualified nodes and does not stop just after finding the destination. With this termination condition, Step 2 does not expand next hops from destination node d if it is already projected, and this is to keep the neighbors of d from being reached through d with unnecessary routes going through d. The iteration runs over the entire nodes until no further expansion can occur and P becomes to include all qualified nodes and this comprehensive iteration constructs newer routes to each node. In Step 2, F'' is a new function based on F' in the single path computation. It not only does the same operation as F' does but also updates the new two variables in the QoS descriptor, n and o. For this new operation, we must assume that the algorithm keeps all paths which are the results of the algorithm and F'' checks if the link between i and its neighbor j has been already included in any previously computed paths. If the link is found in the previously computed paths, $D(j)$ expanded from $D(i)$ through the link (i, j) becomes to have $D_o(i) + 1$ for its $D_o(j)$. Otherwise, $D_n(j)$ becomes $D_n(i) + 1$. The values of o and n are used in Step 1 which determines which route is newer and more preferable in terms of hop counts. Although the algorithm prefers newer paths compared to previously computed paths, it does not mean that the algorithm finds longer paths. Instead, it looks for a path more disjoint from previously computed paths, but if several new paths exist and their degree of being disjoint are the same, the shortest one among them is selected. This is achieved by the new condition in Step 1, **if** $D'_o(i) > D_o(i)$ **or** $(D'_o(i) = D_o(i)$ **and** $D'_n(i) > D_n(i))$ **then** $P \leftarrow D(i)$. This process can be more clearly illustrated as in Fig. 2.

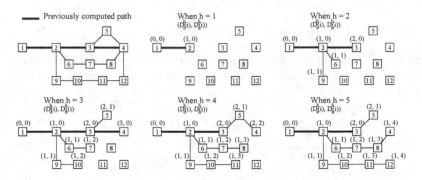

Fig. 2. The Multiple Path Algorithm Illustration.

The sample network in Fig. 2 is assumed to have a previously computed path (i.e., the thick line), and all the nodes are reachable with given QoS constraints all satisfied (i.e., $f_Q(D(i)) = 1$ for all $i = \{1, \ldots, 12\}$), and the multiple path algorithm is running again to find the second path for destination node 4. When $h = 3$, the destination is found through the old path, but the algorithm does not stop here as explained earlier. When $h = 4$, the preceding node of the destination is changed to node 5 because the path through node 5 is newer than the path found at $h = 3$. When $h = 5$, the destination becomes reachable through another newer path which is more preferable than the one found at $h = 4$ in terms of being more disjoin. When $h = 6$, it is not depicted in the figure though, the third new path is verified through node 12, but it is less preferable than the path found at $h = 5$ since both paths have the same number of old links (i.e., $D_o = 1$) in their paths but the lastly found path has more hops. Thus, the path found at $h = 5$ is kept and the preceding node of d remains the same. All these determinations are carried out by the new condition in Step 1.

In addition, if the algorithm were to find alternate paths by removing all the links previously found (i.e., the links in thick lines in Fig. 2), its subsequent runs would not be able to find alternate paths at all since link $(1, 2)$ is actually a *bridge* which makes the graph two separate ones if it is removed, and explicitly finding bridges in a graph prior to actual path computation usually takes extra time and becomes harder as the complexity of the topology increases. Therefore, such a naive approach may not sufficiently provide multiple paths with which network services become robust in unreliable network conditions. On the other hand, the multiple path computation algorithm we defined here is certainly able to intelligently find maximally disjoint multiple paths without explicitly finding bridges first.

3   Path Management Schemes

In this section, we make use of the path computation algorithms and define path management schemes which determine how many paths are computed for each destination and how calls are allocated to them. These schemes primarily deal with *path sets*. The path sets are collections of paths which are computed with the same constraints. Each path set is associated with a certain destination and a set of multiple QoS constraints, and all the elements in the path set are the results of the path computations for the same destination with the constraints. The number of paths in each path set is determined by the path computation algorithm that the network system uses.

As the first operational aspect of running the path computation algorithms, we consider where QoS constraints are derived from and how they are bounded to corresponding path sets. QoS constraints can be either defined by applications or provided by the network. In the former, the QoS constraints are assumed to be continuous variables. An application can choose and negotiate an arbitrary set of QoS constraints with the network. This allows applications broad flexibility in defining QoS constraints, but makes it difficult for the network to precompute and provision QoS paths. For the network, it would be more convenient to enforce limited sets of QoS constraints, to which the applications must abide. We refer to this pre-definition of constraints *constraint quantization*.

When applications provide their sets of QoS constraints, the constraints are assumed to be non-quantized and a new path set is created whenever a call request comes in. The demand placed on system resources for the path management is proportional to the number of admitted calls. On the other hand, the quantization allows the network to limit the number of possible QoS constraint sets and forces applications to select the one that best fits their traffic characteristics. Therefore, the calls with same quantized constraints and same destination will be aggregated together (a la DiffServ) and use the same path sets. The path set for a certain quantized constraint is opened by the first call, and is shared by all subsequent calls in the same group without path recomputation. This *fast QoS provisioning* is another benefit of the constraint quantization.

In the quantized option, when multiple paths are available, a further choice is available regarding *call allocation*. As a first choice, each call is allocated to a specific path in the set. When the specific call allocation to a single path is exercised, the network system performs *flow-based* load balancing. A second option is to perform *packet-based* load balancing by spreading the packets over the multiple paths regardless of the calls they belong.

As the possible combinations of the above concepts, the following five path management schemes are defined.

3.1   Non-quantized Constraint Flow-Based Single Path (NFSP)

This is the simplest approach of using the single path computation. The constraints are arbitrarily given by applications and each path set has only one path. Thus, each call is allocated specifically to a single QoS path. This is the fundamental QoS routing mechanism used in [5] and the path computation is carried out whenever a new call request comes in. The non-quantization property leads this to having each path set owned by each call. Obviously, this scheme is prone to link failures due to lack of alternate (standby) paths. However, all the packets follow the same path and this produces relatively low jitter and no out-of-sequence packets. With respect to load balancing, no special gain is obtained.

3.2   Quantized Constraint Flow-Based Single Path (QFSP)

Conceptually, this scheme is expected to possess the same properties of NFSP in terms of the network performance such as the call admission rate, the degree of being prone to link failures, the load balancing, etc. But, practically, it is expected to provide the fast QoS provisioning since it comes with the constraint quantization and also system resources are saved and shared for the path management.

3.3   Non-quantized Constraint Packet-Based Multiple Paths (NPMP)

Since this scheme, as the name implies, does not perform the constraint quantization, each call owns their path sets. Therefore, further specific call allocation to a single path in the path set cannot be provided for the reason described earlier, and the packet-based load balancing is the only possibility. Now that this scheme is equipped with the multiple path computation and spreads packets over multiple paths which are *maximally disjoint* from each other, it is highly robust against link failures by simply withdrawing broken paths from the path set and utilizing the rest paths as soon as link failures are detected. The withdrawal is performed transparently to application. The robustness against link failures is expected to remain until all the rest paths in the path set get lost due to their link failures. In addition, the withdrawal of broken paths and switching over to the rest paths will not harm the QoS guarantee since all the paths in the path set were computed with the same QoS constraints and the traffic load has been spread over multiple paths, which could be handled by a single path. Besides, multiple paths will be always kept as many as the network and the system configuration allows. However, if high link failure rate comes along and no more alternate paths are left in the path set, finally the allocated calls get aborted. Although we expect high robustness from this scheme, it inevitably produces higher jitter than the other flow-based schemes since packets follow multiple different paths. Thus, this scheme is expected to be practically efficient when applications need fine granules of the QoS constraints for their specific traffic characteristics and they are not susceptible to relatively high jitter.

3.4   Quantized Constraint Flow-Based Multiple Paths (QFMP)

This scheme quantizes the QoS constraints and allocates calls to specific paths in their path sets. This is the only scheme performing the specific allocation among the multiple path management schemes here. This scheme also benefits the low jitter property with the flow-based call allocation and high robustness against link failures with multiple paths. The reaction to link failures is different from the packet-based schemes like NPMP. When a broken path is detected due to link failures, the allocated calls to the path are switched over to another valid paths in the path set. This redistribution process of the calls to other paths can be executed evenly over the rest valid paths such that the valid paths have the same amount of load. Like other multiple path management schemes, if no more valid paths are left, the allocated calls get aborted. With respect to load balancing, it allocates calls over multiple paths evenly and this leads to a certain type of load balancing with the granule of flows. The benefit of this type of load balancing can be maximized when there are sufficient flows to make use of all paths in the path set, which also means the case where many calls for the same destination arise, but this might not be practical. However, this scheme is expected to be the most efficient approach if many calls for the same destinations arrive in unreliable conditions since this scheme makes each flow follow the same paths resulting in low jitter and the QoS services robust with multiple paths with quantized constraints.

3.5   Quantized Constraint Packet-Based Multiple Paths (QPMP)

Like NPMP, this scheme spreads packets over multiple paths resulting in high robustness against link failures and load balancing with the granule of packets. In addition, this scheme quantizes QoS constraints and lets the same path sets be shared for the same destinations resulting in low resource requirement. The reaction to link failures is the same as in NPMP, and this scheme also inherits inevitably the high jitter generation by the nature of the packet-based call allocation. This scheme must be the most cost-effective approach among the multiple path management schemes in terms of requiring system resources since the constraint quantization relieves the heavy use of resources and the packet-based call allocation simplifies the response to link failures without redistribution of allocated calls.

Three of the above five path management schemes, NPMP, QFMP, and QPMP, run the multiple path computation algorithm and the number of possible multiple paths is quite closely related to network topology. Within the topological limit, we can also bound the maximum number of multiple paths for each path set to avoid superfluous use of resources with marginal gains. For the sake of simplicity, we defined two conditional factors. The first one is just a hard boundary such that the number of multiple paths does not exceed a certain number which presumably produces the maximal gains yet minimizing the system resources. This can be determined by system administrators. The second factor is the end-to-end delay difference d between the longest and the shortest paths in the path set. For instance, if there exists an application which produces data in a certain period of time t and does not want to let packets arrive out of order, we can limit the number of multiple paths with $d \le t$. This is the case of the IP Telephony experiments in Section 4.

4   Simulation Experiments

In this section, we present simulation experiments performed with SENSE[3] to illustrate the various benefits of having multiple paths over single paths. We compare the five path management schemes with respect to fault tolerance and load balancing.

The traffic model in the simulation experiments is IP Telephony described in [5]. The traffic is generated by the application layer; 160 bytes per every 20 ms with "talk" probability of 35 %. Thus, the peak rate of the traffic is 64 Kbps and the average rate about 23 Kbps. The topology for the experiments is 6×6 grid with all the same link capacities of 1.5 Mbps and the same link propagation delays of 1 ms. Call requests arrive periodically at fixed rates and their duration is exactly 60 sec for all cases. The simulation time is 600 sec for all cases. The first set of experiments is to compare the performance of the five path management schemes in unreliable networks. In this scenario, each link in the topology fails and recovers with predefined rates. Their failures are randomly generated by the exponential distribution. The call interarrival time is fixed to 1 sec. Thus, for the entire 600-sec simulation time, exactly 600 calls are generated over the network so that the path management schemes are all fairly compared. Each call involves two random nodes and two independent paths one in each direction.

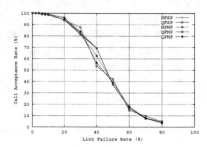

Fig. 3. Call Acceptance Rates of the Five Path Management Schemes over the Unreliable Network.

Fig. 3 shows the call acceptance rate when the link failure rates change. Regardless of the path management schemes, the acceptance rates are quite similar for all cases. That is because all the schemes are running either the single path or the multiple path computation algorithms and that always allows the

---

[3] Network simulator developed at High Performance Internet research group in Network Research Laboratory (NRL): http://www.cs.ucla.edu/NRL/hpi

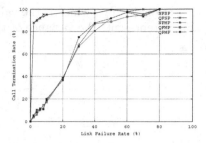

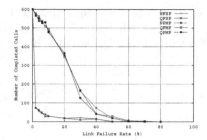

Fig. 4. Call Termination Rates of Accepted Calls Due to Link Failures.

Fig. 5. The Actual Number of Safely Completed Calls without Being Affected by Link Failures.

schemes to find a feasible path if there is one. Thus, the acceptance rates depend primarily on the network capacity of valid paths, not on the path management schemes. Fig. 4 and 5 show the call termination rates due to link failures during the active period of the calls and the actual number of safely completed calls respectively. These figures clearly show that the three multiple path management schemes outperform the two single path management schemes in the face of link failures. As depicted, when calls are allocated to single paths, they become highly vulnerable even at very low failure rates. On the other hand, when calls are allocated to multiple paths, they become quite robust and the gain in terms of fault tolerance is relatively high when the link failure rate ranges between 0 % and 10 %.

The second set of experiments examines load balancing across paths. As described in Section 3, the benefit of using the flow-based call allocation is that all packets follow the same paths. Out-of-sequence packets are avoided and no extra jitter is incurred. However, the flow-based allocation does not take full advantage of load balancing (on a packet by packet basis). In the packet-based allocation, packet spreading causes higher jitter than in the flow-based allocation. In the following load balancing and the jitter experiments, we assume no link failures. We evaluate link load variance across the entire network. We used the following equation to determine the overall load variance.

$$V(t) = \frac{\sum_{i=1}^{L} l_i^2(t)}{L} - \left[\frac{\sum_{i=1}^{L} l_i(t)}{L}\right]^2, \quad V = \frac{\int_0^T V(t)dt}{T}$$

where $l_i(t)$ is the network load of link i monitored at time t and L is the total number of links in the network. As explained, the network topology is 6×6 grid and the total number of links is 60. Each link is full-duplex and the load on a direction is independent from the load of the other direction. Thus, the 60 physical links here are equivalent to 120 unidirectional trunk, which is the value for L. T means the entire simulation time and V represents the variance of the load averaged over the simulation time.

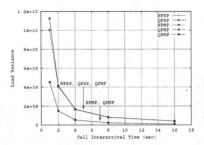
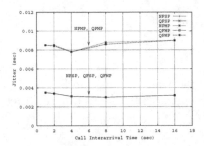

Fig. 6. Load Variance over All Links in the Network.

Fig. 7. Average Jitters of All the Delivered Packets.

Fig. 6 shows the comparisons of the load variances when the path management schemes are applied. As shown, all the three flow-based allocation schemes, NFSP, QFSP, and QFMP, have relatively higher variance than the packet-based ones. This shows the benefit of load balancing by spreading packets over multiple paths. In the case of QFMP, no significant gain is obtained although the path management keeps multiple paths. That is because, as briefly explained in Section 3.4, the scheme allocates in turn calls to single paths in the path sets of their destinations and if there are not many calls for the same destination, this call allocation effect becomes similar to the one of QFSP. This is the reason that at the higher call arrival rate, QFMP shows improvement, but as the call arrival rate decreases, the gain becomes insignificant as shown in the figure.

We also need to consider the jitter that packets experience with the different path management schemes. For this investigation, we also applied to the same simulation configuration of the previous case and measured the average jitter of all packets. We use the definition of jitter that is standardized in the specification of RTCP in [9], which is a running average of end-to-end delay differences of successive paths. Fig. 7 shows the differences of the average jitters caused by the path management schemes. As expected, the packet-based allocation schemes, NPMP and QPMP, have twice as much jitter than the other schemes. Yet, the jitter is still rather modest even for packet based schemes.

As shown through these experiments, the use of multiple paths surely brings significant benefits in response to link failures and in utilizing network links evenly by balancing the load. Especially, these benefits are maximized when multiple paths are efficiently computed as our multiple path algorithm finds *maximally disjoint* (i.e., *minimally overlapped*) multiple paths.

5   Conclusion

We examined the conventional single path computation for QoS support and extended it to compute multiple paths more efficiently in order to improve fault tolerance and load balancing. The multiple path computation algorithm searches for maximally disjoint paths so that the impact of link failures is significantly reduced and links in the network are more evenly utilized by spreading the network load over multiple paths. Still, the multiple paths computed by the algorithm must satisfy given multiple QoS constraints. This source-based computation approach becomes practical in conjunction with the packet-forwarding technology of MPLS such that packets are driven to follow the favorable paths to meet the given QoS constraints even in unreliable network conditions as the experiments proved.

References

1. D. Cavendish and M. Gerla, *Internet QoS Routing using the Bellman-Ford Algorithm*, Proceedings of IFIP Conference on High Performance Networking, Austria, 1998.
2. R. Guerin, A. Orda, and D. Williams, *QoS Routing Mechanisms and OSPF Extensions*, In Proceedings of GLOBECOM'97, Vol. 3, pp. 1903-1908, Phoenix, Arizona.
3. J. Moy, *OSPF Version 2*, RFC 2328, April 1998.
4. E. Rosen, A. Viswanathan, and R. Callon, *Multiprotocol Label Switching Architecture*, RFC 3031, January 2001.
5. A. Dubrovsky, M. Gerla, S. S. Lee, and D. Cavendish, *Internet QoS Routing with IP Telephony and TCP Traffic*, In Proceedings of ICC'00, New Orleans, June 2000.
6. A. Fei and M. Gerla, *Smart Forwarding Technique for Routing with Multiple QoS Constraints*, In Proceedings of GLOBECOM'00.
7. A. W. Brander and M. C. Sinclair, *A Comparative Study of k-Shortest Path Algorithms*, In Proceedings of 11th UK Performance Engineering Workshop, Liverpool, September 1995.
8. D. Eppstein, *Finding the k Shortest Paths*, SIAM Journal on Computing, vol.28, (no.2), SIAM, 1998.
9. H. Schulzrinne, S. Casner, R. Frederick, and V. Jacobson, *RTP: A Transport Protocol for Real-Time Applications*, RFC 1889, January 1996.

On Selection of Paths for Multipath Routing

Srihari Nelakuditi and Zhi-Li Zhang

Deparment of Computer Science & Engineering
University of Minnesota, Minneapolis, MN55455
{srihari,zhzhang}@cs.umn.edu

Abstract. Multipath routing schemes distribute traffic among multiple paths instead of routing all the traffic along a single path. Two key questions that arise in multipath routing are *how many paths are needed* and *how to select these paths*. Clearly, the number and the quality of the paths selected dictate the performance of a multipath routing scheme. We address these issues in the context of the proportional routing paradigm where the traffic is proportioned among a few "good" paths instead of routing it all along the "best" path. We propose a hybrid approach that uses both *globally* exchanged link state metrics — to identify a set of good paths, and *locally* collected path state metrics — for proportioning traffic among the selected paths. We compare the performance of our approach with that of global optimal proportioning and show that the proposed approach yields near-optimal performance using only a few paths. We also demonstrate that the proposed scheme yields much higher throughput with much smaller overhead compared to other schemes based on link state updates.

1   Introduction

It has been shown [21] that *shortest path* routing can lead to unbalanced traffic distribution — links on frequently used shortest paths become increasingly congested, while other links are underloaded. The *multipath* routing is proposed as an alternative to single shortest path routing to distribute load and alleviate congestion in the network. In multipath routing, traffic bound to a destination is split across multiple paths to that destination. In other words, multipath routing uses multiple "good" paths instead of a single "best" path for routing. Two key questions that arise in multipath routing are how many paths are needed and how to find these paths. Clearly, the number and the quality of the paths selected dictate the performance of a multipath routing scheme. There are several reasons why it is desirable to minimize the number of paths used for routing. First, there is a significant overhead associated with establishing, maintaining and tearing down of paths. Second, the complexity of the scheme that distributes traffic among multiple paths increases considerably as the number of paths increases. Third, there could be a limit on the number of explicitly routed paths such as label switched paths in MPLS [16] that can be setup between a pair of nodes. Therefore it is desirable to use *as few paths as possible* while at the same time *minimize the congestion* in the network.

For judicious selection of paths, some knowledge regarding the (global) network state is crucial. This knowledge about resource availability (referred to as *QoS state*) at

L. Wolf, D. Hutchison, and R. Steinmetz (Eds.): IWQoS 2001, LNCS 2092, pp. 170–184, 2001.
© Springer-Verlag Berlin Heidelberg 2001

network nodes, for example, can be obtained through (periodic) information exchange among routers in a network. Because network resource availability changes with each flow arrival and departure, maintaining *accurate* view of network QoS state requires *frequent* information exchanges among the network nodes and introduces both communication and processing overheads. However, these updates would not cause significant burden on the network as long as their frequency is not more than what is needed to convey connectivity information in traditional routing protocols like OSPF [11]. The QoS state of each link could then be piggybacked along with the conventional link state updates. Hence it is important to devise multipath routing schemes that *work well even when the updates are infrequent*.

We propose such a scheme *widest disjoint paths* (wdp) that uses proportional routing — the traffic is proportioned among a few widest disjoint paths. It uses *infrequently* exchanged *global* information for selecting a few good paths based on their long term available bandwidths. It proportions traffic among the selected paths using *local* information to cushion the short term variations in their available bandwidths. This *hybrid* approach to multipath routing adapts at different time scales to the changing network conditions. The rest of the paper discusses what type of global information is exchanged and how it is used to select a few good paths. It also describes what information is collected locally and how traffic is proportioned adaptively.

1.1 Related Work

Several multipath routing schemes have been proposed for balancing the load across the network. The Equal Cost Multipath (ECMP) [11] and Optimized Multipath (OMP) [20,21] schemes perform packet level forwarding decisions. ECMP splits the traffic *equally* among multiple equal cost paths. However, these paths are determined statically and may not reflect the congestion state of the network. Furthermore, it is desirable to apportion the traffic according to the quality of each path. OMP is similar in spirit to our work. It also uses updates to gather link loading information, selects a set of best paths and distributes traffic among them. However, our scheme makes routing decisions at the flow level and consequently the objectives and procedures are different.

QoS routing schemes have been proposed [3,5,10,22] where flow level routing decisions are made based upon the knowledge of the resource availability at network nodes and the QoS requirements of flows. This knowledge is obtained through global link state information exchange among routers in a network. These schemes, which we refer to as global QoS routing schemes, construct a global view of the network QoS state by piecing together the information about each link, and perform path selection based solely on this global view. Examples of global QoS routing schemes are *widest shortest path* [5], *shortest widest path* [22], and *shortest distance path* [10]. While *wdp* also uses link state updates, the nature of information exchanged and the manner in which it is utilized is quite different from global QoS routing schemes. In Section 4, we demonstrate that *wdp* provides higher throughput with lower overhead than these schemes.

Another approach to path selection is to precompute maximally disjoint paths [19] and attempt them in some order. This is static and overly conservative. What matters is not the sharing itself but *the sharing of bottleneck links*, which change with network conditions. In our scheme we dynamically select paths such that they are disjoint *w.r.t* bottleneck links.

The rest of the paper is organized as follows. In Section 2, we introduce the proportional routing framework and describe a global optimal proportional routing procedure (*opr*) and a localized proportional routing scheme *equalizing blocking probability* (ebp). In both these cases, the candidate path set is *static* and large. In Section 2.4, we propose a hybrid approach to multipath routing that selects a few good paths *dynamically* using global information and proportions traffic among these paths using local information. Section 3 describes such a scheme *wdp* that selects widest disjoint paths and proportions traffic among them using *ebp*. The simulation results evaluating the performance of *wdp* are shown in Section 4. Section 5 concludes the paper.

2   Proportional Routing Framework

In this section, we first lay out the basic assumptions regarding the proportional routing framework we consider in this paper. We then present a global optimal proportional routing procedure (*opr*), where we assume that the traffic loads among all source-destination pairs are known. The *opr* procedure gives the least blocking probability that be achieved by a proportional routing scheme. However, it is quite complex and time consuming. We use the performance of *opr* as a reference to evaluate the proposed scheme. We then describe a localized adaptive proportioning approach that uses only locally collected path state metrics and assigns proportions to paths based on their quality. The localized schemes are described in detail in [12,13], a brief summary of which is reproduced here. We then present our proposed hybrid approach to multipath routing that uses global information to select a few good paths and employs localized adaptive proportioning to proportion traffic among these paths.

2.1   Problem Setup

In all the QoS routing schemes considered in this paper we assume that source routing (also referred to as explicit routing) is used. More specifically, we assume that the network topology information is available to all source nodes (e.g., via the OSPF protocol), and one or multiple explicit-routed paths or label switched paths are set up *a priori* between each source and destination pair using, e.g., MPLS [16]. Flows arriving at a source to a destination are routed along one of the explicit-routed paths (hereafter referred to as the *candidate* paths between the source-destination pair). For simplicity, we assume that all flows have the same bandwidth requirement — one unit of bandwidth. When a flow is routed to a path where one or more of the constituent links have no bandwidth left, this flow will be blocked. The performance metric in our study will be the overall blocking probability experienced by flows. We assume that flows from a source to a destination arrive randomly with a Poisson distribution, and their holding time is exponentially distributed. Hence the offered traffic load between a source-destination pair can be measured as the product of the average flow arrival rate and holding time. Given the offered traffic load from a source to a destination, the task of proportional QoS routing is to determine how to distribute the load (i.e., route the flows) among the candidate paths between a source and a destination so as to minimize the overall blocking probability experienced by the flows.

2.2 Global Optimal Proportioning

The global optimal proportioning has been studied extensively in the literature (see [17] and references therein). Here it is assumed that each source node knows the complete topology information of the network (including the maximum capacity of each link) as well as the offered traffic load between every source-destination pair. With the global knowledge of the network topology and offered traffic loads, the *optimal* proportions, for distributing flows among the paths between each source-destination pair, can be computed as described below.

Consider an arbitrary network topology with N nodes and L links. For $l = 1, \ldots, L$, the maximum capacity of link l is $\hat{c}_l > 0$, which is assumed to be fixed and known. The links are unidirectional, i.e., carry traffic in one direction only. Let $\sigma = (s, d)$ denote a source-destination pair in the network. Let λ_σ denote the average arrival rate of flows arriving at source node s destined for node d. The average holding time of the flows is μ_σ. Recall that each flow is assumed to request one unit of bandwidth, and that the flow arrivals are Poisson, and flow holding times are exponentially distributed. Thus the offered load between the source-destination pair σ is $\nu_\sigma = \lambda_\sigma / \mu_\sigma$.

Let $\hat{R}_\sigma$ denote the set of *feasible* paths for routing flows between the pair σ. The global optimal proportioning problem can be formulated [6,7,9] as the problem of finding the optimal proportions $\{\alpha_r^*, r \in \hat{R}_\sigma\}$ where $\sum_{r \in \hat{R}_\sigma} \alpha_r^* = 1$, such that the overall flow blocking probability in the network is minimized. Or equivalently, finding the optimal proportions $\{\alpha_r^*, r \in \hat{R}_\sigma\}$ such that the total carried traffic in the network, $W = \sum_\sigma \sum_{r \in \hat{R}_\sigma} \alpha_r \nu_\sigma (1 - b_r)$ is maximized. Here b_r is the blocking probability on path r when a load of $\nu_r = \alpha_r \nu_\sigma$ is routed through r. Then the set of *candidate* paths R_σ are a subset of feasible paths $\hat{R}_\sigma$ with proportion larger than a negligible value ϵ, i.e., $R_\sigma = \{r : r \in \hat{R}_\sigma, \alpha_r^* > \epsilon\}$. This global optimal proportional routing problem is a constrained nonlinear optimization problem and can be solved using an iterative procedure based on the Sequential Quadratic Programming (SQP) method [4,15].

2.3 Localized Adaptive Proportioning

The optimal proportioning procedure described above requires global information about the offered load between each source-destination pair. It is also quite complex and thus time consuming. We have shown [12] that it is possible to obtain near-optimal proportions using simple localized strategies such as equalizing blocking probability *ebp* and equalizing blocking rate *ebr*. Let $\{r_1, r_2, \ldots, r_k\}$ be the set of k candidate paths between a source destination pair. The objective of the *ebp* strategy is to find a set of proportions $\{\alpha_{r_1}, \alpha_{r_2}, \ldots, \alpha_{r_k}\}$ such that flow blocking probabilities on all the paths are equalized, i.e., $b_{r_1} = b_{r_2} = \cdots = b_{r_k}$, where b_{r_i} is the flow blocking probability on path r_i. On the other hand, the objective of the *ebr* strategy is to equalize the flow blocking rates, i.e., $\alpha_{r_1} b_{r_1} = \alpha_{r_2} b_{r_2} = \cdots = \alpha_{r_k} b_{r_k}$. By employing these strategies a source node can adaptively route flows among multiple paths to a destination, in proportions that are commensurate with the *perceived* qualities of these paths. The perceived quality of a path between a source and a destination is inferred based on locally collected flow statistics: the offered load on the path and the resulting blocking probability of the flows routed along the path.

In this work, we use a simpler approximation to *ebp* that computes new proportions as follows. First, the current average blocking probability $\bar{b} = \sum_{i=1}^{k} \alpha_{r_i} b_{r_i}$ is computed. Then, the proportion of load onto a path r_i is decreased if its current blocking probability b_{r_i} is higher than the average $\bar{b}$ and increased if b_{r_i} is lower than $\bar{b}$. The magnitude of change is determined based on the relative distance of b_{r_i} from $\bar{b}$ and some configurable parameters to ensure that the change is gradual. The mean time between proportion computations is controlled by a configurable parameter θ. This period θ should be large enough to allow for a reasonable measurement of the quality of the candidate paths. The blocking performance of the candidate paths are observed for a period θ and at the end of the period the proportions are recomputed. A more detailed description of this procedure can be found in [14].

2.4   Hybrid Approach to Multipath Routing

The global proportioning procedure described above computes optimal proportions α_r^* for each path r given a feasible path set $\hat{R}_\sigma$ for each source-destination pair σ. Taking into account the overhead associated with setting up and maintaining the paths, it is desirable to minimize the number of candidate paths while minimizing the overall blocking probability. However achieving both the minimization objectives may not be practical. Note that the blocking probability minimization alone, for a fixed set of candidate paths, is a constrained nonlinear optimization problem and thus quite time consuming. Minimizing the number of candidate paths involves experimenting with different combinations of paths and the complexity grows exponentially as the size of the network increases. Hence it is not feasible to find an optimal solution that minimizes both the objectives. Considering that achieving the absolute minimal blocking is not very critical, it is worthwhile investigating heuristic schemes that *tradeoff slight increase in blocking for significant decrease in the number of candidate paths*.

The localized approach to proportional routing is simple and has several important advantages. However it has a limitation that routing is done based solely on the information collected locally. A network node under localized QoS routing approach can judge the quality of paths/links only by routing some traffic along them. It would have no knowledge about the state of the rest of the network. While the proportions for paths are adjusted to reflect the changing qualities of paths, the candidate path set itself remains static. To ensure that the localized scheme adapts to varying network conditions, many feasible paths have to be made candidates. It is not possible to preselect a few good candidate paths statically. Hence it is desirable to *supplement localized proportional routing* with a mechanism that *dynamically selects a few good candidate paths*.

We propose such a hybrid approach to proportional routing where locally collected path state metrics are supplemented with globally exchanged link state metrics. A set of few good candidate paths R_σ are maintained for each pair σ and this set is updated based on the global information. The traffic is proportioned among the candidate paths using local information. In the next section we describe a hybrid scheme *wdp* that selects widest disjoint paths and uses *ebp* strategy for proportioning traffic among them.

3   Widest Disjoint Paths

In this section, we present the candidate path selection procedure used in *wdp*. To help determine whether a path is good and whether to include it in the candidate path set, we define *width* of a path and introduce the notion of *width* of a *set of paths*. The candidate path set R_σ for a pair σ is changed only if it increases the width of the set R_σ or decreases the size of the set R_σ without reducing its width. The widths of paths are computed based on link state updates that carry *average residual bandwidth* information about each link. The traffic is then proportioned among the candidate paths using *ebp*.

A basic question that needs to be addressed by any path selection procedure is what is a "good" path. In general, a path can be categorized as good if its inclusion in the candidate path set decreases the overall blocking probability considerably. It is possible to judge the utility of a path by measuring the performance with and without using the path. However, it is not practical to conduct such inclusion-exclusion experiment for each feasible path. Moreover, each source has to independently perform such trials without being directly aware of the actions of other sources which are only indirectly reflected in the state of the links. Hence each source has to try out paths that are likely to decrease blocking and make such decisions with some local objective that leads the system towards a global optimum.

When identifying a set of candidate paths, another issue that requires attention is the sharing of links between paths. A set of paths that are good *individually* may not perform as well as expected *collectively*. This is due to the sharing of *bottleneck* links. When two candidate paths of a pair share a bottleneck link, it may be possible to remove one of the paths and shift all its load to the other path without increasing the blocking probability. Thus by ensuring that candidate paths of a pair do not share bottleneck links, we can reduce the number of candidate paths without increasing the blocking probability. A simple guideline to enforce this could be that the candidate paths of a pair be mutually disjoint, i.e., they do not share *any* links. This is overly restrictive, since even with shared links, some paths can cause reduction in blocking if those links are not congested. What matters is not the sharing itself but *the sharing of bottleneck links*. While the sharing of links among the paths is *static* information independent of traffic, identifying bottleneck links is *dynamic* since the congestion in the network depends on the offered traffic and routing patterns. Therefore it is essential that candidate paths be *mutually disjoint w.r.t bottleneck links*.

To judge the quality of a path, we define *width* of a path as the the residual bandwidth on its bottleneck link. Let $\hat{c}_l$ be the maximum capacity of link l and ν_l be the average load on it. The difference $c_l = \hat{c}_l - \nu_l$ is the average residual bandwidth on link l. Then the *width* w_r of a path r is given by $w_r = \min_{l \in r} c_l$. The larger its width is, the better the path is, and the higher its potential is to decrease blocking. Similarly we define *distance* [10] of a path r as $\sum_{l \in r} \frac{1}{c_l}$. The shorter the distance is, the better the path is. The widths and distances of paths can be computed given the residual bandwidth information about each link in the network. This information can be obtained through periodic link state updates. To discount short term fluctuations, the *average residual bandwidth* information is exchanged. Let τ be the update interval and u_l^t be the utilization of link l during the period $(t - \tau, t)$. Then the average residual bandwidth at time t, $c_l^t = (1 - u_l^t)\hat{c}_l$. Hereafter without the superscript, c_l refers to the most recently updated value of the average residual bandwidth of link l.

To aid in path selection, we also introduce the notion of *width* for a *set of paths* R, which is computed as follows. We first pick the path r^* with the largest width w_{r^*}. If there are multiple such paths, we choose the one with the shortest distance d_{r^*}. We then decrease the residual bandwidth on all its links by an amount w_{r^*}. This effectively makes the residual bandwidth on its bottleneck link to be 0. We remove the path r^* from the set R and then select a path with the next largest width based on the just updated residual bandwidths. Note that this change in residual bandwidths of links is local and only for the purpose computing the width of R. This process is repeated till the set R becomes empty. The sum of all the widths of paths computed thus is defined as the *width of R*. Note that when two paths share a bottleneck link, the width of two paths together is same as the width of a single path. The width of a path set computed thus, essentially accounts for the sharing of links between paths. The narrowest path, i.e., the last path removed from the set R is referred to as NARROWEST(R).

Based on this notion of width of a path set, we propose a path selection procedure that *adds* a new candidate path only if its inclusion *increases the width*. It *deletes* an existing candidate path if its exclusion *does not decrease* the total width. In other words, each modification to the candidate path set either *improves the width* or *reduces the number* of candidate paths. The selection procedure is shown in Figure 1. First, the load contributed by each existing candidate path is deducted from the corresponding links (lines 2-4). After this adjustment, the residual bandwidth c_l on each link l reflects the load offered on l by all source destination pairs other than σ. Given these adjusted residual bandwidths, the candidate path set R_σ is modified as follows.

The benefit of inclusion of a feasible path r is determined based on the number of existing candidate paths (lines 6-8). If this number is below the specified limit η, the resulting width W_r is the width of $R_\sigma \cup r$. Otherwise, it is the width of $R_\sigma \cup r\backslash$ NARROWEST($R_\sigma \cup r$), i.e., the width after excluding the narrowest path among $R_\sigma \cup r$. Let W^+ be the largest width that can be obtained by adding a feasible path (line 9). This width W^+ is compared with width of the current set of candidate paths. A feasible path is made a candidate if its inclusion in set R_σ increases the width by a fraction ψ (line 10). Here $\psi > 0$ is a configurable parameter to ensure that each addition improves the width by a significant amount. It is possible that many feasible paths may cause the width to be increased to W^+. Among such paths, the path r^+ with the shortest distance is chosen for inclusion (lines 11-13). Let r^- be the narrowest path in the set $R_\sigma \cup r$ (line 14). The path r^- is replaced with r^+ if either the number of paths already reached the limit or the path r^- does not contribute to the width (lines 15-16). Otherwise the path r^+ is simply added to the set of candidate paths (lines 17-18). When no new path is added, an existing candidate path is deleted from the set if it does not change the width (lines 20-22). In all other cases, the candidate path set remains unaffected. It is obvious that this procedure always either increases the width or decreases the number of candidate paths.

```
1.    PROCEDURE SELECT(σ)
2.        For each path r in R_σ
3.            For each link l in r
4.                c_l = c_l + (1 − b_r)ν_r
5.        If |R_σ| < η
6.            W_r = WIDTH(R_σ ∪ r), ∀r ∈ R̂_σ \ R_σ
7.        Else
8.            W_r = WIDTH(R_σ ∪ r \ NARROWEST(R_σ ∪ r)), ∀r ∈ R̂_σ \ R_σ
9.        W⁺ = max_{r ∈ R̂_σ \ R_σ} W_r
10.       If (W⁺ > (1 + ψ) WIDTH(R_σ))
11.           R⁺ = {r : r ∈ R̂_σ \ R_σ, W_r = W⁺}
12.           d⁺ = min_{r ∈ R⁺} d_r
13.           r⁺ = {r : r ∈ R⁺, d_r = d⁺}
14.           r⁻ = NARROWEST(R_σ ∪ r)
15.           If (|R_σ| = η or WIDTH(R_σ ∪ r⁺ \ r⁻) = W⁺)
16.               R_σ = R_σ ∪ r⁺ \ r⁻
17.           Else
18.               R_σ = R_σ ∪ r⁺
19.       Else
20.           r⁻ = NARROWEST(R_σ)
21.           If WIDTH(R_σ \ r⁻) = WIDTH(R_σ)
22.               R_σ = R_σ \ r⁻
23.   END PROCEDURE
```

Fig. 1. The Candidate Path Set Selection Procedure for Pair σ.

It should be noted that though *wdp* uses link state updates it does not suffer from the *synchronization* problem unlike global QoS routing schemes such as *wsp*. There are several reasons contributing to the stability of *wdp*: 1) The information exchanged about a link is its *average* not *instantaneous* residual bandwidth and hence less variable; 2) The traffic is proportioned among few "good" paths instead of loading the "best" path based on inaccurate information; 3) Each pair uses only a few candidate paths and makes only incremental changes to the candidate path set; 4) The new candidate paths are selected for a pair only after deducting the load contributed by the current candidate paths from their links. Due to such adjustment even with link state updates, the view of the network for each node would be different; 5) When network is in a stable state of convergence, the information carried in link state updates would not become outdated and consequently each node would have reasonably accurate view of the network. Essentially the nature of information exchanged and the manner in which it is utilized work in a mutually beneficial fashion and lead the system towards a stable optimal state.

4 Performance Analysis

In this section, we evaluate the performance of the proposed hybrid QoS routing scheme *wdp*. We start with the description of the simulation environment. First, we compare the performance *wdp* with the optimal scheme *opr* and show that *wdp* converges to near-optimal proportions. Furthermore, we demonstrate that the performance of *wdp* is relatively insensitive to the values chosen for the configurable parameters. We then contrast the performance of *wdp* with global QoS routing scheme *wsp* in terms of the overall blocking probability and routing overhead.

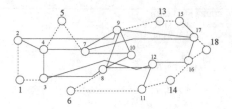

Fig. 2. The Topology Used for Performance Evaluation.

4.1 Simulation Environment

The Figure 4 shows the *isp* topology used in our study. This topology of an ISP backbone network is also used in [1,10]. For simplicity, all the links are assumed to be bidirectional and of equal capacity in each direction. There are two types of links: *solid* and *dotted*. All solid links have same capacity with C_1 units of bandwidth and similarly all the dotted links have C_2 units. The dotted links are the access links and for the purpose of our study their capacity is assumed to be higher than solid links. Otherwise, access links become the bottleneck limiting the impact of multipath routing and hence not an interesting case for our study. Flows arriving into the network are assumed to require one unit of bandwidth. Hence a link with capacity C can accommodate at most C flows simultaneously.

The flow dynamics of the network is modeled as follows (similar to the model used in [18]). The nodes labeled with bigger font are considered to be source (ingress) or destination (egress) nodes. Flows arrive at a source node according to a Poisson process with rate λ. The destination node of a flow is chosen randomly from the set of all nodes except the source node. The holding time of a flow is exponentially distributed with mean $1/\mu$. Following [18], the offered network load on *isp* is given by $\rho = \lambda N\bar{h}/\mu(L_1C_1 + L_2C_2)$, where N is the number of source nodes, L_1 and L_2 are the number of solid and dotted links respectively, and $\bar{h}$ is the mean number of hops per flow, averaged across all source-destination pairs. The parameters used in our simulations are $C_1 = 20$, $C_2 = 30$, $1/\mu = 1$ minute (here after written as just m). The topology specific parameters are $N = 6$, $L_1 = 36$, $L_2 = 24$, $\bar{h} = 3.27$. The average arrival rate at a source node λ is set depending upon the desired load ρ.

The parameters in the simulation are set as follows by default. Any change from these settings is explicitly mentioned wherever necessary. The values for configurable parameters in *wdp* are $\psi = 0.2$, $\tau = 30\ m$, $\theta = 60\ m$, $\xi = 180\ m$. For each pair σ, all the paths between them whose length is at most one hop more than the minimum number of hops is included in the feasible path set $\hat{R}_\sigma$. The amount of offered load on the network ρ is set to 0.55. Each run simulates arrival of $1,000,000$ flows and the results corresponding to the later half the simulation are reported here.

4.2 Performance of *wdp*

In this section, we compare the performance of *wdp* and *opr* to show that *wdp* converges to near-optimal proportions using only a few paths for routing traffic. We also demonstrate that *wdp* is relatively insensitive to the settings for the configurable parameters.

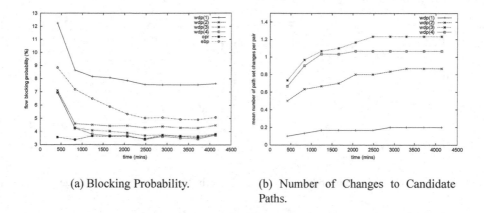

(a) Blocking Probability.

(b) Number of Changes to Candidate Paths.

Fig. 3. Convergence Process of *wdp*.

Convergence. Figure 3 illustrates the convergence process of *wdp*. The results are shown for different values of $\eta = 1 \cdots 4$. Figure 3(a) compares the performance of *wdp*, *opr* and *ebp*. The performance is measured in terms of the overall flow blocking probability, which is defined as the ratio of the total number of blocks to the total number of flow arrivals. The overall blocking probability is plotted as a function of time. In the case of *opr*, the algorithm is run offline to find the optimal proportions given the set of feasible paths and the offered load between each pair of nodes. The resulting proportions are then used in simulation for statically proportioning the traffic among the set of feasible paths. The *ebp* scheme refers to the localized scheme used in isolation for adaptively proportioning across all the feasible paths. As noted earlier all paths of length either minhop or minhop+1 are chosen as the set of feasible paths in our study.

There are several conclusions that can be drawn from Figure 3(a). First, the *wdp* scheme converges for all values of η. Given that the time between changes to candidate path sets, ξ, is 180 *m*, it reaches steady state within (on average) 5 path recomputations per pair. Second, there is a marked reduction in the blocking probability when the number of paths allowed, η, is changed from 1 to 2. It is evident that there is quite a significant gain in using multipath routing instead of single path routing. When the limit η is increased from 2 to 3 the improvement in blocking is somewhat less but significant. Note that in our topology there are at most two paths between a pair that do not share any links. But there could be more than two paths that are mutually disjoint w.r.t bottleneck links. The performance difference between η values of 2 and 3 is an indication that we only need to ensure that candidate paths do not share congested links. However using more than 3 paths per pair helps very little in decreasing the blocking probability. Third, the *ebp* scheme also converges, albeit slowly. Though it performs much better than *wdp* with single path, it is worse than *wdp* with $\eta = 2$. But when *ebp* is used in conjunction with path selection under *wdp* it converges quickly to lower blocking probability using only a few paths. Finally, using at most 3 paths per pair, the *wdp* scheme approaches the performance of optimal proportional routing scheme.

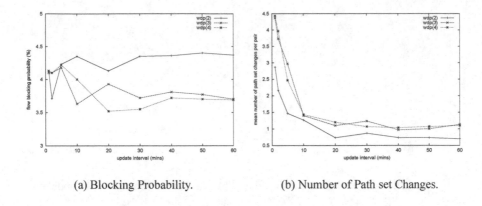

(a) Blocking Probability.   (b) Number of Path set Changes.

Fig. 4. Sensitivity of *wdp* to Update Interval τ.

Figure 3(b) establishes the convergence of *wdp*. It shows the average number of changes to the candidate path set as a function of time. Here the change refers to either addition, deletion or replacement operation on the candidate path set R_σ of any pair σ. Note that the cumulative number of changes are plotted as a function of time and hence a plateau implies that there is no change to any of the path sets. It can be seen that the path sets change incrementally initially and after a while they stabilize. Thereafter each pair sticks to the set of chosen paths. It should be noted that starting with at most 3 minhop paths as candidates and making as few as 1.2 changes to the set of candidate paths, the *wdp* scheme achieves almost optimal performance.

We now compare the average number of paths used by a source-destination pair for routing. Note that in *wdp* scheme η only specifies the maximum allowed number of paths per pair. The actual number of paths selected for routing depends on their widths. The average number of paths used by *wdp* for η of 2 and 3 are 1.7 and 1.9 respectively. The number of paths used stays same even for higher values of η. The *ebp* scheme uses all the given feasible paths for routing. It can measure the quality of a path only by routing some traffic along that path. The average number of feasible paths chosen are 5.6. In case of *opr* we count only those paths that are assigned a proportion of at least 0.10 by the optimal offline algorithm. The average number of such paths under *opr* scheme are 2.4. These results support our claim that *ebp* based proportioning over widest disjoint paths performs almost like optimal proportioning scheme while using fewer paths.

Sensitivity. The *wdp* scheme requires periodic updates to obtain global link state information and to perform path selection. To study the impact of update interval on the performance of *wdp*, we conducted several simulations with different update intervals ranging from 1 *m* to 60 *m*. The Figure 4(a) shows the flow blocking probability as a function of update interval. At smaller update intervals there is some variation in the blocking probability, but much less variation at larger update intervals. It is also clear that increasing the update interval does not cause any significant change in the blocking probability. To study the effect of update interval on the stability of *wdp*, we plotted the average number of path set changes as a function of update interval in Figure 4(b). It shows that the candidate path set of a pair changes often when the updates are frequent. When the update interval is small, the average residual bandwidths of links resemble their instantaneous values, thus highly varying. Due to such variations, paths may appear wider or narrower than they actually are, resulting in unnecessary changes to candidate paths. However, this does not have a significant impact on the blocking performance due to adaptive proportional routing among the selected paths. For the purpose of reducing overhead and increasing stability, we suggest that the update interval τ be reasonably large, while ensuring that it is much smaller than the path recomputation interval ξ. We have also varied other configurable parameters and found that *wdp* is relatively insensitive to the values chosen. For more details, refer to [14].

4.3   Comparison of *wsp* and *wdp*

We now compare the performance of hybrid QoS routing scheme *wdp* with a global QoS routing scheme *wsp*. The *wsp* is a well-studied scheme that selects the widest shortest path for each flow based on the global network view obtained through link state updates. The information carried in these updates is the residual bandwidth at the instant of the update. Note that *wdp* also employs link state updates but the information exchanged is average residual bandwidth over a period not its instantaneous value. We use *wsp* as a representative of global QoS routing schemes as it was shown to perform the best among similar schemes such as shortest widest path (*swp*), shortest distance path (*sdp*). In the following, we first compare the performance of *wdp* with *wsp* in terms of flow blocking probability and then the routing overhead.

Blocking Probability.   Figure 5(a) shows the blocking probability as a function of update interval τ used in *wsp*. The τ for *wdp* is fixed at 30 *m*. The offered load on the network ρ was set to 0.55. It is clear that the performance of *wsp* degrades drastically as the update interval increases. The *wdp* scheme, using at most two paths per pair and infrequent updates with $\tau = 30$ *m*, blocks fewer flows than *wsp*, that uses many more paths and frequent updates with $\tau = 0.5$ *m*. The performance of *wdp* even with a single path is comparable to *wsp* with $\tau = 1.5$ *m*. Figure 5(b) displays the flow blocking probability as a function of offered network load ρ which is varied from 0.50 to 0.60. Once again, the τ for *wdp* is set to 30 *m* and the performance of *wsp* is plotted for 3 different settings of τ: 0.5, 1.0 and 2.0 *m*. It can be seen that across all loads the performance of *wdp* with $\eta = 2$ is better than *wsp* with $\tau = 0.5$. Similarly with just one path, *wdp* performs better than *wsp* with $\tau = 2.0$ and approaches the performance of $\tau = 1.0$ as the load increases. It is also worth noting that *wdp* with two paths rejects significantly fewer flows than with just one path, justifying the need for multipath routing.

It is interesting to observe that even with a single path and very infrequent updates *wdp* outperforms *wsp* with frequent updates. There are several factors contributing to the superior performance of *wdp*. First, it is the nature of information used to capture the link state. The information exchanged about a link is its *average* not *instantaneous* residual bandwidth and hence less variable. Second, before picking the widest disjoint

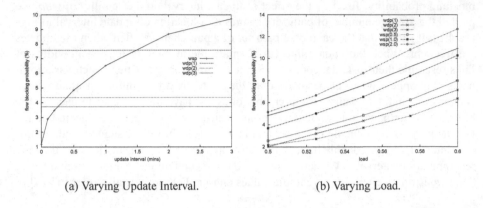

(a) Varying Update Interval.   (b) Varying Load.

Fig. 5. Performance Comparison of *wdp* and *wsp*.

paths, the residual bandwidth on all the links along the current candidate path are adjusted to account for the load offered on that path by this pair. Such a *local adjustment* to the global information makes the network state appear differently to each source. It is as if each source receives a customized update about the state of each link. The sources that are currently routing through a link perceive higher residual bandwidth on that link than other sources. This causes a source to continue using the same path to a destination unless it finds a much wider path. This in turn reduces the variation in link state and consequently the updated information does not get outdated too soon. In contrast, *wsp* exchanges highly varying instantaneous residual bandwidth information and all the sources have the same view of the network. This results in mass synchronization as every source prefers *good* links and avoids *bad* links. This in turn increases the variance in instantaneous residual bandwidth values and causes route oscillation[1]. The *wdp* scheme, on the other hand, by selecting paths using both local and global information and by employing *ebp* based adaptive proportioning delivers stable and robust performance.

Routing Overhead. Now we compare the amount of overhead incurred by *wdp* and *wsp*. This overhead can be categorized into per flow routing overhead and operational overhead. We discuss these two separately in the following.

The *wsp* scheme selects a path by first pruning the links with insufficient available bandwidth and then performing a variant of Dijkstra's algorithm on the resulting graph to find the shortest path with maximum bottleneck bandwidth. This takes at least $O(E \log N)$ time where N is the number of nodes and E is the total number of links in the network. Assuming precomputation of a set of paths R_σ to each destination, to avoid searching the whole graph for path selection, it still need to traverse all the links of these precomputed paths to identify the widest shortest path. This amounts to an

---

[1] Some remedial solutions were proposed in [1,2] to deal with the inaccuracy at a source node. However, the fundamental problem remains and the observations made in this paper still apply.

overhead of $O(L_\sigma)$, where L_σ is the total number of links in the set R_σ. On the other hand, in *wdp* one of the candidate paths is chosen in a weighted round robin fashion whose complexity is $O(\eta)$ which is much less than $O(L_\sigma)$ for *wsp*.

Now consider the operational overhead. Both schemes require link state updates to carry residual bandwidth information. However the frequency of updates needed for proper functioning of *wdp* is no more than what is used to carry connectivity information in traditional routing protocols such as OSPF. Therefore, the average residual bandwidth information required by *wdp* can be piggybacked along with the conventional link state updates. Hence, *wdp* does not cause any additional burden on the network. On the other hand, the *wsp* scheme requires frequent updates consuming both network bandwidth and processing power. Furthermore *wsp* uses too many paths. The *wdp* scheme uses only a few preset paths, thus avoiding per flow path setup. Only admission control decision need to be made by routers along the path. The other overheads incurred only by *wdp* are periodic proportion computation and candidate path computation. The proportion computation procedure is extremely simple and costs no more than $O(\eta)$. The candidate path computation amounts to finding η widest paths and hence its worst case time complexity is $O(\eta N^2)$. However, this cost is incurred only once every ξ period. Considering both the blocking performance and the routing cost, we proclaim that *wdp* yields much higher throughput with much lower overhead than *wsp*.

5 Conclusions

The performance of multipath routing hinges critically on the number and the quality of the selected paths. We addressed these issues in the context of the proportional routing paradigm, where the traffic is proportioned among a few good paths instead of routing it all along the best path. We proposed a hybrid approach that uses both global and local information for selecting a few good paths and for proportioning the traffic among the selected paths. We presented a *wdp* scheme that performs *ebp* based proportioning over widest disjoint paths. A set of widest paths that are disjoint *w.r.t* bottleneck links are chosen based on globally exchanged link state metrics. The *ebp* strategy is used for adaptively proportioning traffic among these paths based on locally collected path state metrics. We compared the performance of our *wdp* scheme with that of optimal proportional routing scheme *opr* and shown that the proposed scheme achieves almost optimal performance using much fewer paths. We also demonstrated that the proposed scheme yields much higher throughput with much smaller overhead compared to other link state update based schemes such as *wsp*.

References

1. G. Apostolopoulos, R. Guerin, S. Kamat, S. Tripathi, "Quality of Service Based Routing: A Performance Perspective", ACM SIGCOMM, September 1998.
2. G. Apostolopoulos, R. Guerin, S. Kamat, S. Tripathi, "Improving QoS Routing Performance under Inaccurate Link State Information", ITC'16, June 1999.

3. Shigang Chen, Klara Nahrstedt, "An Overview of Quality-of-Service Routing for the Next Generation High-Speed Networks: Problems and Solutions", IEEE Network Magazine, Special Issue on Transmission and Distribution of Digital Video, Vol. 12, No. 6, November-December 1998, pp. 64-79.
4. T. Coleman and Y. Li, "An Interior, Trust Region Approach for Nonlinear Minimization Subject to Bounds," in *Journal on Optimization, Vol. 6*, pp. 418–445, SIAM, 1996.
5. R. Guerin, S. Kamat, A. Orda, T. Przygienda, D. Williams, "QoS Routing Mechanisms and OSPF Extensions", *Work in Progress*, Internet Draft, March 1997.
6. F.P. Kelly, "Routing in Circuit-Switched Networks: Optimization, Shadow Prices and Decentralization", Advances in Applied Probability 20, 112-144, 1988.
7. F.P. Kelly, "Routing and capacity Allocation in Networks with Trunk Reservation", Mathematics of Operations Research, 15:771-793, 1990.
8. F.P. Kelly, "Dynamic Routing in Stochastic Networks", In *Stochastic Networks*, ed. F.P. Kellyand R.J. Williams, Springer-Verlag, 169-186, 1995.
9. F.P. Kelly, "Fixed Point Models of Loss Networks," J. Austr. Math. Soc., Ser. B, 31, pp. 204-218, 1989.
10. Q. Ma, P. Steenkiste, "On Path Selection for Traffic with Bandwidth Guarantees", IEEE ICNP, October 1997.
11. J. Moy, "OSPF Version 2", Request For Comments 2328, Internet Engineering Task Force, April 1998.
12. S. Nelakuditi, Z-L. Zhang, R.P. Tsang, "Adaptive Proportional Routing: A Localized QoS Routing Approach", IEEE INFOCOM'00, March 2000.
13. S. Nelakuditi, S. Varadarajan, and Z-L. Zhang, "On Localized Control in Quality-of-Service Routing", October 2000. Submitted to IEEE Transactions on Automatic Control, Special Issue on Systems and Control Methods for Communication Networks.
14. S. Nelakuditi, and Z-L. Zhang, "On Selection of Paths for Multipath Routing", Techincal Report, Department of Computer Science, University of Minnesota, April 2001.
15. M. Powell, "A Fast Algorithm for Nonlinear Constrained Optimization Calculations," in *Numerical Analysis, ed. G.A. Watson, Lecture Notes in Mathematics, Vol 630*, Springer Verlag, 1978.
16. E. Rosen, A. Viswanathan, and R. Callon, "Multi-Protocol Label Switching Architecture", *work in progress*, Internet Draft draft-ietf-mpls-arch-06.txt, August 1999.
17. K.W. Ross, "Multiservice Loss Models for Broadband Telecommunication Networks", Springer-Verlag, 1995.
18. A. Shaikh, J. Rexford, K. Shin, "Evaluating the Overheads of Source-Directed Quality-of-Service Routing", ICNP, October 1998.
19. N. Taft-Plotkin, B. Bellur, and R. Ogier, "Quality-of-Service Routing using Maximally Disjoint Paths", IWQOS, June 1999.
20. C. Villamizar, "MPLS Optimized Multipath (MPLS-OMP)", Internet Draft, February 1999.
21. C. Villamizar, "OSPF Optimized Multipath (OSPF-OMP)", Internet Draft, February 1999.
22. Z. Wang, J. Crowcroft, "Quality-of-Service Routing for Supporting Multimedia Applications", IEEE JSAC, September 1996

Preferential Treatment of Acknowledgment Packets in a Differentiated Services Network

Konstantina Papagiannaki[1], Patrick Thiran[2,3],
Jon Crowcroft[1], and Christophe Diot[2]

[1] Computer Science Department, UCL, Gower street, London, WC1E 6BT, UK
{D.Papagiannaki,jon}@cs.ucl.ac.uk
[2] Sprint ATL, 1 Adrian Court, Burlingame, CA 94010, USA
cdiot@sprintlabs.com
[3] ICA - DSC, EPFL, CH-1015 Lausanne, Switzerland
Patrick.Thiran@epfl.ch

Abstract. In the context of networks offering Differentiated Services (DiffServ), we investigate the effect of acknowledgment treatment on the throughput of TCP connections. We carry out experiments on a testbed offering three classes of service (Premium, Assured and Best-Effort), and different levels of congestion on the data and acknowledgment path. We apply a full factorial statistical design and deduce that treatment of TCP data packets is not sufficient and that acknowledgment treatment on the reverse path is a necessary condition to reach the targeted performance in DiffServ efficiently. We find that the optimal marking strategy depends on the level of congestion on the reverse path. In the practical case where Internet Service Providers cannot obtain such information in order to mark acknowledgment packets, we show that the strategy leading to optimal overall performance is to copy the mark from the respective data packet into returned acknowledgement packets, provided that the affected service class is appropriately provisioned.

1 Introduction - Motivation

There have been several proposals for implementing scalable service differentiation in the Internet. Such architectures achieve scalability by avoiding per-flow state in the core and by moving complex control functionality to the edges of the network. A specific field in the IP header (the DS field) is used to convey the class of service requested. Edge devices perform sophisticated classification, marking, policing (and shaping operations, if required). Core devices forward packets according to the requirements of the traffic aggregate they belong to [3].

Within this framework, several schemes have been proposed, such as the "User Share Differentiation (USD)" [2], the "Two-Bit Differentiated Services" architecture [14], and "Random Early Drop with In and Out packets" (RIO) [5]. Preferential treatment can be provided to flow aggregates according to policies defined per administration domain. All those schemes (along with the work of the Differentiated Services Working Group of the IETF [3]) refer to unidirectional flows.

L. Wolf, D. Hutchison, and R. Steinmetz (Eds.): IWQoS 2001, LNCS 2092, pp. 187–201, 2001.
© Springer-Verlag Berlin Heidelberg 2001

The Transmission Control Protocol (TCP) is the dominant data transport protocol in the Internet [4], [13]. TCP is a bi-directional transport protocol, which uses 40-bytes acknowledgment packets (ACKs) for reliable data transmission and to control its sending rate. Data packet losses are detected through duplicate or lack of acknowledgments on the reverse direction. Such an event is followed by a reduction in transmission rate, as a reaction to possible congestion on the forward path.

Modeling TCP shows that TCP flows suffer when the path serving their ACKs is congested [12]. Therefore, marking data packets belonging to connections facing congestion on their reverse path may not prove adequate for a connection to reach its performance target, due to ACK losses on the congested reverse path. We are interested in identifying ways in which data and acknowledgment packets have to be marked so that connections with different levels of congestion on forward and reverse paths can still achieve their performance goals.

In this paper, we evaluate, through a thoroughly devised experimental plan, TCP performance in a Differentiated Services network featuring congestion on forward and/or reverse paths. We build a testbed that offers three classes of service, namely Premium, Assured and Best-Effort [14]. We examine cases when forward and/or reverse paths are congested, and vary the marks carried by data and acknowledgment packets. We quantify the effect that each one of those factors has on TCP throughput. Our goals are (i) to assess whether such an effect exists, and (ii) to identify the optimal marking strategy for the acknowledgments of both Premium and Assured flows.

To the best of our knowledge, this problem has been addressed only by simulation in [11]. This latter study simulated a dumb-bell topology to investigate the effect of marking acknowledgment packets, and analyzed the behavior of a single flow, for which multiple marking schemes were applied. Our study diverts from [11] in that: (i) we use a testbed instead of a simulator, (ii) we use a more complex network topology with one level of aggregation, which features combinations of congested/uncongested forward/reverse paths for each class of service, (iii) we try out all possible combinations of data - acknowledgment packet markings for all classes of service, (iv) we identify which factors influence TCP throughput in our experiments and quantify their effect, and (v) we propose optimal acknowledgment marking strategies which lead to better Premium and Assured throughput regardless of the level of congestion on the network.

The organization for the rest of the paper is as follows. Section 2 explains the experimental plan and the associated statistical model that we have adopted in order to quantify the effect of the congestion levels and acknowledgment marking strategies on the throughput of the Premium and Assured flows. Section 3 describes the actual testbed on which the measurements were carried out, according to the plan elaborated in Section 2. Section 4 provides the analysis of the collected data for Premium and Assured flows and discusses which factors (or interaction thereof) influences throughput the most. Section 5 identifies the optimal acknowledgment marking strategies in networks where congestion can be predicted. In practice, this may not be possible, so Section 6 proposes a suboptimal strategy which is independent of the network congestion and achieves throughput values close to the optimal ones found in Section 4. We conclude the paper with a summary of our main results, and we discuss whether marking ACKs would be practical in a Differentiated Services network.

2   Experimental Design and Methodology

In the statistical design of experiments, the outcome is called the *response variable*, the variables or parameters that affect the response variable are called *factors*; the values that a factor can take are called *levels*; the repetitions of experiments are called *replications* [9] [10].

In our case, we are interested in two *response variables*, namely the throughput of the Premium flows, henceforth denoted by y, and the throughput of the Assured flows, denoted by y'.

We will study the influence of four *factors*, which are:

- the *marking of the ACKs for the Premium flows*, denoted by P,
- the *marking of the ACKs for the Assured flows*, denoted by A,
- the *marking of the ACKs for the Best-Effort flows*, denoted by B,
- the *existence or absence of congestion* on the forward and/or reverse path, denoted by C.

Each of these four factors can take three *levels*, which are as follows:

- for each factor P, A and B, the levels will be p, a and b, based on whether the acknowledgment packet of the corresponding flow is marked as Premium, Assured or Best-Effort respectively,
- for C, we will distinguish the three following levels: f (the forward path is congested, but not the reverse path), r (the reverse path is congested, but not the forward path) and t (both forward and reverse paths are congested). We chose not to consider the rather trivial case where both forward and reverse paths are not congested, since in this case none of the other factors (marking strategies) will affect the response variables (throughputs of Assured and Premium flows).

The experiments consist in measuring throughputs of Premium and Assured flows under all possible combinations of the four factors P, A, B and C. Since each factor has three levels, there are $3^4 = 81$ possible combinations, and the design is called a full 3^4 design [9]. Each experiment will be *replicated* three times, so that a total of 243 experiments will be carried out. The throughputs achieved by flows during each experiment are not independent; the throughput of a Premium flow depends on the number of flows sharing the allocated Premium capacity, while the throughput of an Assured flow depends on the number of Assured and Best-Effort flows sharing the same path. Each one of those experiments is uniquely defined by a combination of five letters i, j, k, l, m ($i, j, k \in \{p, a, b\}$, $l \in \{f, r, t\}$, $m \in \{1, 2, 3\}$), where the first four are the corresponding level of factor $P = i$, $A = j$, $B = k$ and $C = l$, and where $R = m$ refers to one of the three replications. For example, y_{aapt3} will denote the throughput of the Premium flow measured on the third experiment ($R = 3$) with Premium and Assured acknowledgments marked as Assured ($P = A = a$), with Best-Effort acknowledgments marked as Premium ($B = p$), and with both forward and reverse paths congested ($C = t$).

A 3^4 experimental design with 3 replications assumes the following model for the response variable of the Premium flows, for all 243 possible combinations of the indices i, j, k, l, m [10]:

$$y_{ijklm} = x^0 + x_i^P + x_j^A + x_k^B + x_l^C + x_{ij}^{PA} + x_{ik}^{PB} + \dots$$
$$+ x_{kl}^{BC} + x_{ijk}^{PAB} + \dots + x_{jkl}^{ABC} + x_{ijkl}^{PABC} + \epsilon_{ijklm} \qquad (1)$$

where the terms of the right hand side are computed as follows [9] [10]:

- $x^0 = \sum_{i,j,k,l,m} y_{ijklm}/243$ is the average response (throughput) over the 243 experiments,
- $x_i^P = \sum_{j,k,l,m} y_{ijklm}/81 - x^0$ is the difference between the throughput averaged over the 81 experiments taken when factor P takes level i, with $i \in \{p, a, b\}$, and the average throughput x^0. It is called the *main effect* due to factor P at level i. A similar expression holds for x_j^A, x_k^B and x_l^C,
- $x_{ij}^{PA} = \sum_{k,l,m} y_{ijklm}/27 - x_i^P - x_j^A + x^0$ is called the *effect of (2-factor) interaction* of factors P at level i and A at level j. A similar expression holds for the five other pairs of factors,
- $x_{ijk}^{PAB}, \dots, x_{jkl}^{ABC}$, and x_{ijkl}^{PABC} are the *effects of (3- and 4-factors) interaction*, and are computed using similar expressions,
- ϵ_{ijklm} represents the experimental error in the mth experiment (residual), $1 \le m \le 3$, which is the difference between the actual value of y_{ijklm} and its estimate computed as the sum of all the above terms.

The model for the throughput of the Assured flows is identical, with all terms in (1) denoted with a prime to distinguish them from the variables linked to Premium flows.

The importance of each factor, and of each combination of factors, is measured by the proportion of total variation in the response that is explained by the considered factor, or by the considered combination of factors [9]. These percentages of variation, which are explicited in the Appendix of the full version of this paper [15], are used to assess the importance of the corresponding effects and to trim the model so as to include the most significant terms.

Model (1) depends upon the following assumptions: (i) The effects of various factors are additive, (ii) errors are additive, (iii) errors are independent of the factor levels, (iv) errors are normally distributed, and (v) errors have the same variance for all factor levels [9]. The model can be validated with two simple "visual tests": (i) the normal quantile-quantile plot (Q-Q plot) of the residuals ϵ_{ijkml}, and (ii) the plot of the residuals ϵ_{ijklm} against the predicted responses y_{ijklm}, y'_{ijklm}. If the first plot is approximately linear and the second plot does not show any apparent pattern, the model is considered accurate [9]. If the relative magnitude of errors is smaller than the response by an order of magnitude or more, trends may be ignored.

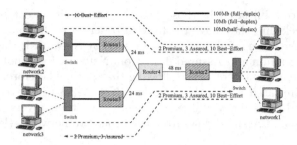

Fig. 1. Experimental Setup. The network consists of 3 ISP networks interconnecting at a Network Access Point. Each network generates a predefined traffic mix and directs its flows to selected networks so that different forward/reverse path congestion levels are achieved.

3 Network Setup

In this section, we describe the way we design the network testbed, which allows us to realize the experimental plan described in Section 2.

3.1 Designing the Testbed Topology

The four factors we identified in the previous section are: factors P, A, and B, denoting the marks of the packets acknowledging Premium, Assured and Best-Effort packets respectively, and factor C, denoting whether forward and/or reverse TCP paths are congested. The different levels of factors P, A, and B can be easily realized, since they correspond to differential packet marking. Factor C is more problematic. We want to be able to test all levels in a single network, with a limited number of experiments.

By the term "congestion", we describe the condition of a link where contention for resources leads to queue build-ups and possible packet losses. Clearly, the more flows a link serves, the more likely it is that this link will reach a state of congestion. Under this assumption, we implement congestion on specific links using different numbers of flows on different links.

Our goal is our network to offer all three levels of congestion, we wish to investigate, for both Premium and Assured flows. A dumb-bell topology cannot implement all three levels f, r and t of factor C for both Premium and Assured flows, even if we assume that the two types of flows are isolated. For instance, we can test levels f and r by having the connecting link congested in one direction and not the other, but we cannot simultaneously test level t, which would require the connecting link to be congested in both directions.

Therefore our network must feature a minimum of four routers. A "Y"-topology, such as the one depicted in Figure 1, is capable of offering all three levels of factor C for both Premium and Assured flows in a single network.

If Network 2 and Network 3 initiate two Premium, three Assured, and ten Best-Effort flows towards Network 1, then the forward path for both Premium and Assured flows initiated in those two networks will be congested, because of the bottleneck link they have to share in order to reach Network 1. On the other hand, if the same number of flows is issued by Network 1, then the forward path of both Premium and Assured flows from Network 1 will not be congested, due to the limited number of flows following that path. Therefore, Premium and Assured flows initiated within the bounds of Network 1 will have no congestion on their forward path. They will face congestion on their reverse path, because of the Premium and Assured flows from the other two networks (level r of factor C).

Flows generated within the bounds of Networks 2 and 3 face congestion on their forward path. Existence and absence of congestion on the reverse TCP paths will give us the other two levels for factor C. More specifically, if we congest the path leading to Network 3, and leave the path to Network 2 uncongested, then we are capable of realizing levels f and t for the flows initiating in Network 2, and Network 3 respectively.

We achieve that by directing the Premium, and Assured flows of Network 1 towards Network 3, and the Best-Effort flows of Network 1 towards Network 2. In that way, the path from Network 1 to Network 2 accommodates the Best-Effort traffic of Network 1 and the acknowledgment traffic of Network 2. Thus, Premium and Assured flows from Network 2 face congestion on their forward and no congestion on their reverse path (level f of factor C). On the other hand, the path from Network 1 to Network 3 is loaded with both the Premium, and Assured traffic of Network 1, as well as the acknowledgment traffic of all 15 flows of Network 3. In other words, flows generated by Network 3 will face congestion on both forward and reverse paths (level t of factor C).

Table 1. Routes Realizing the Different Levels of Factor C for Premium and Assured Flows.

Origin → Destination Network / factor C	Premium flows	Assured flows
Network2 → Network1	on forward path only (f)	on forward path only (f)
Network3 → Network1	on both paths (t)	on both paths (t)
Network1 → Network3	on reverse path only (r)	on reverse path only (r)
Network1 → Network2	*This path does not carry any Premium or Assured flows.*	

One may wonder why the reverse path of flows from Network 3 to Network 1 is congested, while the forward path from Network 1 to Network 3 is not, since they actually follow the same route. This can be explained by the difference in size between packets flowing on both directions: packets on the reverse path of flows from Network 3 to Network 1 are acknowledgment packets (ACKs), which are much smaller in size than the data packets flowing on the forward path from Network 1 to Network 3. The queuing time of ACKs can therefore be quite large compared to their processing time when data packets have to be served ahead of them, making the reverse path for flows from Network 3 to Network 1 congested, even if few data packets are present in the buffer. On the other hand, the delay of a data packet is mostly due to transmission and not to queuing (since not many data packets occupy that buffer), so that the forward path from Network 1 to Network 3 appears not congested to these flows. The resulting topology is presented in Fig. 1.

Table 1 displays the level of congestion for each class of service at each network. For example, a Premium flow originating at Network 3 will have both its forward and reverse paths congested, whereas an Assured flow originating at Network 2 will face congestion on its forward path only, because the path from Network 1 to Network 2 is lightly loaded.

3.2 Testbed Configuration

The testbed derived from the topology described in the previous section (Fig. 1) consists of four routers, Linux PCs with kernel version 2.2.10 that supports differentiated services [1]. Delay elements have been added on the links interconnecting the four routers, so that the bandwidth-delay product is large enough for the TCP flows to be able to open up their windows and reach their steady state (maximum window size is set to 32KB). Each network features two or three HP-UX workstations. Long-lived TCP Reno flows are generated using the "netperf" tool [8]. They last 10 minutes and send 512 bytes packets. Netperf reports the achieved throughput by each flow for the whole duration of the experiment (there is no warm-up period). The choice of long-lived flows was made so as to avoid transient conditions, and to focus on TCP flows in their steady state, where acknowledgment packet marking will have the greatest impact.

End hosts in our testbed did not have the capability of marking packets, and therefore edge routers perform policing, marking, classification and appropriate forwarding. The packets are policed based on their source and destination addresses using a token bucket. The Assured service aggregate is profiled with 2.4 Mbps and any excess traffic is demoted to Best-Effort. The Premium flows are profiled with 1 Mbps each, and any excess traffic is dropped at the ingress (shaping should be performed by the customer). There is no special provision for Best-Effort traffic, except from the fact that we have configured the routers in such a way so that it does not starve.

Each outgoing interface is configured with a Class Based Queue (CBQ) [7] consisting of a FIFO queue for Premium packets and a RIO queue for Assured and Best Effort packets. The RIO parameters are $35/50/0.1$ (min_{th}, max_{th}, max_p) for the OUT packets and $55/65/0.05$ for the IN packets. We chose those values so that: (i) the minimum RED threshold is equal to 40% of the total queue length, as recommended in [6], (ii) the maximum threshold for OUT packets is much lower than the one for IN packets to achieve higher degree of differentiation between Assured and Best-Effort packets, as suggested in [5], and (iii) the achieved rate for each one of the three classes of service is close to the profiled one.

Lastly, in order to enable cost-efficient analysis of all the possible scenarios, we assume that all networks implement the same acknowledgment marking strategy. We believe that this assumption has little influence on the results we observe.

4   Analysis of Experimental Results

We now apply the methodology outlined in Section 2 on the collected experimental results and identify the most important factors for Premium and Assured flows. Due to lack of space, we refrain from presenting the raw experimental results in this paper, and encourage the reader to refer to the full version of this paper [15].

4.1   Analysis of Variance (ANOVA)

The effects of various factors and their interactions can be obtained utilizing the methodology described in Section 2. It is really important to understand the complexity behind the relations among the four factors. Acknowledging the packets of a certain class of service with specific marks does not only influence the throughput of the flows belonging to that service class, but it also affects the flows belonging to the class, which is going to be utilized by the acknowledgment packets.

Table 2. Analysis of Variance table for Premium flows. Effects that account for more than 1.5% of total variation are represented in bold. Notations between parentheses are detailed in [15].

Component	Percentage of Variation
Main effects (total)	48.69%
Congestion (SSC/SST)	**31.23%**
Premium ACKs (SSP/SST)	**10.43%**
Assured ACKs (SSA/SST)	**4.42%**
BE ACKs (SSB/SST)	**2.6%**
First-order interactions (total)	37.17%
Premium ACKs - Congestion (SSPC/SST)	**32.26%**
Assured ACKs - Congestion (SSAC/SST)	0.59%
BE ACKs - Congestion (SSBC/SST)	1.11%
Premium ACKs - Assured ACKs (SSPA/SST)	**2.04%**
Premium ACKs - BE ACKs (SSPB/SST)	0.86%
Assured ACKs - BE ACKs (SSAB/SST)	0.19%

Table 3. Analysis of Variance table for Assured flows. Effects that account for more than 1.5% of total variation are represented in bold. Notations between parentheses are detailed in [15].

Component	Percentage of Variation
Main effects (total)	89.51%
Congestion (SSC/SST)	**86.38%**
Premium ACKs (SSP/SST)	0.54%
Assured ACKs (SSA/SST)	0.8%
BE ACKs (SSB/SST)	**1.77%**
First-order interactions (total)	8.89%
Premium ACKs - Congestion (SSPC/SST)	0.21%
Assured ACKs - Congestion (SSAC/SST)	**7.86%**
BE ACKs - Congestion (SSBC/SST)	0.16%
Premium ACKs - Assured ACKs (SSPA/SST)	0.5%
Premium ACKs - BE ACKs (SSPB/SST)	0.09%
Assured ACKs - BE ACKs (SSAB/SST)	0.04%

Table 2 details the percentages of variation apportioned to the main effects, and to first-order interactions for Premium flows, which amount to 85% of the total variation. Second and third-order interactions turn out to be negligible. Respective results specific to the Assured flows are presented in Table 3. From Table 3, we can see that the role of congestion is much more important for Assured flows than for Premium flows. Furthermore, the main effects and the first-order interactions are adequate to justify 98.4% of the existing variation.

Dropping all interactions that explain less than 1.5% variation as negligible, and using model (1), Premium and Assured throughput can be described with the following formulae:

$$y_{ijklm} = x^0 + x_i^P + x_j^A + x_k^B + x_l^C + x_{ij}^{PA} + x_{il}^{PC} + \epsilon_{ijklm}, \tag{2}$$

$$y'_{ijklm} = x'^0 + x'^B_k + x'^C_l + x'^{AC}_{jl} + \epsilon'_{ijklm}, \tag{3}$$

for $i, j, k \in \{p, a, b\}$, $l \in \{f, r, t\}$, and $1 \leq m \leq 3$.

4.2   Visual Tests

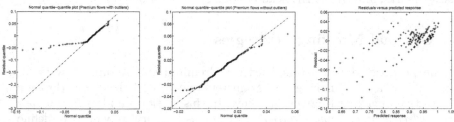

Figure 2. Normal Q-Q plot for residuals.

Figure 3. Normal Q-Q plot for residuals (no outliers).

Figure 4. Residuals versus predicted response.

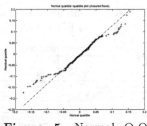
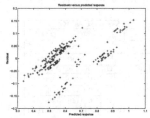

Figure 5. Normal Q-Q plot for residuals

Figure 6. Residuals vs. predicted response

To test whether (2), and (3) are indeed valid models, we perform the tests described at the end of Section 2. Figure 2 presents the Q-Q plot between the residuals and a normal distribution. Since the plot is not linear, we cannot claim that the errors in our analysis are normally distributed. However, if we take out the 20 outliers (out of 243 values), then the new Q-Q plot displays a linear relation between the two distributions (Figure 3). Investigating the data, we see that those 20 outlying values do not occur for the same experiment, and therefore the error introduced by ignoring them does not challenge the validity of our analysis. We suspect that those outlying values are due to SYN packets getting lost at the beginning of the experiments, when all the flows start simultaneously. A flow will normally need 6 seconds (TCP Reno timeout) to recover from such a loss, delaying its transmission while other flows increase their sending rates.

The second visual test is the plot for the residuals ϵ_{ijklm} versus the predicted response y_{ijklm}, and is presented in Figure 4. No apparent trends appear in the data. This validates our model, and allows us to conclude that the throughput of a Premium flow in a Differentiated Services network is mostly affected by 1) congestion on forward/reverse paths, 2) the interaction between congestion and the Premium acknowledgment marks, and 3) the Premium acknowledgment marks themselves.

The visual tests for the Assured flows are presented in Figures 5, and 6, and also confirm the accuracy of model (3) for the Assured throughput. Therefore, an Assured flow is mainly affected by 1) congestion on its forward and reverse path,

2) the interaction between the congestion level and the Assured acknowledgment packet marks, and 3) the BE acknowledgment packet marks.

5   Optimal Marking Strategies

In the previous section we modeled the Premium and Assured flow throughput utilizing a small number of parameters and proved that TCP throughput is sensitive to acknowledgment marking. In this section we identify the *optimal acknowledgment marking strategy* for each class of service, which we define as the marking algorithm which results in the highest throughput values for the class of service in consideration.

Table 4. Confidence Intervals for Premium Flow Analysis.

Congestion		Assured ACKs	
x_r^C	(-0.0538, -0.0442)	x_p^A	(-0.072, 0.0347)$^\alpha$
x_f^C	(0.0135, 0.0232)	x_a^A	(-0.0447, 0.0619)$^\alpha$
x_t^C	(0.0257, 0.0354)	x_h^A	(-0.0433, 0.0634)$^\alpha$
Premium ACKs		**Best-Effort ACKs**	
x_p^P	(-0.0076, 0.0020)$^\alpha$	x_p^B	(-0.0676, 0.0391)$^\alpha$
x_a^P	(0.0212, 0.0309)	x_a^B	(-0.0475, 0.0592)$^\alpha$
x_h^P	(-0.0281, -0.0184)	x_h^B	(-0.0449, 0.0618)$^\alpha$
Premium ACKs - Congestion		**Premium - Assured ACKs**	
x_{pr}^{PC}	(0.040, 0.0537)	x_{pp}^{PA}	(-0.0233, -0.0096)
x_{ar}^{PC}	(0.0162, 0.0299)	x_{pa}^{PA}	(-0.0002, 0.0134)$^\alpha$
x_{br}^{PC}	(-0.0768, -0.0631)	x_{pb}^{PA}	(0.003, 0.0167)
x_{pf}^{PC}	(-0.0309, -0.0172)	x_{ap}^{PA}	(0.0072, 0.0209)
x_{af}^{PC}	(-0.0197, -0.006)	x_{aa}^{PA}	(-0.0129, 0.0007)$^\alpha$
x_{bf}^{PC}	(0.0301, 0.0438)	x_{ab}^{PA}	(-0.0149, -0.0012)
x_{pt}^{PC}	(-0.0296, -0.0159)	x_{bp}^{PA}	(-0.0044, 0.0092)$^\alpha$
x_{at}^{PC}	(-0.0169, -0.0032)	x_{ba}^{PA}	(-0.0074, 0.0062)$^\alpha$
x_{bt}^{PC}	(0.026, 0.0397)	x_{bb}^{PA}	(-0.0086, 0.005)$^\alpha$

α: indicates that the effect is not significant

Table 5. Confidence Intervals for Assured Flow Analysis.

Congestion		Best-Effort ACKs	
$x_r^{'C}$	(0.1648, 0.1892)	$x_p^{'B}$	(-0.0074, 0.0169)$^\alpha$
$x_f^{'C}$	(-0.1047, -0.0804)	$x_a^{'B}$	(-0.0361, -0.0118)
$x_t^{'C}$	(-0.0966, -0.0722)	$x_h^{'B}$	(0.007, 0.0314)
Assured ACKs - Congestion			
$x_{pr}^{'AC}$	(0.0224, 0.0569)		
$x_{ar}^{'AC}$	(0.0185, 0.053)		
$x_{br}^{'AC}$	(-0.0927, -0.0583)		
$x_{pf}^{'AC}$	(-0.0374, -0.003)		
$x_{af}^{'AC}$	(-0.0365, -0.0021)		
$x_{bf}^{'AC}$	(0.0223, 0.0567)		
$x_{pt}^{'AC}$	(-0.0367, -0.0022)		
$x_{at}^{'AC}$	(-0.0337, 0.0007)$^\alpha$		
$x_{bt}^{'AC}$	(0.0187, 0.0532)		

α: indicates that the effect is not significant

In order to evaluate whether a factor has a positive or negative effect, we must first provide confidence intervals for the effects identified as important in the previous section. Those confidence intervals can be computed using t-values read at the number of degrees of freedom associated with the errors. Due to lack of space we refer to [9] for the techniques used to compute those intervals and the degrees of freedom associated with each one of the components in our analysis. The obtained 90% confidence intervals for Premium and Assured flows are presented in Tables 4 and 5 respectively.

If the interval contains the value 0, then the effect of the corresponding component is not statistically significant. For the rest, an interval with positive values indicates higher than average throughput, while an interval with negative values indicates lower than average throughput.

5.1   Results Specific to Premium Flows

In Section 4 we showed that the throughput of a Premium flow is mostly affected by congestion (31.23%), acknowledgments to Premium packets (10.43%), acknowledgments to Assured packets (4.42%), the interaction between acknowledgments to Premium packets and congestion (32.26%), and the interaction between ACKs directed to Premium and Assured packets (2.04%). We can further derive from Table 4 that:

1. **effect of factor C**: the average throughput of a Premium flow suffers when the reverse path is congested (the confidence interval for $C = r$ lies in the negative), and reaches high values when the forward path is also congested ($C = f$, $C = t$).
   Therefore, marking acknowledgment packets is especially important when the reverse path is congested ($C = r$). In this case, marking data packets does not affect the performance since the forward path is lightly utilized. If congestion exists on the forward path or on both forward and reverse paths ($C = f$, or $C = t$), it seems that the protection of data packets on the forward path is capable of making up for the lost (or delayed) acknowledgment packets.
2. **effect of factor PC**: when a Premium flow suffers from congestion on the reverse path, then acknowledgment packets have to be marked as Premium ($PC = pr$). Best-Effort acknowledgment packets in this case are not adequate so that Premium flows reach their performance goal ($PC = br$). Lastly, if no congestion is present on the reverse path, then even Best-Effort acknowledgment marking is capable of offering a sustained rate to Premium flows ($PC = bf$, and $PC = bt$ offer higher than the average throughput x^0).
3. **effect of factor P**: regardless of congestion, the Premium flows achieve better throughput when their acknowledgments are marked as Assured ($P = a$). In this case, acknowledgment traffic exceeding the agreed-upon profile can still get transmitted as Best-Effort,
4. **effect of factor PA**: Premium flows perform poorly when both Premium and Assured flows are acknowledged with Premium packets (confidence interval is negative when $PA = pp$).

Consequently, the strategy leading to higher Premium throughput values, is to acknowledge Premium data packets with Assured packets. If the reverse TCP path is congested, though, then Premium ACKs have to be marked as Premium.[1]

5.2   Results Specific to Assured Flows

We have seen in Section 4 that Assured flows' throughput is mostly affected by congestion (86.38%), the interaction between acknowledgments to Assured packets and congestion (7.86%), and acknowledgments to Best-Effort packets (1.77%). From the confidence intervals presented in Table 5, we can further derive that:

1. **effect of factor C**: Assured flows perform poorly when their forward path is congested ($C = f, C = t$).

---

[1]   In our experiments we had not specifically provisioned for ACKs and therefore we would expect slightly different results for point 3, and 4 in case we had.

The Assured service is a statistical service which is contracted to face less loss than BE in case of congestion. Therefore, its throughput is affected by congestion on the forward path much more than Premium flows (a minimum throughput service).

2. **effect of factor AC**: if only the reverse path is congested $(C = r)$, then flows should protect their ACKs by subscribing them to a provisioned class of service. If only the forward path is congested $(C = f)$, it is better if flows receive ACKs that do not belong to a rate-limited class of service. For the third case $(C = t)$, when both paths are congested, we see that it is better if Assured flows send their ACKs as BE. We believe that this result stems from the fact that the forward path contains an aggregation point which makes flows from networks 2 and 3 behave in similar ways.

3. **effect of factor B**: lastly, it is better if Best-Effort traffic does not utilize the Assured service class for its ACKs; in such a case Assured flows have to compete with Assured-marked ACKs for resources (indeed other measurements taken throughout the experiments show that such a combination leads to the largest number of retransmissions for Assured flows).

Therefore, the optimal strategy for Assured flows is to mark ACKs as Premium or Assured in networks with congested reverse paths, and as BE in networks where the reverse TCP paths are lightly utilized.

6   Practical and Efficient Marking Strategies

In the previous section, we identified the optimal acknowledgment marking strategies for Premium and Assured flows. Those strategies depend on the level of congestion on forward and reverse paths, and should therefore be specific to particular networks. It is however difficult, if not impossible, to predict the level of congestion on the reverse path (especially since this path may change in time due to routing), and marking ACKs depending on their source and destination pair imposes a rather large overhead.

In this section, we identify a sub-optimal acknowledgment marking strategy, which still leads to higher Premium and Assured throughputs, but which no longer depends on the network specifics. We then show that this marking strategy achieves throughput close to the optimal values obtained in the previous section, and outperforms the "best-effort marking" used by default.

Sections 4 and 5 showed that in cases of congestion on the reverse TCP path, ACKs have to be protected. In such cases, Premium and Assured flows benefit when their acknowledgment packets belong to a provisioned class of service. More specifically, we know from Section 5 that: (i) regardless of congestion, Premium flows perform the best when their ACKs are marked as Premium (effect of P on Premium throughput), (ii) Premium flows acknowledged with Premium packets perform better when Assured flows do not use the Premium service class for their ACKs (effect of PA on Premium throughput), and (iii) Assured flows perform better when Best-Effort flows receive BE ACKs (effect of B on Assured throughput). Therefore, marking strategies that are practical while maintaining a good performance for both Premium and Assured flows, are the ones described by $P = p$, $B = b$ and $A = a$ or $A = b$.

Throughput values achieved by Assured flows under those schemes and different levels of congestion are similar except when the reverse TCP path is congested. For the latter case, Assured acknowledgements lead to higher Assured throughput.

Therefore, a practical and yet optimal strategy seems to be the one where each flow receives acknowledgments in the same class of service. Such a strategy can be easily implemented by a network, if TCP implementations are modified so that acknowledgment packets copy the mark of the packet, they acknowledge. In this section we evaluate the performance of this particular marking strategy by comparing it with the optimal strategy identified in Section 4, and the default strategy, where all ACKs are marked as BE. Figure 7 displays the average throughput achieved by Premium and Assured flows under those three acknowledgment marking strategies.

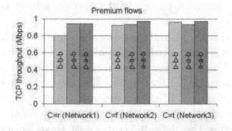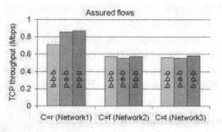

Figure 7. Premium and Assured flow throughput under 3 acknowledgment marking strategies (the default strategy, the strategy where each acknowledgment packet copies the mark of the data packet, and the optimal). Strategies are described by the combination of factors P-A-B.

From Figure 7 we see that the optimal acknowledgement marking strategy may improve performance by 20% over the throughput achieved when ACKs are marked as BE (Assured flows in Network 1 achieve 20% higher throughput when all the flows are acknowledged with Premium packets - case p-p-p - rather than with Best-Effort packets - case b-b-b). Furthermore, we see that the marking strategy, where each flow receives ACKs belonging to the same class as the data packets, performs well in most cases and independently of the level of congestion on the forward and reverse paths for both Premium and Assured flows.

7   Conclusions - Discussion

Carrying out experiments in a Differentiated Services network, offering three classes of service (Premium, Assured and Best-Effort), we have investigated the effect of acknowledgment packet marking, and congested/uncongested forward/reverse paths, onto TCP throughput. We have studied and quantified the effect of the identified factors onto the throughput of the provisioned classes of service, and we have identified the levels of those factors that lead to optimal throughput values.

The analysis of the collected results shows that TCP throughput is sensitive to congestion on the reverse TCP path, thereby confirming the results obtained by modeling the TCP behavior in asymmetric networks, with slow or congested reverse TCP paths [12].

Consequently, bi-directionality of traffic must be taken into account in Service Level Specifications and special provision must be made for bi-directional flows in a Differentiated Services network. Protection of TCP data packets alone will not ensure better performance to flows, requesting a better than Best-Effort service.

Our results indicate that in cases of congested reverse TCP paths, Premium and Assured flows have to be acknowledged by preferentially treated packets. Nevertheless, there is no single best marking strategy, which leads to optimal performance for both Premium and Assured flows. Furthermore, the optimal marking strategy for each provisioned class of service requires explicit knowledge of the level of congestion on the reverse path; a piece of information which is really hard to obtain.

As a consequence, we have investigated sub-optimal marking strategies, which still lead to high throughput values for the two provisioned classes of service, and are independent of the network specifics. We have proven that the marking strategy, which fulfills those requirements and is easily implementable, is the one where acknowledgment packets carry the marks of their respective data packets. We have shown that this strategy leads to performance comparable to the optimal marking strategy, and outperforms the default strategy of Best-Effort ACKs.

How to offer better than Best-Effort services over a single network is still an open, and active area of research. No specific recommendations have been made by the appropriate bodies and investigation of the performance achieved by flows, when they request preferential treatment, is still under investigation. Furthermore, provisioning of Differentiated Services networks remains an interesting problem, still to be solved.

This paper has shown that in the case of TCP flows, provisioning of forward paths is not adequate for TCP flows to reach their target rate. Reverse paths have to be provisioned as well, and acknowledgment packets have to be appropriately marked. In other words, network providers offering Differentiated Services do not only have to draw agreements with the providers handling their egress traffic, but also have to draw agreements with the providers returning ACKs for the TCP flows initiating within their bounds. We have to notice, however, that the additional provisioning that has to be done due to the acknowledgment traffic will be rather limited, because of the limited size of the acknowledgment packets. More specifically, in our experiments our provisioned traffic classes managed to reach their target levels with differential acknowledgment marking without special provisioning of the reverse paths.

References

1. W. Almesberger. Linux Network Traffic Control - Implementation Overview. http://icawww1.epfl.ch/linux-diffserv/, April 1999.
2. A. Basu and Z. Wang. A Comparative Study of Schemes for Differentiated Services. Technical report, Bell Laboratories, Lucent Technologies, July 1998.
3. S. Blake, D. Black, M. Carlson, E. Davies, Z. Wang, and W. Weiss. An Architecture for Differentiated Services. Request for Comments (Proposed Standard) 2475, Internet Engineering Task Force, October 1998.
4. K. Claffy. The Nature of the Beast: Recent Traffic Measurements from an Internet Backbone. In *INET'98*, 1998.
5. D. D. Clark and W. Fang. Explicit Allocation of Best-Effort Packet Delivery Service. *ieanep*, 6(4):362–373, August 1998.
6. S. Floyd and V. Jacobson. Random early detection gateways for congestion avoidance. *IEEE/ACM Transactions on Networking*, 1(4):397–413, August 1993.
7. S. Floyd and V. Jacobson. Link-sharing and resource management models for packet networks. *IEEE/ACM Transactions on Networking*, 3(4), August 1995.
8. Information Networks Division, Hewlett-Packard Company. Netperf: A Network Performance Benchmark. http://www.netperf.org, February 1995.
9. Raj Jain. *The art of computer systems performance analysis: techniques for experimental design, measurement, simulation, and modeling*. John Wiley, New York, 0-471-50336-3, 1991.
10. Peter W. M. John. *Statistical Design and Analysis of Experiments*. Society for Industrial and Applied Mathematics, 3600 University City Science Center, Philadelphia, PA 19104-2688, 0-89871-427-3, 1998.
11. S. Köhler and U. Schäfer. Performance Comparison of Different Class-and-Drop Treatment of Data and Acknowledgements in DiffServ IP Networks. Technical Report 237, University of Würzburg, August 1999.
12. T. V. Lakshman, U. Madhow, and B. Suter. TCP/IP Performance with Random Loss and Bidirectional Congestion. *IEEE/ACM Transactions on Networking*, 1998.
13. S. McCreary and K. Claffy. Trends in Wide Area IP Traffic Patterns - A View from Ames Internet Exchange. In *ITC'00*, Monterey, September 2000.
14. K. Nichols, V. Jacobson, and L. Zhang. A Two-bit Differentiated Services Architecture for the Internet. Internet Draft, Internet Engineering Task Force, May 1999. Work in progress.
15. K. Papagiannaki, P. Thiran, J. Crowcroft, and C. Diot. Preferential Treatment of Acknowledgment Packets in a Differentiated Services Network. Technical Report TR10-ATL-041001, Sprint Advanced Technology Laboratories, April 2001.

A Quantitative Model for the Parameter Setting of RED with TCP Traffic

Thomas Ziegler[1], Christof Brandauer[2], and Serge Fdida[3]

[1] FTW, Telecommunications Research Center Vienna, Maderstr.1, 1040 Vienna, Austria
Thomas.Ziegler@ftw.at
[2] Salzburg Research, J. Haringerstr. 5, 5020 Salzburg, Austria
Christof.Brandauer@salzburgresearch.at
[3] Université Pierre et Marie Curie, Laboratoire Paris 6, 75015 Paris, France
Serge.Fdida@lip6.fr

Abstract. This paper systematically derives a quantitative model how to set the parameters of the RED queue management algorithm as a function of the scenario parameters bottleneck bandwidth, round-trip-time, and number of TCP flows. It is shown that proper setting of RED parameters is a necessary condition for stability, i.e. to ensure convergence of the queue size to a desired equilibrium state and to limit oscillation around this equilibrium. The model provides the correct parameter settings, as illustrated by simulations and measurements with FTP and Web-like TCP flows in scenarios with homogeneous and heterogeneous round trip times.

1 Introduction

Although the RED (Random Early Detection) queue management algorithm [1] is already well known and has been identified as an important building block for Internet congestion control [2], parameter setting of RED is still subject to discussion. Simulations investigating the stability of RED with infinite-length (i.e. FTP-like) TCP flows [3][4][5] show that achieving convergence of RED's average queue size (*avg*) between the *minth* and *maxth* thresholds depends on adequate parameter setting. Our observation in [6] is that improper setting of RED parameters may cause the queue size to oscillate heavily around an equilibrium point (the queue average over infinite time intervals) outside the desired range from *minth* to *maxth*. High amplitude oscillations are harmful as they cause periods of link under utilization when the instantaneous queue size equals zero followed by periods of frequent "forced packet-drops" [7] when the average queue size exceeds *maxth* or the instantaneous queue approaches the total buffer size. Forced packet drops are in contradiction to the goal of early congestion detection and decrease the performance of ECN [8], attempting to avoid packet loss and to provide increased throughput for low-demand flows by decoupling congestion notification from dropping packets. In case of WRED [9] or RIO [10] oscillations may cause poor discrimination among in-profile and out-of-profile packets in a DiffServ environment. When the average queue size decreases below the maximum queue size threshold for out-of-profile packets the out-packets may enter the queue. Subsequently, the average queue size increases again and in-packets may be dropped with high probability.

However, the intention of this paper is not to investigate RED's performance decrease due to oscillation around an unfavorable equilibrium point but to build a

---

* This work is partly sponsored by the IST project Aquila.

L. Wolf, D. Hutchison, and R.Steinmetz (Eds.): IWQoS, LNCS 2092, pp. 202-216, 2001.

quantitative model for RED parameter setting in case of TCP flows to provide convergence (i.e. desirable behavior) of the queue. Regarding the kind of convergence we may hope to achieve with RED and infinite-length TCP flows it is important to mention that we do not mean convergence to a certain queue size value in the mathematical sense. Contrary, we define convergence very loosely as "achieving a state of bounded oscillation of the queue size around an equilibrium point of ($minth$+$maxth$)/2 so that the amplitude of the oscillation of the average queue size is significantly smaller than the difference between $maxth$ and $minth$ and the instantaneous queue size remains greater than zero and smaller than the total buffer size".

Realistic Internet traffic consists mainly of short-living (Web-like) TCP flows and not FTP-like traffic, thus it might be stated that a model based on infinite length TCP flows is unrealistic. We argue, however, that a quantitative model for RED parameter setting based on Web-like TCP traffic would be definitely too sophisticated to build. Thus our approach is to build a model for infinite length flows and then investigate if the model provides also proper parameter setting for short living TCP flows. Speaking in other words, our model is based on the commonly used control-theoretic approach of stabilizing a complex, non-linear dynamic system in the steady state case (infinite length flows) as a pre-requirement for desirable behavior in the dynamic-state case (Web flows).

As shown qualitatively in related papers [6][11][12], the RED parameters relevant for stability are the maximum drop probability ($maxp$) determining the equilibrium point, the minimum difference between the queue size thresholds ($maxth$-$minth$) required to keep the amplitude of the oscillation around the equilibrium point sufficiently low and the queue weight (wq). The RED parameters exhibit interdependencies demanding for a systematical approach to derive their values. As a first step, a quantitative model how to set $maxp$ dependent on the bottleneck bandwidth, RTT distribution, and the number of flows is derived in section 4. Subsequently, taking into account the $maxp$ and wq models (see section 2 for wq), an empirical model for the setting of $maxth$-$minth$ is derived in section 5. This model gives a lower bound for $maxth$-$minth$ as a function of the bottleneck bandwidth, RTT distribution and number of flows to avoid extensive oscillation of the queue size around the equilibrium point. Finally, after coming back to parameter interdependencies and combining the models for $maxth$-$minth$, $maxp$ and wq into a non-linear system of equations in section 6, simulations and measurements with FTP-like and Web-like TCP traffic are performed to evaluate the model in section 7.

Abbreviations used throughout the paper:
- B: buffer size at bottleneck in packets
- C: capacity of bottleneck link in Mbps
- L: bottleneck-capacity in mean packets per second
- D: delay of bottleneck link in ms
- N: number of flows

2   Related Research on RED Modelling and Parameter Setting

Existing publications discussing the setting of RED's $minth$, $maxth$ and $maxp$ parameters give either rules of thumb and qualitative recommendations [1][3][4][12][13][14] or quantitative models assuming infinite time averages [11] which are not capable of modelling the oscillatory behavior of the RED queue.

[12] proposes a simple quantitative model how to set wq based on RED's response to a unit-step input signal. In [11] the length of the TCP period of window increase and decrease *(I)* and an upper bound of one RTT for RED's queue-sampling interval (δ) is derived using the TCP model proposed in [15]. It is found that setting the RED averaging interval equal to *I* results in a good compromise between the opposing goals of maintaining the moving average close to the long term average and making the moving average respond quickly to a change in traffic conditions. RED's averaging interval equals the TCP period *I* if the queue weight is computed as follows:

$$wq = 1 - a^{\delta/I},$$

where a is a constant parameter in the order of 0.1, providing a quantitative model for the setting of wq.

Independently and simultaneously to the present paper (see [6] for an earlier version) a control theoretic approach to model the parameter setting of RED has been made in [16]. This paper is based on a linearization of the TCP model published in [17].

Additionally, [18] compares the end-to-end performance of RED variants and Drop-Tail. [19] shows that RED-like mechanisms tend to oscillate in the presence of two-way TCP traffic. [20] investigates the tuning of RED parameters to minimize flow-transfer times in the presence of Web traffic.

3   Simulation Settings

Simulations are performed with the *ns* network simulator [21] using the topology shown in figure 1. All 500Mbps links employ Drop Tail queue management; buffer sizes at access links are set sufficiently high to avoid packet loss. Thus packets are discarded solely at the bottleneck link from router1 to router2.

The link between router1 and router2 uses RED queue management. RED is operated in packet mode only as simulations in [6] and [12] show the same results regarding the convergence behavior of the queue with RED in byte and packet mode. The "mean packet size" parameter of RED is set to 500 bytes.

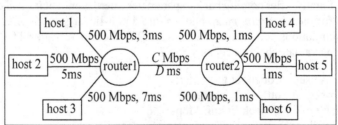

Fig. 1. Simulated Network.

Unless otherwise noted, N TCP flows start at random times between 0 and 10 seconds of simulation time. Hosts at the left hand side of the simulated network act as sources, hosts at the right hand side act as sinks. Host i starts $N/3$ TCP flows to host $3+i$. TCP data senders are of type Reno and New-Reno. Packet sizes are uniformly distributed with a mean of 500 bytes and a variance of 250 bytes.

In all queue size over time figures shown in this paper the average queue size is plotted with a bold line, the instantaneous queue size is plotted with a thin line. The unit of the x-axis is seconds, the unit of the y-axis is mean packet sizes. Simulations last for 100 seconds. However, only the last 30 seconds are plotted in queue size over time figures as we are rather interested in the steady-state than the startup behavior of the system.

4   Determining the Equilibrium Point

The following simulation shows that *maxp* has to be set as a function of the aggressiveness of traffic sources in order to enable convergence of the average queue to an equilibrium point between *minth* and *maxth*. The aggressiveness of an aggregate of TCP flows is inversely proportional to its per-flow bandwidth*RTT product[1], defined as $C*RTT/N$. Thus we can for instance increase the bottleneck capacity C and leave *maxp* constant to illustrate the drift of the equilibrium point.

Constant parameters: $D = 100ms$, $N = 100$; *maxp* is set to 0.1 as recommended in [13]. Lacking models for *maxth, minth* and *wq* these parameters have to be set reasonably based on experience from former simulations and rules of thumb for simulations in this chapter. The buffer size B is set to the bandwidth*RTT product of the scenario or 40 packets, whichever is higher; *maxth* = 2B/3, *minth* = *maxth*/4. These settings result in a difference between *maxth* and *minth* sufficiently high to avoid high amplitude oscillations, and a setting of *wq* to make the average queue size track the instantaneous queue properly as evaluated in earlier simulations and shown also in figure 2.

Simulation	C	minth	maxth	B	wq
1	2	29	114	171	0.0042
2	10	143	571	857	0.00063
3	50	714	2857	4286	0.000035

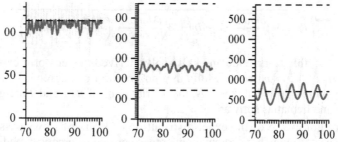

Fig. 2. Simulations 1-3, inst. and Average Queue Size over Time, maxp = 1/10.

As the increase of C causes the per-flow bandwidth*RTT product to increase, the aggressiveness of the TCP flows and thereby the average queue size decreases. In simulation 1 a constant *maxp* of 0.1 (and thus the RED drop probability) is too small, causing convergence of *avg* to *maxth* and thus frequent forced packet drops. In simulation 2 *maxp* is well chosen given the specific scenario, thus the queue's equilibrium point is in-between *minth* and *maxth*. In simulation 3 a *maxp* of 0.1 is too high causing *avg* to oscillate around *minth* resulting in suboptimal link utilization as the queue is often empty.

---

[1] The per-flow bandwidth*RTT product can also be considered as the long term average TCP window of flows.

4.1   A Model for maxp

We present a steady-state analysis resulting in a quantitative model on how to set *maxp* as a function of C, N and *RTT*. As shown in [15], the rate R_i of a TCP flow i in units of pkt/s as a function of the loss probability (p), the round trip time (RTT_i) and retransmission time-out time (T_i) can be modelled by the following expression:

$$R_i = \frac{1}{RTT_i \cdot \sqrt{\frac{2bp}{3}} + T_i \cdot min\left(1, 3 \cdot \sqrt{\frac{3bp}{8}}\right) p(1 + 32p^2)} \tag{1}$$

The constant b in the above equation denotes the number of packets received at the TCP data receiver to generate one ACK. With a delayed ACK TCP data receiver b would be equal to 2. For the moment, we do not use delayed ACKs, hence b is set to one.

For any aggregate of N TCP flows the long term average rate equals the link capacity:

$$L = \sum_{i=1}^{N} R_i \tag{2}$$

Substituting R_i by eq. 1 we get a function of p, L, N, a distribution of round trip times and retransmission time-out times. Our goal is to make the average queue size converge at *(minth+maxth)/2*, which corresponds to a drop probability of *maxp/2*. Substituting p with *maxp/2*, we can derive the optimum setting of *maxp* as a function of L, *RTT* and N:

$$L = \sum_{i=1}^{N} \frac{1}{RTT_i \cdot \sqrt{\frac{bmaxp}{3}} + T_i min\left(1, 3\sqrt{\frac{3bmaxp}{16}}\right)\left(\frac{maxp}{2} + 4maxp^3\right)} \tag{3}$$

We are not able to provide an analytically derived closed-form expression for *maxp* as solving (3) for *maxp* results in a polynomial of degree seven. However, numerical solution provided that L, N, the distributions of *RTT* and T are given is feasible with a mathematics software like [22].

ISPs are not aware of the exact distribution of RTTs of flows passing a RED queue. The best we may hope to obtain in order to provide a practical model is a histogram, grouping flows into m RTT classes where each RTT class j has n_j flows and homogeneous round trip times. For such a model, eq. (3) can be rewritten as

$$L = \sum_{j=1}^{m} \frac{n_j}{RTT_j \cdot \sqrt{\frac{bmaxp}{3}} + T_j min\left(1, 3\sqrt{\frac{3bmaxp}{16}}\right)\left(\frac{maxp}{2} + 4maxp^3\right)} \tag{4}$$

For derivation of *maxp* in subsequent simulations we have set RTT_j and T_j as follows:

$$RTT_j = 2d_j + (minth+maxth)/(2L)$$

$$T_j = RTT_j + 2(minth+maxth)/L \tag{5}$$

The term *(minth+maxth)/(2L)* matches the average queueing delay at the bottle-neck, d_j denotes the total propagation delay of RTT class *j*. As implemented by TCP, *T* is computed as the *RTT* plus four times the variance of the *RTT*, approximated by the average queueing delay at the bottleneck.

The following paragraphs repeat the simulations at the beginning of this section but with *maxp* adapted according to the model:

Simulation	4	5	6
1/maxp	4.15	30	618

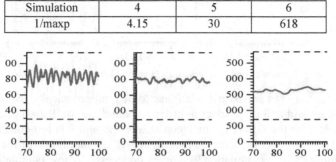

Fig. 3. Simulations 4-6, inst. and Average Queue Size over Time, maxp Adapted.

Figure 3 shows that adapting *maxp* according to equation 4 makes the queue size converge close to *(minth+maxth)/2*.

For the remainder of the paper, *maxp* is set according to the model proposed in this section. Having a model for *maxp* we may from now on additionally use the model how to set the *wq* parameter as explained in [11] and section 2. Note that the model for the setting of *wq* parameter is based on eq. 1 and thus requires the drop probability and RTT as an input. As for the derivation for the *maxp* model we may approximate the drop probability for the *wq* model as *maxp/2*; for the RTT we refer to equation 5. As the *ns* RED implementation computes the average queue size at each packet arrival the δ parameter is set to the inverse of the link capacity in pack-ets. The constant *a* is set to 0.01 instead of 0.1 (see section 2) resulting in a slightly higher queue weight parameter and thus marginally shorter memory in RED's queue averaging compared to [11]. Simulations show evidence that choosing parameters for the setting of *wq* as described above results in the average queue size properly tracking the instantaneous queue size and avoids possible oscillations by choosing *wq* too small (i.e. too long memory).

5   Determining the Amplitude of the Queue Size Oscillation

As TCP is volume controlled, the buffer requirements of an aggregate of TCP flows solely depends on the product of bandwidth and RTT, no matter how the individual values for bandwidth or RTT are set in a scenario. Simulations 7-9 successively increase *D* to vary the bandwidth*RTT product, showing the dependency of *maxth-minth* on bandwidth and RTT. Constant parameters: *C* = 20Mbps, *minth* = 40, *maxth* = 300, *B* = 400. *N* is adapted such that the per flow bandwidth*RTT product stays roughly constant.

Simulation	D	N	1/maxp	wq
7	1ms	18	42	0.0025
8	50ms	67	37.5	0.0015
9	300ms	317	36	0.00017

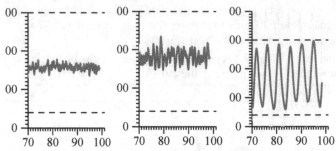

Fig. 4. Simulations 7-9, inst. and Average Queue Size over Time

The difference between *maxth* and *minth* has to be a monotonically increasing function of the bandwidth*RTT product in order to bound the oscillation of the average queue size. In case the difference between *maxth* and *minth* is left constant and the bandwidth*RTT product is increased (see figure 4), the amplitude of the oscillation may be improperly high (simulation 9) or even unnecessarily low (simulation 7).

Simulations 10-12 investigate the behavior of the RED queue in case of a varying number of TCP flows. Constant parameters: C = 20Mbps, D = 100ms, *minth* = 40, *maxth* = 300, B = 400.

Simulation	N	1/maxp	wq
10	100	49	0.0004
11	250	10.3	0.00065
12	300	8	0.0007

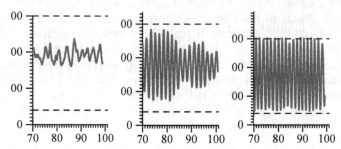

Fig. 5. Simulations 10-12, inst. and Average Queue Size over Time.

According to our model for *maxp*, an increase in the number of TCP flows requires setting *maxp* higher in order to keep the equilibrium point close to *(minth+maxth)/2* (see parameter table for figure 5). Increasing *maxp* and leaving *maxth-minth* constant means increasing the slope of RED's drop probability function as defined by *maxp/ (maxth-minth)*. A steeper drop probability function means that small changes in TCP congestion windows, in other words small changes in the queue size, cause high changes in RED's drop probability. In response to a high change in drop probability, TCP flows drastically alter their congestion windows causing the oscillatory behavior shown in figure 5, simulation 12. Thus, to avoid these oscillations, we conclude that an increase of *maxp* (due to an increase in N) must be accompanied by an increase of *maxth-minth* in order to bound the slope of the drop probability function.

5.1   Model for maxth-minth and Homogeneous RTTs

Employing the qualitative insights from the introduction of this section and the models already derived for *maxp* and *wq* to make the queue converge to *(maxth+minth)/2* we use an empirical approach to quantitatively model the required difference between *maxth* and *minth* as a function of the bottleneck capacity and round-trip-time product *(C*RTT)*, and the number of flows *(N)*. For the moment we only consider the case of homogeneous RTTs (section 5.2 generalizes the model to the heterogeneous RTT case).

The goal of our model is to determine the difference between *maxth* and *minth* such that the amplitude of the oscillation around the equilibrium point is kept constant at a desirable value of *(maxth-minth)/4*.[2] This should happen independently of the *C*RTT* and *N* input parameters. Our approach towards generation of such a model is to perform simulations over a broad range of input parameters and thereby manually adjust *maxth-minth* until the above condition is met. The resulting cloud of points in the 3-dimensional *(C*RTT, N, maxth-minth)* space can then be approximated numerically by a closed-form expression, providing the desired quantitative model for *maxth-minth*.

In all simulations the total buffersize *(B)* is set to 3*maxth/2, sufficiently high to avoid losses due to buffer overflow; *minth* equals *(maxth-minth)/3;* packet sizes have a mean of 500 bytes, see section 3. For other simulation settings we refer to section 3.

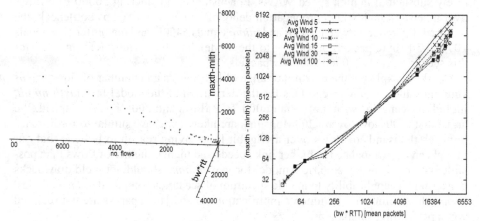

Fig. 6. Simulation results for maxth-minth (in mean packets) as a function of the bandwidth*RTT product (in mean packets) and the number of flows.

Figure 6a shows the resulting cloud of points where each point represents a proper value for *maxth-minth* which has been found by simulation using our empirical approach. Figure 6b illustrates how figure 6a has been created and ameliorates its readability. Six sets of simulations have been conducted for creation of figure 6a, each consisting of 10 simulations having identical average per-flow windows (as defined by *C*RTT/N*) and being represented by one curve in figure 6b.

The distribution of points in figure 6a and the curves in figure 6 show that *maxth-minth* can be considered as almost linearly dependent on the bandwidth*RTT product and the number of flows. Thus the cloud of points in figure 6a may be approxi-

---

[2] The choice of *(maxth-minth)/4* for the oscillation of the average queue size reflects a reasonable compromise between having stability margins for a bounded oscillation and trying to keep queueing delay and buffer requirements as small as possible.

mated by a linear least-square fit, yielding *maxth-minth* as a linear function in two variables:

$$maxth - minth = c1 \cdot C \cdot RTT + c2 \cdot N + c3, \tag{6}$$

with the values of $c1 = 0.02158$, $c2 = 0.567$, $c3 = 85$ for an average packet size of 500 bytes. We have repeated our empirical approach for packet sizes of 250, 1000, and 1500 bytes and found analogous approximations as different packet sizes did not change the linear shape of the cloud of points; see table 1 for the resulting constants.

avg. packet size	c1	c2	c3
250	0.02739	0.7324	17
500	0.02158	0.5670	85
1000	0.01450	0.3416	46
1500	0.01165	0.09493	85

Table 1: Constants for *maxth-minth* Model.

Note that eq. 6 gives a lower bound; simulations show that the amplitude of the oscillation decreases further if *maxth-minth* is set higher than suggested by eq. 6. Although eq. 6 results in settings of *maxth-minth* significantly smaller than the bandwidth*delay product (compare to [12]), the buffer requirements of RED with TCP traffic are definitely substantial in high speed WANs. For instance, considering 50000 FTP flows over a network path with a propagation delay of 100ms and a 5Gbps bottleneck, the required difference between *maxth* and *minth* equals 54000, and the buffer size equals 107000 500 byte packets according to the model. The bandwidth*RTT product for this scenario equals 250000 500 byte packets.

We have exploited the maximum range of link speeds and number of flows the *ns* simulator allows on high-end PCs for the derivation of the model for *maxth-minth*, including scenarios with more than 6000 TCP flows and 200Mbps link speed. We argue that 6000 flows over a 200Mbps link are likely to behave similar to for instance hundred thousands of flows over a Gigabit link, giving reason to believe that our model can be extrapolated to higher link speeds and higher number of flows than possible with ns. Thus the empirical model for *maxth-minth* should not yield drawbacks concerning its applicability to a wide spectrum of scenarios compared to an analytical approach. Additionally, common simplifications in analytical papers are not required when performing simulations.

5.2   Incorporating Heterogeneous RTTs

Equation 6 provides a model how to set *maxth-minth* as a function of the bottleneck bandwidth, number of flows and the RTT, assuming that all flows have homogeneous (i.e. equal) RTTs. In order to extend the model to the heterogeneous RTT case, we propose to compute the RTT quantity in eq. 6 as a weighted average (*aRTT*) of the per-flow RTTs. A TCP flow's share of the bottleneck capacity and the bottleneck queue is inversely proportional to its RTT. Flows utilizing a higher portion of the bottleneck resources have greater influence on the queue dynamics than flows utilizing a smaller portion of the bottleneck resources. It is thus reasonable to use a flow's share of the link capacity for computation of the weights for *aRTT*.

Similar to section 4.1, we assume a histogram of m RTT classes, where each RTT class j has n_j flows and homogeneous round trip time RTT_j. The rate R_j of a TCP flow joining class j can be computed according to eq. 1, section 4.1. Again, we approximate the drop probability required for the TCP model as *maxp/2*. The average RTT, weighted by the rate of flow aggregates can be computed as follows:

$$aRTT = \sum_{j=1}^{m} \frac{n_j R_j}{L} RTT_j \tag{7}$$

Finally, equation 6 has to be rewritten as

$$maxth - minth = c1 \cdot C \cdot aRTT + c2 \cdot N + c3 \tag{8}$$

6   Parameter Dependencies and Model Assembly

RED parameter settings are partially interdependent. Thus it is important to take these dependencies into account when designing a model. The *maxp* parameter determines the equilibrium point of the queue size and depends on *C, N* and *RTT. RTT* means that *maxp* depends on queueing delay and thus on the setting of *maxth* and *minth*. However, the equilibrium point (as it is the *infinite* time average), does not depend on how accurately RED's average queue size follows the instantaneous queue size. Thus *maxp* can be derived independently of the setting of *wq* (compare section 4.1).

The *wq* parameter, on the other hand, depends on the number of flows, the bottleneck bandwidth, the drop probability (and thus on *maxp*) and the RTT (and thus implicitly on *maxth* and *minth*). Thus we can compute the *wq* parameter having the model for *maxp* and assuming fixed settings for *maxth* and *minth*.

The setting of *maxth-minth* depends on *maxp, wq, RTT*, the bottleneck bandwidth and the number of flows. The dependency on *maxp* and *wq* is taken into account by our empirical model for *maxth-minth* as the *maxp* and *wq* models are used as an input.

Deriving the model for *maxth-minth* yields an equation for *maxth-minth* as a function of *C*RTT* and *N*. Although *maxp* and *wq* have been taken into account for the derivation of the model, they do not appear directly in the equation (compare to eq. 6) due to the empirical approach of deriving it. Thus, assuming homogeneous RTTs, we may first compute *maxth-minth* and then compute *maxp* knowing how to set *maxth* and *minth* when applying the model to compute the RED parameters for scenario defined by *C*RTT* and *N*. Finally *wq* can be computed knowing *maxp, maxth* and *minth*.

For the following reason it is not possible anymore to consider the computation of *maxp* and *maxth-minth* as independent from each other in the case of heterogeneous RTTs: As explained above the setting of *maxp* depends on *maxth* and *minth*. As shown in section 5.2, the setting of the *aRTT* quantity and thus the setting of *maxth-minth* depends on *maxp*. As a consequence, equation (8) and equation (4) have to be combined to a system of non-linear equations which can be solved numerically for *maxp, minth,* and *maxth*. Finally, knowing how to set *maxp, minth,* and *maxth* for computation of the average length of the TCP period *I, wq* can be computed as explained in section 2.

The quantitative model for RED parameter setting has been implemented in the *maple* mathematics software [22] and can be easily used via the Web, see [23].

7   Evaluation of the Model

7.1   FTP-Like Traffic

More than 200 simulations with arbitrary points in the (C*RTT,N) space, with the
TCP delayed ACKs option enabled and disabled and with homogeneous and heterogeneous RTTs have been conducted to evaluate the model for FTP-like TCP flows
(see also [6]). In all these simulations the queue size converges between *minth* and
maxth as desired. This paper only shows a small subset of these simulations as the
new insights the reader may gain are restricted.

Simulation	C	D	N	minth	maxth	B	1/maxp	wq
13	1	0.01	3	31	125	187	280	0.002540
14	50	0.1	313	176	703	1054	35	0.000172
15	150	0.25	2163	1064	4255	6383	36	2.383e-05

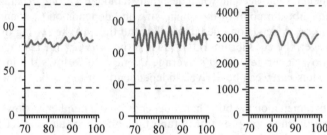

Fig. 7. Simulations 13-15, inst. and Average Queue Size over Time.

Figure 7 shows that the parameter settings proposed by the model are appropriate for
the selected scenarios.

A model partially developed empirically by simulation should not only be evaluated by simulation. Thus we have performed extensive (more than 50) measurements
with various points in the (C*RTT,N) space using a topology as illustrated in figure 8.

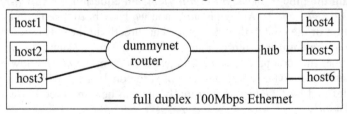

Fig. 8. Measured Network.

The propagation delay of all links can be assumed to be equal zero as all equipment is
located in one lab; the packet processing time in nodes has been determined as negligible. TCP data senders are located at host1-3, TCP data-receivers at hosts 4-6. All
hosts are Pentium III PCs, 750Mhz, 128 MB RAM and run the ttcp tool to generate
the FTP-like bulk-data TCP flows. Host operating system is Linux running kernel version 2.2.16 implementing TCP SACK with an initial congestion window of two segments and delayed ACKs. The initial congestion window property of the Linux TCP
implementation makes sources somewhat more aggressive compared to the simulations.

The "dummynet router" is another Pentium III PC, 750Mhz, 128MB RAM running the FreeBSD operating system version 3.5. This PC runs the dummynet tool [24] for emulation of a bottleneck link (specified by bandwidth C, delay D and an MTU of 500 bytes) and the RED queue management algorithm. Note that dummynet operates only on packets flowing in direction host1-3 to host4-6. In the reverse direction dummynet is transparent. Congestion and packet drops happen solely due to RED at dummynet's emulated bottleneck link. The bottleneck capacity is chosen sufficiently small to make the influence of collisions at the Ethernet layer on the measurement results negligible. Additionally, we have verified carefully that all PCs provide sufficient computational power in order to avoid influencing the measurement results by side-effects.

Measurement	C	D	N	minth	maxth	B	1/maxp	wq
1	0.5	50	10	33	132	197	69	0.003
2	3	100	200	76	303	455	5.9	0.00159
3	10	50	50	48	193	290	41.3	0.000966
4	10	300	200	118	471	706	57	0.000178

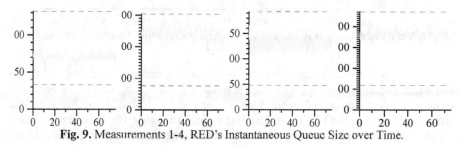

Fig. 9. Measurements 1-4, RED's Instantaneous Queue Size over Time.

Figure 9 has been created by dummynet dumping the instantaneous queue size on the harddisk every time a packet arrives at the RED queue. As illustrated in figure 9, the measurements confirm simulations in verifying the correctness of the model. Measurements 2 and 4 exhibits a somewhat higher equilibrium point, which can be explained by the TCP model (equation 1) becoming less accurate for very high drop probabilities (this effect has also been observed in simulations) and the aggressive kind of TCP implemented in the used Linux OS (initial congestion window equals 2 segments).

7.2 Web Traffic

This section simulates Web-like TCP SACK traffic, employing a model for HTTP 1.0 as described in [25] and being shipped with *ns-2*. We simulate U Web client-server connections. A client downloads a page consisting of several objects from the server, is idle for some think time and downloads the next page. Between getting two subsequent objects, the client waits for some inter-object time. According to [25] all random variables of the model are Pareto distributed. Table 2 shows the parameters of the distributions.

parameter of pareto distribution	think time	objects per page	inter object time	object size
mean	50 ms	4	0.5 ms	12 Kbyte
shape	2	1.2	1.5	1.2

Table 2: Parameters for Web Model.

The N parameter (number of active TCP flows) required as an input to the RED parameter model is estimated by the mean number of active flows over the entire simulation time. Comparing the number of Web-sessions and the average number of flows in the subsequent table, we can observe a non linear dependency of traffic load (as indicated by N) on the number of Web Sessions U.

Constant parameters: $C = 10$Mbps, $D = 0.1$s, packet size = 500 bytes.

Simulation	N	minth	maxth	B	wq	1/maxp	U
16	50	56	224	336	0.000371	113	900
17	250	97	388	581	0.000748	9.7	1500
18	1000	250	999	1498	0.000478	4.2	1600

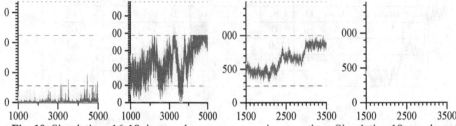

Fig. 10. Simulations 16-18, inst. and average queue size over time. Simulation 18, number of active flows over time.

Simulations 16 to 18 show scenarios with low, medium and high load due to different Web traffic demands. For all scenarios, the queue size curves depend to a great extent on the traffic demand created by the Web-sessions which is highly variable over time. This is illustrated in the two rightmost part-figures of figure 10, showing the queue size and the number of active flows (an indicator for the instantaneous traffic demand) over time for simulation 18. The queue size curve tracks the number of active flows curve closely.

In the lightly loaded scenario (simulation 16) the queue hardly exceeds *minth* due to the low traffic demand, thus RED has no influence on system dynamics. Simulations 17 and 18 represent scenarios with sufficient load to fill the buffer dependent on the variations in instantaneous traffic demands. Of course we can not expect convergence of the queue with bounded oscillation between *minth* and *maxth* similar to the case of FTP-like flows for such a scenario. However, the RED model proposes reasonable parameter settings as the RED queue buffers most traffic bursts between *minth* and *maxth*.

A comparison of the number of flows and the queue size over time curve for simulation 18 explains why the queue size may deviate extensively from the desired value of *(minth+maxth)/2* and proofs that this deviation happens due to the variability in load and not due to an inaccuracy of the RED model. Up to a simulation time of about 2400 seconds the real number of active flows is smaller than 1000 (the value of the N parameter used as input to the model) thus the queue size is below the desired value. After 2800 seconds the real number of flows exceeds N making the queue approach *maxth*. From 2400 to 3000 seconds of simulation time, however, the real number of active flows is approximately equal to N and the queue size is close to the desired value. Thus we can conclude that the model provides proper parameter settings as long as we are able to estimate the input parameters to the model with sufficient accuracy (see section 8 for a remark). This has been observed also in other simulations with Web traffic.

8   Conclusions

We find that the RED parameters relevant for stability are the maximum drop probability (*maxp*) determining the equilibrium point (i.e. the queue average over infinite time intervals), the minimum difference between the queue size thresholds (*maxth-minth*) required to keep the amplitude of the oscillation around the equilibrium point sufficiently low and the queue weight (*wq*). The major part of this paper is dedicated to the systematical derivation of a quantitative model for the setting of the RED parameters mentioned above to achieve stability of the queue size. The model is suitable for heterogeneous RTTs and assumes TCP traffic. The stability of the system does not solely depend on the dynamics of the RED drop-function and end-to-end congestion control but also on the time scales of the arrival-process. It would be, however, too complex to build a quantitative model with realistic Web-like TCP traffic, thus we focus on infinite life-time, bulk-data TCP flows for derivation of the model.

Note that there exist other constraints for RED parameter setting which are not considered in this paper as they are not relevant for system stability but which are nevertheless important (e.g. setting *minth* as a trade-off between queueing delay and burst tolerance). For a discussion of these constraints we refer to [13].

The model is evaluated by measurements with bulk-data TCP traffic and simulations with bulk-data and Web-like TCP traffic, showing that RED parameters are set properly and that deriving the RED parameter model under the simpler steady state conditions (infinite length flows) provides good parameter settings not only for the steady state case but also for the dynamic state case (realistic Web-like flows). Web traffic, however, has a highly variable demand thus it is difficult to determine a good, fixed value for an estimator of the traffic load (average number of active flows) which is required as an input parameter to *any* model proposing parameter settings for active queue management mechanisms. As a consequence, we can not expect the queue to stay always within some desired range even though the model for parameter setting itself may be correct. Showing the correctness of our model for Web traffic, we have demonstrated by simulation that if we are able to feed our RED model with sufficiently accurate input parameters, stability (i.e. bounded oscillation of the average queue size between *minth* and *maxth*) can be achieved for Web traffic. In a realistic environment, however, it is questionable whether ISPs are capable of retrieving sufficiently accurate information about the RTT of flows and the average number of flows traversing a router. Additionally, the number of flows is highly variable thus finding a good fixed value for e.g. the number of flows may not be possible. Simple extensions to RED may help diminishing dependencies on the scenario parameters and thus provide performance benefits compared to the original version of RED. For instance, gentle RED decreases the dependency on traffic variability in terms of the number of flows due to its smoother drop function, as shown in [18]. However, no matter whether the gentle or the original version of RED is used, the model proposed in this paper can provide useful insights how to set active queue management parameters.

Acknowledgements

The authors are grateful to Eduard Hasenleithner, Ferdinand Graf and Matthäus Gruber from Salzburg University of Applied Sciences for building the testbed and supporting the measurements.

References

1. S. Floyd, V. Jacobson, "Random Early Detection Gateways for Congestion Avoidance", IEEE/ACM Transactions on Networking, August 1993
2. B. Braden, D. Clark et al., "Recommendations on Queue Management and Congestion Avoidance in the Internet", RFC 2309, April 1998
3. W. Feng, D. Kandlur, D. Saha, K.G. Shin, " A self-configuring RED gateway", Proc. of IEEE Infocom 1998
4. W. Feng et al., "Techniques for eliminating Packet Loss in congested TCP/IP Networks", University of Michigan CSE-TR-349-97, November 1997
5. T.J. Ott, T.V. Lakshman, L.H. Wong, "SRED: Stabilized RED", Proc. of IEEE Infocom 1998
6. T. Ziegler, S.Fdida, C. Brandauer, "Stability Criteria of RED with TCP Traffic", Tech. Report, unpublished, August 1999, http://www-rp.lip6.fr/publications/production.html
7. S. Floyd, K. Fall, K. Tieu, "Estimating arrival rates from the RED packet drop history", March 1998, unpublished, http://www.aciri.org/floyd/end2end-paper.html
8. K.K. Ramakrishnan, S. Floyd, "A Proposal to add Explicit Congestion Notification (ECN) to IP", RFC2491, January 1999
9. Cisco Web pages, http://www.cisco.com/warp/public/732/netflow/qos_ds.html
10. D. Clark, "Explicit Allocation of Best Effort Packet Delivery Service", http://www.ietf.org/html.charters/diffserv-charter.html
11. V. Firoiu, M. Borden, " A Study of active Queue Management for Congestion Control", Proc. of IEEE Infocom, Tel Aviv, 2000
12. V. Jacobson, K. Nichols, K. Poduri, "RED in a different Light", Draft Technical Report, Cisco Systems, Sept. 1999
13. S. Floyd, "Discussions on setting RED parameters", http://www.aciri.org/floyd/red.html , Nov. 1997
14. C. Villamizar, C. Shong, " High performance TCP in ANSNET", Computer Communication Review, V.24. N.5, October 1994, pp. 45-60
15. J. Padhye et al.,"Modeling TCP Troughput: A simple Model and its empirical Validation", Proceedings of ACM SIGCOMM, August 1998
16. C. Hollot et al., "A Control Theoretic Analysis of RED", To appear in proceedings of IEEE Infocom 2001, Tel Aviv, Israel
17. V. Misra, W. Gong, D. Towsley, A Fluid-based Analysis of a Network of AQM Routers Supporting TCP Flows with an Application to RED", SIGCOMM 2000
18. G. Iannacone, C. Brandauer, T. Ziegler, C. Diot, Serge.Fdida, M. May, "Comparison of Tail Drop and AQM Performance for bulk-data and web-like Internet traffic", to appear in proceedings of ISCC 2001, July 2001, Hammameth, Tunesia
19. T. Ziegler, S. Fdida, C. Brandauer, B. Hechenleitner, "Stability of RED with two-way TCP Traffic", IEEE ICCCN, Las Vegas, Oct. 2000
20. Mikkel Christiansen, Kevin Jeffay, David Ott, F. Donelson Smith, "Tuning RED for Web traffic", Proc. of ACM SIGCOMM 2000
21. NS Simulator Homepage, http://www.isi.edu/nsnam/ns/
22. Maple home page, http://www.mit.edu/afs/athena.mit.edu/software/maple/www/home.html
23. RED model homepage, www.salzburgresearch.at/~cbrand/REDmodel
24. L. Rizzo, "An embeddded Network Simulator to support Network Protocols' Development", Proceedings of Tools'97, St. Malo, France, June 1997
25. A. Feldmann, A. Gilbert, P. Huang, W. Willinger, "Dynamics of IP traffic: A Study of the Role of Variability and the Impact on Control", Proceedings of ACM SIGCOMM'99, pages 301-313, 1999

Evaluation of the QoS Offered by PRTP-ECN - A TCP-Compliant Partially Reliable Transport Protocol

Karl-Johan Grinnemo[1] and Anna Brunstrom[2]

[1] Ericsson Infotech AB, Lagergrens gata 4
SE-651 15 Karlstad, Sweden
Karl-Johan.Grinnemo@ein.ericsson.se
[2] Karlstad University, Dept. of Computer Science
SE-651 88 Karlstad, Sweden
Anna.Brunstrom@kau.se

Abstract. The introduction of multimedia in the Internet imposes new QoS requirements on existing transport protocols. Since neither TCP nor UDP comply with these requirements, a common approach today is to use RTP/UDP and to relegate the QoS responsibility to the application. Even though this approach has many advantages, it also entails leaving the responsibility for congestion control to the application. Considering the importance of efficient and reliable congestion control for maintaining stability in the Internet, this approach may prove dangerous. Improved support at the transport layer is therefore needed. In this paper, a partially reliable transport protocol, PRTP-ECN, is presented. PRTP-ECN is a protocol designed to be both TCP-friendly and to better comply with the QoS requirements of applications with soft real-time constraints. This is achieved by trading reliability for better jitter characteristics and improved throughput. A simulation study of PRTP-ECN has been conducted. The outcome of this evaluation suggests that PRTP-ECN can give applications that tolerate a limited amount of packet loss significant reductions in interarrival jitter and improvements in throughput as compared to TCP. The simulations also verified the TCP-friendly behavior of PRTP-ECN.

1 Introduction

Distribution of multimedia traffic such as streaming media over the Internet poses a major challenge to existing transport protocols. Apart from having demands on throughput, many multimedia applications are sensitive to delays and variations in those delays [28]. In addition, they often have an inherent tolerance for limited data loss [13].

The two prevailing transport protocols in the Internet today, TCP [22] and UDP [21], fail to meet the the QoS requirements of streaming-media and other applications with soft real-time constraints. TCP offers a fully reliable transport service at the cost of increased delay and reduced throughput. UDP, on the other

L. Wolf, D. Hutchison, and R. Steinmetz (Eds.): IWQoS 2001, LNCS 2092, pp. 217–230, 2001.

hand, introduces virtually no increase in delay or reduction in throughput, but provides no reliability enhancement over IP. In addition, UDP leaves congestion control to the discretion of the application. If misused, this could impair the stability of the Internet.

In this paper, we present a novel transport protocol, Partially Reliable Transport Protocol using ECN (PRTP-ECN), which offers a transport service that better complies with the QoS requirements of applications with soft real-time requirements. PRTP-ECN is a receiver-based, partially reliable transport protocol that is implemented as an extension to TCP. Thus it is able to work within the existing Internet infrastructure. It employs a congestion control mechanism which to a large extent corresponds to the one used in TCP. A simulation evaluation suggests that by trading reliability for latency, PRTP-ECN is able to offer a service with significantly reduced interarrival jitter, and increased throughput and goodput as compared to TCP. In addition, the evaluation implies that PRTP-ECN is TCP-friendly, something which may not be the case with some RTP/UDP solutions.

The rest of this paper is organized as follows. Section 2 discusses related work. In Section 3, we give a brief overview of the design principles behind PRTP-ECN. Next, in Section 4, the design of the simulation experiment is described. The results of the simulation experiment are discussed in Section 5. Finally, in Section 6, we summarize the major findings, and indicate further areas of study.

2   Related Work

PRTP-ECN builds on the work of a number of researchers. The feasibility of using retransmission-based, partially reliable error control schemes to address the QoS requirements of digital continuous media in general, and interactive voice in particular was demonstrated by Dempsey [11]. Based on his findings, he introduced two new retransmission schemes: Slack Automatic Repeat Request (S-ARQ) [9] and Partially Error-Controlled Connection (PECC) [10]. The principle behind the S-ARQ technique is to extend the buffering strategy at the receiver to handle jitter in such a way that a retransmission could be done without violating the time limit imposed by the application. In contrast to S-ARQ, PECC does not involve any modifications to the playback buffer. Instead, it modifies the retransmission algorithm so that retransmission of lost packets only occur in those cases it could be done without violating the latency requirements of the application. PECC was incorporated into the Xpress Transport Protocol [29], a protocol designed to support a variety of applications ranging from real-time embedded systems to multimedia distribution. Even though PRTP-ECN and PECC share some similarities, they differ from each other in that PRTP-ECN considers congestion control, something which is not done by PECC.

Extensive work on using partially reliable and partially ordered transport protocols to offer a service better adapted to the QoS needs of streaming media applications has been conducted at LAAS-CNRS [12,24], and at the University of Delaware [4,5]. Their work resulted in the proposal of a new transport protocol, Partial Order Connection (POC) [3,2]. The POC approach to realize a partially reliable service combines a partitioning of the media stream into objects, with the notion of reliability classes. An application designates individual objects as needing different levels of reliability, i.e., reliability classes are assigned at the object level. By introducing the object concept, and letting applications specify their reliability requirements on a per-object basis, POC offers a very flexible transport service. However, PRTP-ECN is significantly easier to integrate with current Internet standards. An early version of POC was considered as an extension to TCP, but required extensive rework of the TCP implementation [7]. In addition, POC needs to be implemented at both the sender and the receiver side, while PRTP-ECN only involves the receiver side.

Three examples of partially reliable protocols utilizing the existing Internet infrastructure are: Cyclic-UDP [27], VDP (Video Datagram Protocol) [6], and MSP (Media Streaming Protocol) [15]. Cyclic-UDP works on top of UDP, and supports the delivery of prioritized media units. It uses an error correction strategy that makes the probability of successful delivery of a media unit proportional to the unit's priority. The CM Player [26], which is part of an experimental video-on-demand system at University of California at Berkeley, employs Cyclic-UDP for the transport of video streams between the video-on-demand server and the CM Player client. VDP is more or less an augmented version of RTP [25]. It is specifically designed for transmission of video, and uses an application-driven retransmission scheme. MSP, a successor to VDP, uses a point-to-point client-server architecture. A media session in MSP comprises two connections. One UDP connection used for the media transfer, and one TCP connection for feedback control. Based on the feedback, the sender starts dropping frames from the stream, taking into account the media format. In contrast to these protocols, PRTP-ECN is a general transport protocol, not aimed at a particular application domain. Furthermore, PRTP-ECN uses the same congestion control mechanism as TCP does; the same congestion control mechanism which has proven successful in the Internet for years. To this comes that all these three protocols involves the sender, which is not the case for PRTP-ECN. By involving the sender, these protocols only lend themselves to small-scale deployments and homogeneous environments, which is not the case for PRTP-ECN.

3   Overview of PRTP-ECN

As already mentioned, PRTP-ECN is a partially reliable transport protocol. It is implemented as an extension to TCP, and only differs from TCP in the way it handles packet losses. PRTP-ECN needs only to be employed at the receiver side. At the sender side, an ECN-capable TCP is used.

PRTP-ECN lets the QoS requirements imposed by the application govern the retransmission scheme. This is done by allowing the application to specify the parameters in a retransmission decision algorithm. The parameters let the application directly prescribe an acceptable packet-loss rate, and indirectly affect the interarrival jitter, throughput, and goodput. By relaxing on the reliability, the application receives less interarrival jitter and improved throughput and goodput.

PRTP-ECN works identical to TCP as long as no packets are lost. Upon detection of a packet loss, PRTP-ECN must decide whether the lost data is needed to ensure the required reliability level imposed by the application. This decision is based on the success-rate of previous packets. In PRTP-ECN, the success-rate is measured as an exponentially weighted moving average over all packets up to and including the last one received. This weighted moving average is called the *current reliability level*, $crl(n)$, and is defined as:

$$crl(n) = \frac{\sum_{k=0}^{n} af^{n-k} * p_k * b_k}{\sum_{k=0}^{n} af^{n-k} * b_k} , \tag{1}$$

where n is the last packet received, af, is the weight or *aging factor*, and b_k denotes the number of bytes contained in packet k. The variable, p_k, is binary valued. If the kth packet was successfully received, then $p_k = 1$, otherwise $p_k = 0$.

The QoS requirements imposed on PRTP-ECN by the application translates into two parameters in the retransmission scheme: af and rrl. The *required reliability level*, rrl, is the reference in the the feedback control system made up of the data flow between the sender and the receiver, and the flow of acknowledgements in the reverse direction. As long as $crl(n) \geq rrl$, dropped packets need not be retransmitted and are therefore acknowledged. If an out-of-sequence packed is received and $crl(n)$ is below rrl, PRTP-ECN acknowledges the last in-sequence packet, and waits for a retransmission. In other words, PRTP-ECN does the same thing as TCP does in this situation.

There is, however, a problem with acknowledging lost packets. In TCP, the retransmission scheme and the congestion control scheme are intertwined. An acknowledgement not only signals successful reception of one or several packets, but also indicates that there is no noticeable congestion in the network between the sender and the receiver. PRTP-ECN decouples these two schemes by using the TCP portions of ECN (Explicit Congestion Notification) [23].

The only requirement imposed on the network by PRTP-ECN is that the TCP implementation on the sender side has to be ECN-capable. It does not engage intermediary routers. In the normal case, ECN enables direct notification of congestion, instead of indirectly via missing packets. It engages both the IP and TCP layers. Upon incipient congestion, a router sets a flag, the Congestion Experienced bit (CE), in the IP header of arriving packets. When the receiver of a packet finds that the CE bit has been set, it sets a flag, the ECN-Echo flag, in the TCP header of the subsequent acknowledgement. Upon reception of an acknowledgement with the ECN-Echo flag set, the sender halves its congestion window and performs fast recovery. However, PRTP-ECN does not involve intermediate routers, and correspondingly does not need the IP parts of ECN. It only employs the ECN-Echo flag to signal congestion. When an out-of-sequence packet is acknowledged, the ECN-Echo flag is set in the acknowledgement. When receiving the acknowledgement, the sender will throttle its flow, but refrain from re-sending any packet.

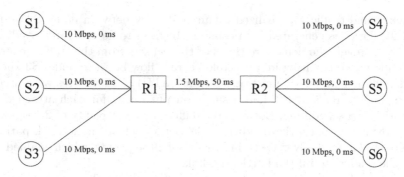

Fig. 1. Network Topology.

4   Description of Simulation Experiment

The QoS offered by PRTP-ECN as compared to TCP was evaluated through simulation, examining potential improvements in average interarrival jitter, average throughput, and average goodput. In addition, we investigated whether PRTP-ECN connections are TCP-friendly and fair against competing flows.

4.1   Implementation

We used version 2.1b5 of the ns2 network simulator [19] to conduct the simulations presented in this paper. The TCP protocol was modeled by the FullTcp agent, while PRTP-ECN was simulated by PRTP, an agent developed by us.

The FullTcp agent is similar to the 4.4 BSD TCP implementation [20,30]. This means, among other things, that it uses a congestion control mechanism similar to TCP Reno's [1]. However, SACK [18] is not implemented in FullTcp. The PRTP-ECN agent, PRTP, inherits most of its functionality from the FullTcp agent. Only the retransmission mechanism differs between FullTcp and PRTP.

4.2   Simulation Methodology

The network topology that was used in the simulation study is depicted in Figure 1. There were three primary factors in the experiment:

1. protocol used at node S4,
2. traffic load, and
3. starting times for the flows emanating from nodes S1 and S2.

   Two FTP applications attached to nodes S1 and S2 sent data at a rate of 10 Mbps to receivers at nodes S4 and S5. The FTP applications were attached to TCP agents. Node S4 accommodated two transport protocols: TCP and PRTP-ECN. Both protocol agents used an initial congestion window of two segments [1]. All other agent parameters were assigned their default values.

Background traffic was realized by an UDP flow between nodes S3 and S6. The UDP flow was generated by a constant bitrate traffic generator residing at node S3. However, the departure times of the packets from the traffic generator were randomized, resulting in a variable bit rate flow between nodes S3 and S6. In all simulations, a maximum transfer unit (MTU) of 1500 bytes were used.

The routers, R1 and R2, had a single output queue for each attached link and used FCFS scheduling. Both router buffers had a capacity of 25 segments, i.e. approximately twice the bandwidth-delay product of the network path. All receivers used a fixed advertised window size of 20 segments, which enabled each one of the senders to fill the bottleneck link.

The traffic load was controlled by setting the mean sending rate of the UDP flow to a fraction of the nominal bandwidth on the R1-R2 link. Tests were run for seven traffic loads: 20%, 60%, 67%, 80%, 87%, 93%, and 97%. These seven traffic loads corresponded approximately to the packet-loss rates: 1%, 2%, 3%, 5%, 8%, 14%, and 20% in the reference tests, i.e. the tests where TCP was used at node S4. Tests were run for eight PRTP-ECN configurations (see Section 4.3), and each test was run 40 times to obtain statistically significant results.

In all simulations, the UDP flow started at 0 s, while three cases of start times for the FTP flows were studied. In the first case, the flow between nodes S1 and S4 started at 0 s, and the flow between nodes S2 and S5 started at 600 ms. In the second case, it was the other way around, i.e. the flow between nodes S1 and S4 started at 600 ms, and the flow between nodes S2 and S5 started at 0 s. Finally, in the last case both flows started at 0 s. Each simulation run lasted 100 s.

4.3   Selection of PRTP-ECN Configurations

As already explained in Section 3, the service provided by PRTP-ECN depends on the values of the parameters af and rrl. From now on, we call an assignment of these parameters, a PRTP-ECN configuration.

The PRTP-ECN configurations used in the simulation experiment were selected based on their tolerance for packet losses. Since the packet-loss frequency a PRTP-ECN configuration tolerates depends on the packet-loss pattern, we defined a metric, the *allowable steady-state packet-loss frequency*, derived from the particularly favorable scenario in which packets are lost at all times when PRTP-ECN allows it, i.e. all times when $crl \geq rrl$. Simulations suggest that the allowable packet-loss frequency in this scenario approaches a limit, f_{loss}, as the total number of sent packets, n, reaches infinity, or more formally stated:

$$f_{loss} \stackrel{\text{def}}{=} \lim_{n \to \infty} \frac{loss(\sigma^n)}{n} , \tag{2}$$

where σ^n denotes the packet sequence comprising all packets sent up to and including the nth packet, and $loss(\sigma^n)$ is a function that returns the number of lost packets in σ^n. Considering that packet losses almost always happen in less favorable situations, the allowable steady-state packet-loss frequency could be seen as a rough estimate of the upper bound of the packet-loss frequency tolerated by a particular PRTP-ECN configuration.

We selected seven PRTP-ECN configurations which had allowable steady-state packet-loss frequencies ranging from 2% to 20%. Since our metric, the allowable steady-state packet-loss frequency, did not capture all aspects of a particular PRTP-ECN configuration, care was taken to ensure that the selection was done in a consistent manner. Of the eligible configurations for a particular packet-loss frequency, we always selected the one having the largest aging factor. If there were several configurations having the same aging factor, we consistently selected the configuration having the largest required reliability level. In Table 1 below, the selected PRTP-ECN configurations are listed. As seen from the table, the allowable steady-state packet-loss frequencies for the selected PRTP-ECN configurations roughly corresponds with the packet-loss rates in the reference tests (see Section 4.2).

Table 1. Selected PRTP-ECN Configurations.

Configuration Name	Steady-State Packet-Loss Frequency	af	rrl
PRTP-2	0.02	0.99	0.97
PRTP-3	0.03	0.99	0.96
PRTP-5	0.05	0.99	0.94
PRTP-8	0.08	0.99	0.91
PRTP-11	0.11	0.99	0.88
PRTP-14	0.14	0.99	0.85
PRTP-20	0.20	0.99	0.80

4.4   Performance Metrics

This subsection provides definitions of the performance metrics studied in the simulation experiment.

Average interarrival jitter: The average interarrival jitter is the average variation in delay between consecutive deliverable packets in a flow [8].

Average throughput: The average throughput of a flow is the average bandwidth delivered to the receiver, *including* duplicate packets [16].

Average goodput: The average goodput of a flow is the average bandwidth delivered to the receiver, *excluding* duplicate packets [14].

Average fairness: The average fairness of a protocol on a link is the degree to which the utilized link bandwidth has been equally allocated among contending flows. A metric commonly used for measuring fairness is *Jain's fairness index* [17]. For n flows, with flow i receiving a fraction, b_i, on a given link, the fairness of the allocation is defined as:

$$Fairness\ index \overset{\text{def}}{=} \frac{\left(\sum_{i=1}^{n} b_i\right)^2}{n\left(\sum_{i=1}^{n} b_i^2\right)} . \tag{3}$$

TCP-friendliness A flow is said to be TCP-friendly if its arrival rate does not exceed the arrival rate of a conformant TCP connection under the same circumstances [14]. In this simulation study, we make use of the TCP-friendliness test presented by Floyd and Paxson [14]. According to their test, a flow is TCP-friendly if the following inequality holds for its arrival rate:

$$\vartheta \leq \frac{1.5\sqrt{2/3}\lambda}{RTT\sqrt{p_{loss}}} \ , \tag{4}$$

where ϑ is the arrival rate of the flow in Bps, λ denotes the packet size in bytes, RTT is the minimum round-trip time in seconds, and p_{loss} is the packet-loss frequency.

5   Results

In the analysis of the simulation experiment, we performed a TCP-friendliness test, and calculated the average interarrival jitter, the average throughput, and the average goodput for the flow between nodes S1 and S4. In addition, we calculated the average fairness in each run. We let the mean, taken over all runs, be an estimate of a performance metric in a test. Of the three primary factors studied in this experiment, the starting times of the two FTP flows were found to have marginal impact on the results. For this reason, we focus our discussion on one of the three cases of starting times: the one in which the FTP flow between the nodes S1 and S4 started 600 ms after the flow between nodes S2 and S5. However, it should be noted that the conclusions drawn from these tests, also applies to the tests in the other two cases.

In order to make comparisons easier, the graphs show interarrival jitter, throughput and goodput for the PRTP-ECN configurations relative to TCP, i.e. the ratios between the metrics obtained for the PRTP-ECN configurations and the metrics obtained for TCP are plotted. As a complement to the graphs, Tables 2, 3, and 4 show the estimates of the metrics together with their 99%, two-sided, confidence interval for a select of traffic loads.

Our evaluation of the jitter characteristics of PRTP-ECN gave very promising results. As can be seen from the graph in Figure 2, and from the tables, the PRTP-ECN configurations decreased the interarrival jitter as compared to TCP. At low traffic loads, the reduction was about 30%, but for packet-loss rates in the neighborhood of 20%, the reduction was in some cases as much as 68%. The confidence intervals show that the improvements in interarrival jitter obtained from using PRTP-ECN are statistically significant. They also show that by using a properly configured PRTP-ECN configuration, not only could the interarrival jitter be decreased, but also the variations in the interarrival jitter could become more predictable.

Considering the importance of jitter for streaming media, the suggested reduction in interarrival jitter could make PRTP-ECN a viable alternative for such applications. For example, could a video broadcasting system that tolerates high

Table 2. Performance Metrics for Tests where the Traffic Load Was 20%.

	Jitter (ms)	Throughput (bps)	Goodput (bps)	Fairness Index
TCP	27.52 ± 1.48	579315.40 ± 17992.27	578781.40 ± 17980.79	0.99 ± 0.0027
PRTP-2	18.89 ± 0.75	662325.40 ± 14828.90	662082.40 ± 14832.24	0.98 ± 0.0061
PRTP-3	18.60 ± 0.73	664734.40 ± 13948.62	664467.40 ± 13951.00	0.98 ± 0.0055
PRTP-5	18.05 ± 0.74	669018.40 ± 14384.92	668883.40 ± 14394.67	0.98 ± 0.0058
PRTP-8	17.30 ± 0.60	680175.40 ± 12561.11	680046.40 ± 12565.65	0.98 ± 0.0053
PRTP-11	17.31 ± 0.55	677904.40 ± 12587.62	677778.40 ± 12586.03	0.98 ± 0.0058
PRTP-14	16.99 ± 0.68	683313.40 ± 14350.01	683193.40 ± 14350.01	0.98 ± 0.0057
PRTP-20	17.10 ± 0.58	681975.40 ± 13046.69	681855.40 ± 13046.69	0.98 ± 0.0058

Table 3. Performance Metrics for Tests where the Traffic Load Was 67%.

	Jitter (ms)	Throughput (bps)	Goodput (bps)	Fairness Index
TCP	101.78 ± 3.56	237156.40 ± 5479.40	235920.40 ± 5483.22	1.00 ± 0.0016
PRTP-2	72.57 ± 2.83	268659.40 ± 6627.15	268248.40 ± 6630.07	0.98 ± 0.0050
PRTP-3	63.47 ± 2.52	284100.40 ± 6706.80	283797.40 ± 6724.92	0.98 ± 0.0066
PRTP-5	51.78 ± 1.44	309105.40 ± 5812.80	308934.40 ± 5814.24	0.95 ± 0.0098
PRTP-8	49.99 ± 0.84	312003.40 ± 4110.89	311871.40 ± 4107.60	0.94 ± 0.0072
PRTP-11	49.91 ± 1.13	312624.40 ± 5633.09	312483.40 ± 5640.09	0.94 ± 0.0091
PRTP-14	49.75 ± 1.03	312618.40 ± 5070.95	312471.40 ± 5064.95	0.94 ± 0.0083
PRTP-20	49.85 ± 0.98	311787.40 ± 4778.42	311652.40 ± 4783.02	0.94 ± 0.0077

Table 4. Performance Metrics for Tests where the Traffic Load Was 97%.

	Jitter (ms)	Throughput (bps)	Goodput (bps)	Fairness Index
TCP	737.09 ± 112.25	39741.00 ± 4520.74	39093.00 ± 4405.56	0.92 ± 0.053
PRTP-2	722.55 ± 77.09	37811.76 ± 3710.43	37391.76 ± 3663.80	0.95 ± 0.033
PRTP-3	701.42 ± 96.97	38213.52 ± 3493.51	37847.52 ± 3464.48	0.95 ± 0.030
PRTP-5	582.70 ± 56.08	42669.00 ± 3433.95	42393.00 ± 3419.70	0.93 ± 0.042
PRTP-8	485.24 ± 36.03	46736.76 ± 2880.10	46556.76 ± 2864.18	0.91 ± 0.045
PRTP-11	425.21 ± 44.98	48396.16 ± 4511.87	48237.16 ± 4512.41	0.88 ± 0.046
PRTP-14	329.73 ± 28.22	53304.00 ± 3137.76	53184.00 ± 3131.41	0.86 ± 0.050
PRTP-20	242.80 ± 11.79	58173.00 ± 2853.42	58059.00 ± 2850.96	0.83 ± 0.047

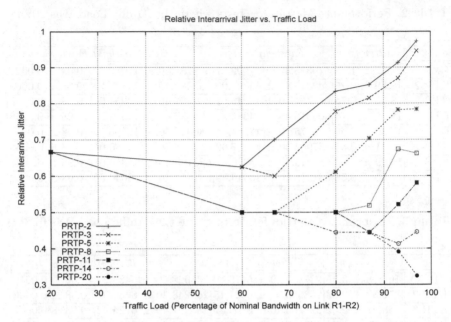

Fig. 2. Interarrival Jitter vs. Traffic Load.

packet-loss rates, theoretically decrease its playback buffer significantly by using PRTP-ECN.

Also our evaluation of the throughput and goodput of PRTP-ECN gave positive results. As is evident from Figures 3 and 4, and the tables, a significant improvement in both throughput and goodput were obtained using PRTP-ECN. For example, an application accepting a 20% packet-loss rate could increase its throughput, as well as its goodput, with as much as 48%. However, also applications that only tolerate a few percents packet-loss rate could experience improvements in throughput and goodput with as much as 20%. From the confidence intervals, it follows that the improvements in throughput and goodput were significant, and that PRTP-ECN could provide a less fluctuating throughput and goodput than TCP. A comparison of the throughputs and goodputs for PRTP-ECN and TCP also suggest that PRTP-ECN is better to utilize the bandwidth than TCP. However, this has not been statistically verified.

Recall from Section 4.2 that a traffic load approximately corresponds to a particular packet-loss rate. Taking this into account when analyzing the results, it may be concluded that a PRTP-ECN configuration had its optimum in both relative interarrival jitter, relative throughput, and relative goodput when the packet-loss frequency was almost the same as the allowable steady-state packet-loss frequency. This is a direct consequence of the way we defined the allowable steady-state packet loss frequency. At packet-loss frequencies lower than the allowable steady-state packet-loss frequency, the gain in performance was limited by the fact that not so many retransmissions had to be done in the first place.

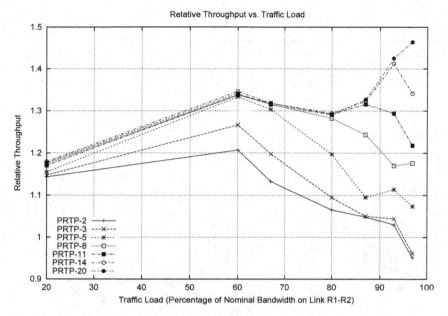

Fig. 3. Throughput vs. Traffic Load.

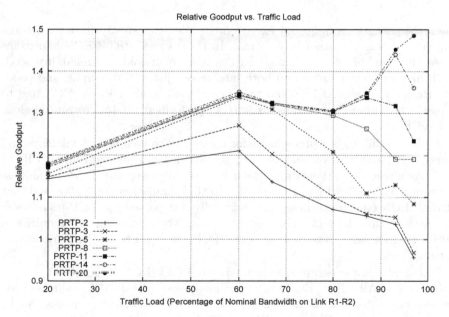

Fig. 4. Goodput vs. Traffic Load.

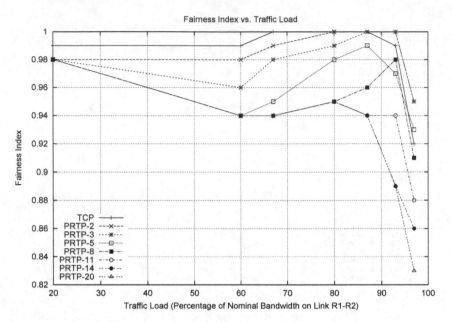

Fig. 5. Fairness Index vs. Traffic Load.

When the packet-loss frequency exceeded the allowable steady-state packet-loss frequency, it was the other way around. In these cases, PRTP-ECN had to increase the number of retransmissions in order to uphold the reliability level, which had a negative impact on both interarrival jitter, throughput, and goodput. However, it should be noted that even in cases when PRTP-ECN had to increase the number of retransmissions, it performed far less retransmissions than TCP.

TCP-friendliness is a prerequisite for a protocol to be able to be deployed on a large scale, and it was therefore an important design consideration for PRTP-ECN. As already mentioned in Section 4.4, we employed the TCP-friendliness test proposed by Floyd and Paxson [14] in this simulation study. We said that a PRTP-ECN configuration was TCP-friendly if it passed the TCP-friendliness test in more than 95% of the simulation runs. The reason for not requiring a 100% pass-frequency, was that not even TCP managed to be TCP-friendly in all runs.

Our simulation experiment suggested that PRTP-ECN is indeed TCP-friendly. All PRTP-ECN configurations passed the TCP-friendliness test. We also computed the fairness index [17] for each simulation run. As can be seen in the graph in Figure 5, PRTP-ECN is reasonable fair. However, since PRTP-ECN gave better throughput than TCP, it follows from the definition of the fairness index that it must be lower for PRTP-ECN than for TCP.

6   Conclusions and Future Work

This paper presented a TCP-compliant, partially reliable transport protocol, PRTP-ECN, that addresses the QoS requirements of applications with soft real-time constraints. PRTP-ECN was built as an extension to TCP, and trades reliability for reduced interarrival jitter and improved throughput and goodput. Our simulation evaluation suggest that PRTP-ECN is able to offer a service with significantly reduced interarrival jitter, and increased throughput and goodput as compared to TCP. Our simulations also found PRTP-ECN to be TCP-friendly. In all tests, PRTP-ECN passed the TCP-friendliness test proposed by Floyd et al. [14]. In a broader perspective, our simulations illustrate the trade-off existing between different QoS parameters, and how it could be exploited to get a general transport service that better meets the need of a particular application. Furthermore, it also demonstrates the appropriateness of letting the application control the QoS trade-offs to be made, but letting a general transport protocol be responsible for carrying it out. Thereby making it possible to enforce congestion control on all flows. Something of major importance for the stability of the Internet.

In the near future, a more extensive simulation experiment of PRTP-ECN will be conducted. This study will not only treat the steady-state behavior of PRTP-ECN, but also its transient behavior. We are also working on an implementation of PRTP-ECN. Once the implementation is completed an experimental study will be initiated.

References

1. M. Allman, V. Paxson, and W. Stevens. TCP congestion control. *RFC 2581*, April 1999.
2. P. Amer, T. Connolly, C. Chassot, and M. Diaz. Partial order transport service for multimedia applications: Reliable service. *2nd High Performance Distributed Computing Conference, Spokane, Washington*, July 1993.
3. P. Amer, T. Connolly, C. Chassot, and M. Diaz. Partial order transport service for multimedia applications: Unrelaible service. *International Networking Conference (INET'93), San Fransisco, CA*, August 1993.
4. P. D. Amer, C.Chassot, T. Connolly, M. Diaz, and P. Conrad. Partial order transport service for multimedia and other applications. *IEEE/ACM Transactions on Networking*, 2(5), October 1994.
5. Paul Amer, Philip Conrad, Edward Golden, Sami Iren, and Armando Caro. Partially-ordered, partially-reliable transport service for multimedia applications. In *Proc. Advanced Telecommunications/Information Distribution Research Program Annual Conference*, pages 215–220, College Park, MD, January 1997.
6. Z. Chen, S. Tan, R. Campbell, and Y. Li. Real time video and audio in the world wide web. *Proc. Fourth International World Wide Web Conference*, 1995.
7. T. Connolly, P. Amer, and P. Conrad. RFC 1693: An extension to TCP: Partial order service, November 1994.
8. J. Davidson and J. Peters. Voice over IP fundamentals. *Cisco Press*, March 2000.

9. B. Dempsey, J. Liebeherr, and A. Weaver. A delay-sensitive error control scheme for continuous media communications. *Second IEEE Workshop on the Architecture and Implementation of High Performance Communication Subsystems (HPCS '93)*, September 1993.

10. B. Dempsey, W. Strayer, and A. Weaver. Adaptive error control for multimedia data transfers. *International Workshop on Advanced Communications and Applications for High Speed Networks*, pages 279–289, March 1992.

11. B. J. Dempsey. Retransmission-based error control for continuous media. *PhD thesis, University of Virginia*, 1994.

12. M. Diaz, A. Lopez, C. Chassot, and P. Amer. Partial order connections: a new concept for high speed and multimedia services and protocols. *Annales des telecoms*, 49(5-6), 1994.

13. D. Wijesekera et al. Experimental evaluation of loss perception in continuous media. *ACM Multimedia Journal*, 7:486–499, July 1999.

14. Sally Floyd and Kevin Fall. Promoting the use of end-to-end congestion control in the internet. *IEEE/ACM Transactions on Networking*, 7(4):458–472, August 1999.

15. C. Hess. Media streaming protocol: An adaptive protocol for the delivery of audio and video over the internet. *Master's Thesis, Graduate College of Computer Science at the University of Illinois at Urbana-Champaign*, 1998.

16. R. Jain. The art of computer systems performance analysis. *John Wiley & Sons, Inc.*, 1991.

17. R. Jain, D. Chiu, and W. Hawe. A quantitative measure of fairness and discrimination for resource allocation in shared computer systems. *Technical Report, Digital Equipment Corporation, DEC-TR-301, 1984*, 1984.

18. M. Mathis, J. Mahdavi, S. Floyd, and A. Romanow. RFC 2018: TCP selective acknowledgment options, October 1996.

19. Steven McCanne and Sally Floyd. UCB/LBNL/VINT Network Simulator - ns (version 2). `http://www-mash.CS.Berkeley.EDU/ns/`, April 1999.

20. M. McKusick, K. Bostic, and M. Karels. The design and implementation of the 4.4BSD operating system. *Addison-Wesley*, May 1996.

21. J. Postel. RFC 768: User datagram protocol, August 1980. Status: STANDARD.

22. J. Postel. RFC 793: Transmission control protocol, September 1981. Status: STANDARD.

23. K. Ramakrishnan and S. Floyd. RFC 2481: A proposal to add explicit congestion notification (ECN) to IP, January 1999.

24. L. Rojas-Cardenas, L. Dairaine, P. Senac, and M. Diaz. An adaptive transport service for multimedia streams. *IEEE International Conference on Multimedia Computing and Systems*, June 1999.

25. H. Schulzrinne, S. Casner, R. Frederick, and V. Jacobson. RFC 1889: RTP: A transport protocol for real-time applications, January 1996.

26. B. Smith. Implementation techniques for continuous media systems and applications. *Ph.D. Thesis, University of California, Berkeley*, 1993.

27. Brian C. Smith. Cyclic-UDP: A priority-driven best-effort protocol. *Unpublished manuscript*, May 1994.

28. R. Steinnetz. Human perception of jitter and media synchronization. *IEEE Journal on Selected Areas in Communications*, 14(2):61–72, February 1996.

29. Timothy W. Strayer, Bert J. Dempsey, and Alfred C. Weaver. XTP – The xpress transfer protocol. *Addison-Wesley Publishing Company*, 1992.

30. G. Wright and W. Stevens. TCP/IP illustrated, volume 2: The implementation. *Addison-Wesley*, December 1999.

GAME Based QoS Provisioning in Multimedia Wideband CDMA Networks

Mohamed Moustafa[1], Ibrahim Habib[1], and Mahmoud Naghshineh[2]

[1] The City University of New York, Department of Electrical Engineering
New York, NY 10031, USA
moustafa@ieee.org, eeiwh@ee-mail.engr.ccny.cuny.edu
[2] IBM Watson Research Center,Hawthorne, NY 10532, USA
mahmoud@us.ibm.com

Abstract. In this paper, we present a novel scheme that maximizes the number of admissible mobiles while preserving quality of service and resources in wideband CDMA networks. We propose the manipulation of two main control instruments in tandem when previous work has focused on handling them separately: transmitter power level and bit rate. The active component of this scheme is the Genetic Algorithm for Mobiles Equilibrium (GAME). The base station measures each mobile received signal quality, bit rate and power. Accordingly, based on an evolutionary computational model, it recommends the calculated optimum power and rate vector to everyone. Meanwhile, a standard closed loop power control command is maintained to facilitate real time implementation. The goal here is to achieve an adequate balance between users. Thereof, each mobile can send its traffic with a suitable power to support it over the different path losses. In the mean time, its battery life is being preserved while limiting the interference seen by neighbors. Consequently, more mobiles can be handled. A significant enhancement in cell capacity, signal quality and power level has been noticed through several experiments on combined voice, data and video services.

1 Introduction

Wireless communications have become a very widely discussed research topic in recent years. Lately, extensive investigations have been carried out into the application of Code Division Multiple Access (CDMA) system as an air interface multiple access scheme for the International Mobile Telecommunications System 2000 (IMT-2000). These third generation (3G) mobile radio networks will support wideband multimedia services at bit rates as high as 2 Mbps, with the same quality as fixed networks [9]. Users in a mobile multimedia system have different characteristics; consequently, their resource requirements differ. Some of the most important user characteristics are variable quality of service (QoS) requirements, and variable bit rate (VBR) during operation. In a multimedia system, different services are provided such as voice, data, or video or a combination of some of these. Consequently, the users have different requirements in terms

L. Wolf, D. Hutchison, and R. Steinmetz (Eds.): IWQoS 2001, LNCS 2092, pp. 235–249, 2001.

of bandwidth, maximum bit rate, bit error rate (BER), and so on. The system should process all these requirements, calculate the necessary resource allocation for each connection, and manage resources accordingly. Since some of the multimedia services have VBR, the resource allocation to each user will vary in time. The system should monitor the instantaneous allocation per user and, if necessary, police the existing connections when they exceed the allocated resources. In a CDMA network, resource allocation is critical in order to provide suitable Quality of Service for each user and achieve channel efficiency [12]. Many QoS measures, including Bit Error Rate, depend on the received bit energy-to-noise density ratio E_b/N_o given by

$$\left(\frac{E_b}{N_o}\right)_i = \frac{G_{bi}P_i/R_i}{\left(\sum_{j\neq i}^M G_{bj}P_j + \eta\right)/W} \tag{1}$$

where W is the total spread spectrum bandwidth occupied by the CDMA signals. G_{bi} denotes the link gain on the path between mobile i and its base b. η denotes background noise due to thermal noise contained in W and M is the number of mobile users. The transmitted power of mobile i is P_i which is usually limited by a maximum power level as

$$0 \leq P_i \leq P_i^{max} \quad \forall i \in [1, M] \tag{2}$$

R_i is the information bit rate transmitted by mobile i. This rate is bounded by the value; R_i^P the peak bit rate, designated in the traffic contract once this user has been admitted into the system.

$$0 \leq R_i \leq R_i^P \quad \forall i \in [1, M] \tag{3}$$

An increase in the transmission power of a user increases its E_b/N_o, but increases the interference to other users, causing a decrease in their E_b/N_os. On the other hand, an increase in the transmission rate of a user deteriorates its own E_b/N_o. Controlling powers and rates of the users therefore amounts to directly controlling the QoS that is usually specified as a pre-specified target E_b/N_o value (Θ). It can also be specified in terms of the outage probability, defined as the probability that the E_b/N_o falls below Θ.

Power control is a means primarily designed to compensate for the loss caused by propagation and fading. There have been numerous papers published addressing power control problems in CDMA cellular systems. The Interim Standard 95 (IS-95) [10] has two different power control mechanisms. In the uplink, both open loop (OLPC) and fast closed loop power control (CLPC) are employed. In Open loop scheme, the measured received signal strength from the base station is used to determine the mobile transmit power. OLPC has two main functions: it adjusts the initial access mobile transmission power and compensates large abrupt variations in the path loss attenuation. Since the uplink and downlink fading processes are not totally correlated, the OLPC cannot compensate for the uplink fading. To account for this independence CLPC is used. In this mode, the base station measures the received E_b/N_o over a 1.25 ms period, and compares that to the target Θ. If the received $E_b/N_o < \Theta$, a '0' is generated to instruct the mobile to increase its power, otherwise, a '1' is generated to instruct the mobile to decrease its power. These commands instruct the mobile to adjust transmitter power by a predetermined amount, usually 1 dbm.

Previous work has focused on finding adequate power levels that maximize the received bit energy-to-noise density ratio E_b/N_o where the transmission rate of each user is fixed [13]. Recently, [4] presented an algorithm for controlling mobiles transmitter power levels following their time varying transmission rates.

In this study, we propose a scheme (GAME-C) for controlling mobiles transmitter power and bit rate concurrently. Our objective is to maximize the number of possible users while preserving QoS and resources. GAME-C is composed of two components: GAME and CLPC. The Genetic Algorithm for Mobiles Equilibrium (GAME) is the main control authority that recommends power and rate. We adopted also the use of CLPC to make real time implementation affordable. The remainder of the paper is organized as follows. The proposed scheme is described in section 2. Experiments with some numerical results are presented in section 3 illustrating the deployment of the proposed GA. In section 4, we conclude.

2   GAME-C

In this section, we detail the proposed protocol. We start by describing the mobile-base station interactions. Next, the optimization problem is formulated followed by the genetic algorithm solution. Our analysis applies to the uplink (mobile to base), which we assume is orthogonal to the downlink, and can be treated independently. We concentrate on the uplink as it is generally accepted that its performance is inferior to that of the downlink.

2.1   Mobile-Base Interface

The main signaling messages, interchanged between the mobile station (MS) and the base station (BS) in the proposed scheme, are depicted in fig. 1. Initially, during a call setup, MS and BS negotiate the terms of a traffic contract. This contract includes some traffic descriptors as well as some parameters representing the required quality of service (QoS). Four items of this contract are needed by GAME: P_i^{max}, Θ, R^P and R^G. P_i^{max} is the maximum power the MS can transmit. This value is acting as a constraint on the power level that can be recommended by the BS. Θ, the required received E_b/N_o, represents the QoS of the call since it is directly related to the connection bit error rate (BER). R^P is the peak bit rate to be generated by the mobile application. The last item in the traffic contract is, R^G, the guaranteed bit rate. It indicates the maximum bit rate that the BS will guarantee to the mobile. Any traffic above R^G and below R^P can be accepted or rejected by GAME according to the cell load and congestion status. In fact, R^G has also a direct relationship with the maximum tolerable delay for this type of traffic. The more the bearable delay the less R^G is.

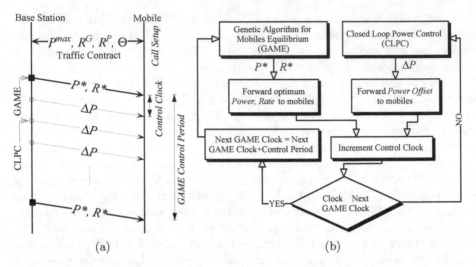

Fig. 1. GAME-C Protocol: (a) Signaling: Every control clock, the CLPC sends 1 bit ΔP to the MS. GAME is triggered every control period and then forwards to MS optimum power P^* and bit rate R^*. (b) High level Data Flow Diagram.

In brief, since GAME is going to manipulate bit rate, it requests this parameter to know the upper limit on any proposed rate cut.

Immediately, after the end of this call setup phase, and every *control period*, the BS triggers the GAME who tries to find the optimum power P^* and the optimum bit rate R^* for each mobile. Optimal solution is in the sense that each user gets merely enough power, P^*, to fulfill its Θ with the maximum possible rate, $R^* \geq R^G$. Therefore, each MS preserves its battery life and always has a guaranteed QoS. Meanwhile, the BS can admit the maximum number of calls since each one is generating the minimum interference to others. In the mean time, every *control clock* (1.25 ms), the BS controls each MS power through CLPC that generates the binary ΔP based on the received E_b/N_o. If the received $E_b/N_o < \Theta$, a "$\Delta P = 0$" is generated to instruct the mobile to increase its power, otherwise, a "$\Delta P = 1$" is generated to instruct the mobile to decrease its power. These commands instruct the mobile to adjust transmitter power by a predetermined amount, usually 1 dbm [10].

Figure 1(b) depicts the main flow of the proposed scheme. Every control clock, the BS increments its clock counter and test the control period maturity. If it is the GAME turn to optimize P and R then the BS activates the genetic algorithm and forwards the result (P^*, R^*) to each MS. Otherwise, the standard CLPC is in charge of advising the mobiles to offset their power levels by the pre-specified amount.

2.2   Problem Formulation

Thus, the objective here is to find a nonnegative power $\underline{P} = [P_1, P_2, \ldots, P_M]$ and $\underline{R} = [R_1, R_2, \ldots, R_M]$ vectors within some boundaries and maximize the function F that can be proposed initially as

$$F = \sum_{i=1}^{M} F_i^E$$

where F_i^E is a threshold function defined for user i as

$$F_i^E = \begin{cases} 1 & \text{if } E_b/N_o \geq \Theta \\ 0 & \text{otherwise} \end{cases} \tag{4}$$

This means to maximize the number of users that have their signal qualities above the minimum requirement, Θ. However, this objective function is incomplete since it does not give preferentiality to solutions that use less power. Hence, while limiting the $P - R$ search space to solutions that maximize the number of QoS satisfied mobiles through F_i^E, minimizing P is essential. Since low mobile transmitter power means little interference to others and long battery life, we then proposed the power objective component, F_i^P, that gives credit to solutions that utilize low power and punishes others using high levels.

$$F_i^P = 1 - \frac{P_i}{P_i^{max}} \tag{5}$$

Consequently, we modified our main objective function F to reflect this power preference as

$$F = \sum_{i=1}^{M} F_i^E + \frac{1}{M} \sum_{i=1}^{M} F_i^E F_i^P$$

The reason of multiplying F_i^E by F_i^P is to prevent those users who have failed their QoS qualification from contributing to the objective score.

   Another goal is to fulfill every user bandwidth request. Each call is guaranteed a specific bandwidth, R^G, according to the traffic contract. However, a user should not be prevented from getting higher rate if there is a chance. Thus, from bandwidth point of view, the R search space should avoid values below the baseline R^G while encouraging solutions to go as high as possible, below the upper bound R^P. This bandwidth objective is represented by

$$F_i^R = \begin{cases} (R_i - R_i^G) / (R_i^P - R_i^G) & \text{if } R_i^G \leq R_i \leq R_i^G \\ 0 & \text{otherwise} \end{cases} \tag{6}$$

Accordingly, the final main objective function becomes

$$F = \sum_{i=1}^{M} F_i^E + \frac{1}{M} \sum_{i=1}^{M} F_i^E \left(F_i^P + F_i^R \right) \tag{7}$$

Notice also that F proposed in (7) can overcome what is known as the Near-Far effect [7] that usually happens when a mobile MS1 near the BS uses power P1 in excess of its need and as a result prohibits another far one MS2 from reaching its E_b/N_o threshold. However, this case can not happen using F since there exists a solution where $P1' < P1$ and hence higher F_i^P while having the same remaining objectives scores. Therefore, GAME will prefer the solution with lower power. In conclusion, the jointly optimal power and rate is obtained by solving the following optimization problem:

$$\max_{(P^*, R^*) \in \Omega} F(P, R) \tag{8}$$

where F is defined in (7) and the feasible set Ω is subject to power and rate constraints

$$0 \le P_i \le P_i^{max} \text{ and } R_i^G \le R_i \le R_i^P \quad \forall i \in [1, M] \tag{9}$$

2.3   GAME Engine

Genetic algorithms (Gas), as described in [3], are search algorithms based on mechanics of natural selection and natural genetics. In every generation, a new set of artificial creatures (chromosomes) is created using bits and pieces of the fittest of the old. Gas use random choice as a tool to guide a search toward regions of the search space with likely improvement. They efficiently exploit historical information to speculate on new search points with expected improved performance.

The **G**enetic **A**lgorithm for **M**obiles **E**quilibrium core is a steady state GA with the ability to stop its evolution after a timeout period being expired. As illustrated in fig. 2, the inputs are measured from the users as their current rate $R(t)$ and power $P(t)$ vectors. Additional information is needed as well in the input like the required signal quality Θ, the maximum possible power P^{max}, and the contract rates R^P, R^G.

The GAME starts by clustering users according to their instantenous bandwidth and QoS requirements. Therefore, mobiles with similar demands can be represented by the same cluster. The purpose of this aggregation is to minimize the dimensionality of the search space. GAME then proceeds by encoding $R(t)$ and $P(t)$ into chromosomes to form the initial population.The chromosome is a binary string of N digits. It encodes the (power, rate) values of all C mobile clusters. Each cluster occupies N_P bits for its power and N_R bits for its rate.

Immediatly after this initialization, the usual steady state genetic algorithm [3] cycle starts. This cycle includes *Evaluation*, *Generation*, and *Convergence* steps. The evaluation part is responsible for giving a score to each chromosome. This score is calculated using a *fitness* function that we propose in 2.2. The generation step keeps producing new chromosomes based on the fittest ones by applying different genetic operators like: crossover and mutation [3]. Finally, the convergence step checks the validity of the stopping criterion.

There are two ways to stop the GAME progress: Convergence or Time-Out. Convergence means that the fittest chromosome, the one with highest fitness

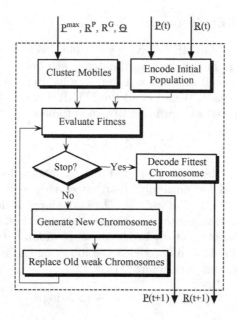

Fig. 2. GAME Engine Data Flow Diagram.

score, has the same fitness value as the average one. This indicates that GAME has saturated and no need to continue evolving. Time-Out case happens when the time slice allotted to GAME to solve the optimization problem has expired and it has to give a solution immediately. This later case usually happens when the control period is too small so the GAME is triggered with high frequency.

Once the GAME stops it decodes the fittest chromosome to power and rate. Consequently, the base station forwards this new $\underline{R}(t+1)$ and $\underline{P}(t+1)$ vectors to the users. According to the fitness function, used to compare the solutions chromosomes, the fittest vectors $\underline{R}(t+1)$ and $\underline{P}(t+1)$ should be within the boundaries (9). In the mean time the bit rate should be as high as possible while the power level is as minimum as possible. This finest solution also should be able to make each user surmounts its required E_b/N_o value (Θ).

3   Experiments

In this section, we describe the experiments conducted to test GAME-C as well as to compare with the standard IS-95 [10] power control method. Notice that only the Closed Loop Power Control (CLPC) was in effect in our comparison since we assumed that the mobiles were stationary at the time GAME is active, so there was no sudden change in link path loss, which is remedied by the open loop control.

3.1   Experimental Environment

Experiments simulating GAME behavior were done over 19 hexagonal cells representing 3 inner rings in cellular coordinates. Each base station was situated at its cell center. Each cell radius was 1 km. Mobile users were distributed uniformly over each cell space. We adopted the ITU-R distance loss model [6]. The link gain, G_{ij}, is modeled as a product of two variables

$$G_{ij} = A_{ij} \times D_{ij}$$

A_{ij} is the variation in the received signal due to shadow fading, and assumed to be independent and log normally distributed with a mean of 0 dB and a standard deviation of 8 dB. The variable D_{ij} is the large-scale propagation loss, which depends on the transmitter and the receiver location, and on the type of geographical environments. Let d_{ij} be the distance in km between transmitter j and receiver i, the ITU-R formula yields the following path loss equation for a typical 3G CDMA system parameters [11].

$$10 \log D_{ij} = -76.82 - 43.75 \log d_{ij} \tag{10}$$

The center frequency is 1975 GHz, antenna heights of the mobile and the base station are 1.5 and 30 m respectively. We assume that 20% of base station coverage area is occupied by buildings. The system bandwidth W can increase up to 20MHz and the background thermal noise density is 174 dbm/Hz. All mobiles maximum transmitter power was set to 1000 mW. The multimedia services that have been investigated through the experiments, summarized in table 3.1, covered many possible applications.

Table 1. Traffic Types Tested in the Simulations.

	Voice	Data	Video
Mean Rate (kbps)	4.5	82	145
R^P (kbps)	9.6	144	1125
R^G (kbps)	8.4	50	844
Θ (dB)	4.2	3.7	5
UMTS Class	A (LDD)	D (UDD)	B (LDD-VBR)

Voice users used in the simulations we following the On-Off model [1]. Talkspurt and silence periods were independent and exponentially distributed with means of 1.0 sec and 1.35 sec respectively. Talk periods generated 9600 bps while 512 bps were generated during silence. This traffic type is classified as Low Delay Data (LDD) or class A in UMTS proposal [2]. We assumed its guaranteed bit rate R^G only 12.5% lower than its peak rate R^P. Data traffic generated 144 Kbps at its peak with average of 82 Kbps and minimum 16 Kbps. This service mapped class D, Unconstrained Delay Data (UDD) and can represents lots of connectionless services including IP, FTP, and e-mail. Its guaranteed level stood at 50 Kbps (65% away from its peak). Mobiles with MPEG encoded video traffic [8] were categorized as class B: Low Delay Data Variable Bit Rate (LDD-VBR). Encoder input is 384x288 pixels with 12-bit color information. A 12 frames GOP pattern (IBBPBBPBBPBB) was generated at 25 frames/sec. Mean bit rate was 145Kbps while the peak rate is 1.125 Mbps. We assumed its guaranteed rate 25% away from the peak. In the course of a simulation run, 27 simulation cycles were conducted. During each cycle, the location of each mobile unit was generated randomly and its initial power was then set to a level proportional to its path loss from the nearest base station. After passing an initial stabilizing time of 300 power control clocks, data were collected for statistics from all mobiles and averaged per service type in the following 30000 clocks. Thus for each trial a total of 810000 observation clocks were collected, which should be sufficient for representative simulation results. The data collected from these simulations were the average mobile transmitter power, bit rate, received E_b/N_o, outage probability and the average cell loading. The outage probability of a mobile unit is defined as $\text{Prob}[E_b/N_o < \Theta]$. Cell loading is defined as the ratio of the number of active users over the maximum allowable number of users and it was approximated [6] as

$$\text{Cell Loading}_b \approx \frac{1}{M} \sum_{i=1}^{M} \frac{G_{ib}P_i}{\sum_{j \neq i}^{M} G_{jb}P_j + \eta} \qquad (11)$$

This relationship illustrates the fact that the system capacity is self-limiting, because the amount of interference is proportional to the number of users in the same or other cells. The loading is a convenient way to refer to the amount of potential capacity being used.

3.2   Capacity

The objective here was to determine the capacity of the described CDMA cellular system while using the proposed GAME-C and compare with the standard CLPC method. We started this experiment by using only one voice user and no data or video users. Gradually, we increased the number of voice users until any mobile drop happened. We then increased the number of data users gradually and try all voice combinations: zero to maximum. Finally, we increased the video users and tested all data and voice mixtures. A mobile drop happened if its E_b/N_o last below its Θ for a continuous 0.5 sec. We adjusted the GAME-C control period to 0.1 sec, that means it will be triggered 10 times each sec. The admissible combinations of users were determined and presented in fig. 3 for IS-95 CLPC and the proposed GAME-C. Figure 3 shows the voice versus data users limits that can coexist while having one and ten video users. A snapshot of these curves demonstrates that GAME-C was able to increase the maximum number of voice and data users. On the average we noticed 17%, 57% and 8% gain in voice, data and video users respectively. As we expected, the gain in the data users was the greatest since they were the most flexible in terms of their guaranteed bit rate that reflected their high delay tolerance. Overall, when

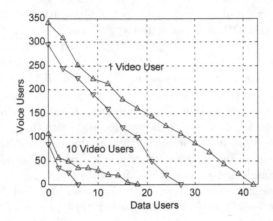

Fig. 3. Admissible Combination of Voice Vs. Data Users: '∇' represents IS-95. '$\triangle$' represents GAME-C with Control Period $T = 0.1$. Upper curves at one Video User. Lower curves at ten Video users.

doing the multidimensional integration to calculate the volume under the surface formed by voice, data, and video axes , CLPC attained 24707 while GAME-C achieved 41299, which is a mere profit of 67%.

3.3   Quality of Service

This experiment aimed to investigate the effect of the proposed scheme on the quality of service offered to users. We used the E_b/N_o and the outage probability to measure the QoS. The average granted bit rate by GAME-C to mobiles was also reported since it is related to the QoS because of its relationship to the delay. In this experiment we varied the number of a specific service users of from 1 to a maximum number. This maximum number is reached once a call dropping case is attained. Each time we collected our data and extract some statistics including the averages that are plotted. We adjusted the GAME-C control period to a reasonable 0.1 sec that is very affordable on current data microprocessors. The experiment was repeated two times once per service type. We tested each traffic type separately to get a clear picture of the effect on each specific class. Results on mixed traffics will be given in the following section.

Outage probability is one of the basic QoS representatives. Figure 4 illustrates this probability versus the number of voice and data users respectively. Unsurprisingly, the outage probability increased with the growing number of users. Each curve expresses the GAME-C QoS superiority by yielding lower outage than the standard CLPC for the same number of users. Notice also the

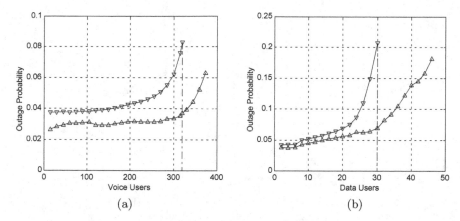

Fig. 4. Outage Probability vs. (a) Voice Users, (b) Data Users. '∇' represents IS-95. '$\triangle$' represents GAME-C with Control Period $T = 0.1$.

vertical dashed line that indicates the maximum tolerable number of users using the standard method. At this peak capacity, the voice outage probability dropped from 0.083 to 0.037 (55% gain), and data outage fell from 0.21 to 0.07 (67% gain) by using GAME-C. Therefore, these curves also confirmed the ability to expand the cell capacity in different types of traffic as illustrated in fig. 3 earlier.

Another important QoS performance measure is the E_b/N_o because of its relationship to BER. Figure 5 provides another proof of GAME-C QoS lead. These figures plot the average E_b/N_o after subtracting the threshold Θ. Almost every time, GAME-C provided its users with extra energy density more than the classic CLPC scheme. It is also clear that this extra QoS faded while increasing the users since the smaller the number the more the room GAME-C has to allocate to mobiles. It is also noteworthy to mention that on the average this extra E_b/N_o was higher than zero every time for both CLPC and GAME-C, and this demonstrates how CLPC is ably performing its job and how hard was it for GAME-C to surpass it.

As can be deduced from (1), there are two ways to boost a mobile E_b/N_o: increasing transmitting power, or decreasing transmitting bit rate. Each solution has advantages and disadvantages. The first solution, increasing a mobile power, is an easy way to enhance QoS but on the downside, it drains the mobile own battery rapidly. In addition, this action increases the interference in the face of other users, which can lead to the drop of their communication links. The disadvantage of the second solution is that a call may be forced to cut its transmitting bit rate to some extent. However, the main advantage is that its effect is self limited, i.e., no other users will be negatively affected but in contrast it may help them by reducing the interference. What GAME-C is trying to do is to find a combined solution within these two extremes that get the most out of

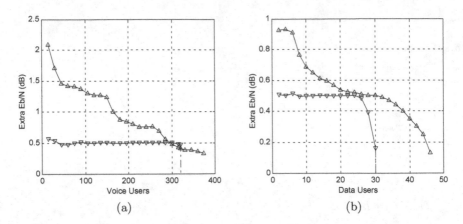

Fig. 5. Average E_b/N_o vs. (a) Voice Users, (b) Data Users. '∇' represents IS-95. '$\triangle$' represents GAME-C with Control Period $T = 0.1$.

their advantages while minimizing their shortcomings. Therefore the proposed scheme increases the power marginally and decreases the bit rate slightly [1] until the resulting power is sufficient to make E_b/N_o tops Θ while introducing reasonable interference to others. We believe that this was the main cause that made GAME-C outperforming the IS-95 CLPC since the later uses only the power as its QoS provisioning tool.

Figure 6 shows the price of this QoS enhancement and capacity stretch. It is some bit rate liberation. As illustrated in the plots, using GAME-C, the rate cut increased with the number of users since more mobiles means more interference, which tweaked the solution towards the bit rate cut side. It is also clear that the amount of bit rate slash depends on the traffic type. For instance, in case of voice traffic, GAME-C managed to keep the bit rate on the average almost the same as requested until the maximum capacity of the CPLC (320 mobiles). Afterward, GAME-C utilized its second option, which is the rate cut, aggressively until it reached its own capacity limit where its two tools had been exhausted: the power reached a high level and the reduced rate was near the guaranteed one R^G.

In order to assess the damage in bit rate caused by GAME-C, we can extract the rate cut at the maximum number of possible users using CLPC. At 320 users, in the case of voice (fig. 6(a)), the average rate drop was 1.2%. Average Rate reduction was 21.2% at 30 users in case of data (fig. 6(b)) services. It is also clear from the plot (fig. 6(b)) that the rate cut in the data users case was the biggest since they had the most flexible delay constraints that translated into the least guaranteed bit rate.

---

[1] GAME-C has the option to reduce bit rate as long as it is higher than the guaranteed rate R^G specified in the traffic contract.

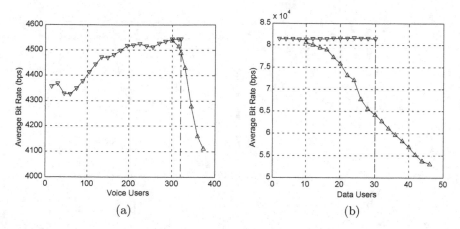

Fig. 6. Average Transmitter Bit Rate vs. (a) Voice Users, (b) Data Users. '∇' represents IS-95. '$\triangle$' represents GAME-C with Control Period $T = 0.1$.

3.4 Power

In the previous experiment, since we targeted to show the GAME-C result on the QoS of each traffic class separately, we did not mix users from different types in the same simulation run.

In this experiment we aimed to study the effect of applying GAME-C on the mobile transmitter power and the cell loading. We combined different services users simultaneously to ensure that the proposed algorithm was able to deal with the mix as good as the solo traffic type. Again, we varied the number of service users from 1 to a maximum number. This maximum number is reached once a call dropping case is attained. Each time we collected the transmitting power as well as the cell loading as defined in (11). We adjusted the GAME-C control period to a reasonable 0.1 sec. The experiment was repeated two times with varying mobiles number from each service type (voice and data) at a time.

As expected, GAME-C was able to reduce mobile power consumption as illustrated in fig. 7. The major reason for that savings is again the bit rate manipulation that gave the base station another degree of freedom in the restricted power allocation problem. It is obvious again, from fig. 7, that data users were the primary beneficiaries of the power reserves. No surprise, since they were the most willing to trade their bit rate by less power and thus adding more users. If we go through the numbers at the maximum CLPC users capacity, we find the following. Power savings were 50% at 45 voice users (fig. 7(a)), and 60% at 15 data users (fig. 7(b)). As seen, again in these plots, the average power consumed rose steadily with the growing number of users. This is natural since the more the mobiles, the more the interference produced, and the more power needed to overcome it.

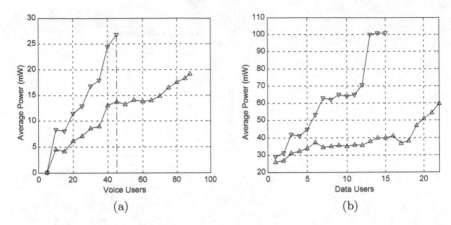

Fig. 7. Average Transmitter Power vs. (a) Voice Users (Video Users=7 and Data Users=13), (b) Data Users (Video Users=6, Voice Users=50). '∇' represents IS-95. '△' represents GAME-C with Control Period $T = 0.1$.

4 Conclusions

In this paper, we introduce the GAME-C scheme for wireless resource management in CDMA networks and applied it to multimedia traffics. It integrates the Genetic Algorithm for Mobiles Equilibrium (GAME) technique with the regular Closed Loop Power Control (CLPC) specified in the standards [10,11,2]. The proposed method includes the control of two main resources in a wireless network: mobile transmitting bit rate and corresponding power level. The main algorithm is to be implemented in the base station that forwards the controlling signals to the mobiles. The basic idea is that all the mobiles have to harmonize their rate and power according to their location, QoS, and density. Actually, GAME uses the bit rate as an additional tool to solve situations where only power control failed because of high interference. Therefore, it trades bit rate for higher QoS, more users capacity, and less transmitting power. This rate reduction is subject also to a maximum, so the resulting one is always above the minimum guaranteed level R^G specified in the traffic contract.

The advantages of using genetic algorithms for optimization are numerous. Parallelism, GAME can be implemented as multiple synchronized threads to take advantage of the full processing power of the used hardware. Evolving nature, GAME can be stopped any moment while having the assurance that the current solution is better than all the previous ones. Scalability, mobiles can be added or removed simply by adjusting the chromosome length and leaving everything else intact. A Small control period T is better than a large one, since it gives GAME-C a chance to watch the situation closely. However, convergence to the optimum or at least a satisfying solution takes some time. Monitoring experiments results reported in the previous section, we noticed that on the average, GAME needed 49 chromosome operations to reach 91% of the optimal answer. Each operation includes chromosomes parents' selection, parents' crossover to generate a child, and mutation to introduce new genes. Given the capabilities of the current microprocessors, we may have a restriction on the maximum frequency of GAME activation. However, GAME-C was able to outperform the standard CLPC with a conservative $T = 0.1$ sec, and even with one full second as control period.

The proposed scheme performed acceptably during the experiments done to test it. The enhancements over the standard CLPC case are substantial. The volume of admissible users within a cell has seen an average expansion of 30%. The outage probability has decreased by an average of 40% with better corresponding signal quality (E_b/N_o). In the mean time, the average power consumption has been saved by 46% while the bit rate, declined on the average by 11% to pay for these enhancements.

The noticeable GAME drawback is that the processing time to solve the optimization problem is seen to be proportional to the number of users. However, we solved this problem by clustering mobile users before generating GAME chromosomes to shorten their lengths. Therefore the cluster chromosome represents a whole group instead of a single user. Current research work is in progress to apply the GAME on the forward link as well as study the performance over a microcell high-rise buildings environment.

References

1. Brady, P. : A Statistical Analysis of On-Off Patterns in 16 conversations. Bell Systems Tech. Journal, 47 (1968) 73–91.
2. ETSI, The ETSI UMTS Terrestrial Radio Access (UTRA) ITU-R RTT Candidate Submission, June 1998.
3. Goldberg, D. : Genetic Algorithms in Search, Optimization and Machine Learning. Addison-Wesley, 1989.
4. Kim, D. : Rate-Regulated Power Control for Supporting Flexible Transmission in Future CDMA Mobile Networks. IEEE Journal on Selected Areas in Communications, 17 (1999) 968–977.
5. Le Gall, D. : MPEG: A Video Compression Standard for Multimedia Applications. Communications of the ACM, April 1991, 47–58.
6. Lee, J., Miller, L. : CDMA Systems Engineering Handbook, Artech House Publishers, Maryland, 1998.
7. Rappaport, S. : Wireless Communications. Prentice Hall, New Jersey, 1996.
8. Rose, O. : Statistical Properties of MPEG Video Traffic and Their Impact on Traffic Modeling in ATM Systems. Inst. Of Comp. Science, University of Wurzburg, Germany. Research Report No 101.
9. Special Issue, IMT-2000: Standards Efforts of the ITU. IEEE Personal Communications, August 1997.
10. TIA/EIA/IS-95: Mobile Station-Base Station Compatibility Standard for Dual-Mode Wideband Spread Spectrum Cellular Systems. Telecommun. Industry Assoc., 1995.
11. TIA/EIA/IS-2000: The cdma2000 ITU-R RTT Candidate Submission, June 1998.
12. Zander, J. : Radio Resource Management in Future Wireless Networks: Requirements and Limitations. IEEE Communications, 35 (1997) August 30–36.
13. Zander, J. : Distributed Cochannel Interference Control in Cellular Radio Systems. IEEE Transactions on Vehicular Technology, 41 (1992) 305–311.

QoS-Aware Adaptive Services
in Mobile Ad-Hoc Networks

Baochun Li

Department of Electrical and Computer Engineering
University of Toronto
bli@eecg.toronto.edu

Abstract. Ad-hoc wireless networks consist of mobile nodes interconnected by multi-hop wireless paths. Unlike conventional wireless networks, ad-hoc networks have no fixed network infrastructure or administrative support. Because of the dynamic nature of the network topology and limited bandwidth of wireless channels, Quality-of-Service (QoS) provisioning is an inherently complex and difficult issue. In this paper, we propose a fully distributed and adaptive algorithm to provide statistical QoS guarantees with respect to *accessibility* of services in an ad-hoc network. In this algorithm, we focus on the optimization of a new QoS parameter of interest, *service efficiency*, while keeping protocol overheads to the minimum. To achieve this goal, we first theoretically derive the lower and upper bounds of service efficiency based on a novel model for group mobility, followed by extensive simulation results to verify the effectiveness of our algorithm.

1 Introduction

Wireless ad-hoc networks are self-created and self-organized by a collection of mobile nodes, interconnected by multi-hop wireless paths in a strictly peer-to-peer fashion. Each node may serve as a packet-level router for its peers in the same network. Such networks have recently drawn significant research attention since they offer unique benefits and versatility with respect to bandwidth spatial re-use, intrinsic fault tolerance, and low-cost rapid deployment. Furthermore, near-term commercial availability of Bluetooth-ready wireless interfaces may lead to the actual usage of such networks in reality. However, the topology of ad-hoc networks may be highly dynamic due to unpredictable node mobility, which makes Quality of Service (QoS) provisioning to applications running in such networks inherently hard. The limited bandwidth of wireless channels between nodes further exacerbates the situation, as message exchange overheads of any QoS-provisioning algorithms must be kept at the minimum level. This requires that the algorithms need to be fully distributed to all nodes, rather than centralized to a small subset of nodes.

Previous work on ad-hoc networks has mainly focused on three aspects: general packet routing [1, 2], power-conserving routing [3], and QoS routing [4]. With respect to QoS guarantees, due to the lack of sufficiently accurate knowledge, both instantaneous and predictive, of the network states, even statistical QoS guarantees may be impossible if the nodes are *highly mobile*. In addition, *scalability* with respect to network size becomes an issue, because of the increased computational load and difficulties in

L. Wolf, D. Hutchison, and R. Steinmetz (Eds.): IWQoS 2001, LNCS 2092, pp. 251–265, 2001.
© Springer-Verlag Berlin Heidelberg 2001

propagating network updates within given time bounds. On the other hand, the users of an ad-hoc network may not be satisfied with pure best-effort services, and may demand at least statistical QoS guarantees. Obviously, scalable solutions to such a contradiction have to be based on models which assume that a subset of the network states is sufficiently accurate.

The objective of this work is to provide statistical QoS guarantees with respect to a new QoS parameter in ad-hoc networks, *service efficiency*, used to quantitatively evaluate the ability of providing the best service coverage with the minimum cost of resources. Our focus is on the generic notion of a *service*, which is defined as a collection of identical *service instances*, each may be a web server or a shared whiteboard. Each instance of service runs in a single mobile node, and is assumed to be critical to applications. Since service instances may be created (by replication) and terminated at run-time, we refer to such a service as an ***adaptive service***.

Since nodes are highly mobile, the ad-hoc network may become partitioned temporarily and subsequently reconnected. In this paper, the subset of nodes is referred to as ***groups***. Based on similar observations, Karumanchi et al. [5] has proposed an update protocol to maximize the availability of the service while incurring reasonable update overheads. However, the groups that it utilized were fixed, pre-determined, and *overlapping* subsets of nodes. Such a group definition may fail to capture the mobility pattern of nodes. Instead, we consider *disjoint sets* of nodes as groups, which are discovered at run-time based on observed mobility patterns. This is preferred in a highly dynamic ad-hoc network. With such group definitions, two critical questions are still not addressed:

- **Group division.** How to divide nodes into groups, so that when the network becomes partitioned, the probability of partitioning along group boundaries is high?
- **Service adaptation.** Assuming the first issue is solved, how can we dynamically create and terminate service instances in each of the groups, so that the *service efficiency* converges to its upper bound with a high probability?

In short, we need an optimal algorithm that maximizes *service efficiency*, i.e., covering the maximum number of nodes with the least possible service instances, especially when the network is partitioned. Obviously, if we assume that node mobility is completely unpredictable, it is impossible to address the issue of group division and service adaptation. We need to have a more constrained and predictable model for node mobility. For this purpose, Hong et al. [6] has proposed a *Reference Point Group Mobility model*, which assumes that nodes are likely to move within *groups*, and that the motion of the *reference point* of each group defines the entire group's motion behavior, including location, speed, direction and acceleration. Since in ad-hoc networks, communications are often within smaller teams which tend to coordinate their movements, the group mobility model is a reasonable assumption in many application scenarios, e.g., emergency rescue teams in a disaster scene or groups of co-workers in a convention. Such a group mobility model was subsequently utilized to derive the *Landmark Routing* protocol [7], which showed its effectiveness to increase scalability and reduce overheads. To further justify the group mobility model, prior research work in the study of the behavioral pattern of wild life [8] has shown extensive grouping behavior in nature, which may be useful as far as ad-hoc sensor networks are concerned.

However, a major drawback of the previously proposed group mobility model was its assumptions that all nodes have prior knowledge of group membership, i.e., they know which group they are in, and that the group membership is *static*. These assumptions have provided an answer to the first unaddressed question, but they are too restrictive and unrealistic. In this work, we relax these assumptions and focus on *dynamic* and *time-varying* group memberships[1] to be detected at run-time by running a distributed algorithm on the nodes, based on *only* local states of each node. With such relaxed assumptions, groups are practically formed and adjusted on-the-fly at run-time, without any prior knowledge about static memberships.

In this paper, our original contributions are the following. First, we provide a mathematical model to rigorously characterize the *group mobility* model using normal probability distributions, based on the intuition proposed in previous work [6]. Second, we define our new QoS parameter of focus, *service efficiency*. Third, based on our definition of group mobility, we theoretically derive lower and upper bounds of the *service efficiency*, which measures the effectiveness of provisioning adaptive services in ad-hoc networks. Fourth, we propose a fully distributed algorithm, referred to as the *adaptive service provisioning algorithm*, to be executed in each of the mobile nodes, so that (1) The group membership of nodes are identified; (2) Service instances are created and terminated dynamically; and (3) message exchange overheads incurred by the algorithm are minimized. Finally, we present our simulation testbed and an extensive collection of simulation results to verify the effectiveness of our distributed algorithm for QoS provisioning.

The remainder of the paper is organized as follows. The mathematical formulation of the group mobility model is given in Section 2. Section 3 shows a theoretical analysis of service efficiency, based on the group mobility model. Section 4 presents the adaptive service provisioning algorithm, in order to identify group membership and manage the adaptive service. Section 5 shows extensive simulation results. Section 6 concludes the paper and discusses future work.

2   Group Mobility Model

The definition of group mobility model given in [6] was intuitive and descriptive, but lacked a theoretical model to rigorously characterize its properties. Furthermore, the model was based on the existence and knowledge of a centralized *reference point* for each group, which characterizes group movements. However, assuming that per-group information such as reference points are known *a priori* to all mobile nodes is unrealistic. For example, when a new node is first introduced to an ad-hoc network, it does not have prior knowledge about the reference points, or even which group it is in.

In this work, we assume that the nodes only have access to its local states, which include its distance to all its neighboring nodes, derived from the physical layer. With this assumption, the group mobility model needs to be redefined so that it is characterized based on fully distributed states, e.g., distances between nodes, rather than the

---

[1] Strictly speaking, we need to impose some restrictions on the degree of dynamics with respect to group membership changes. It may not exceed the frequency of running the adaptive service provisioning algorithm.

availability of a reference point. Intuitively, nodes within the same group tends to have a high probability of keeping stable distances from each other.

In this paper, we assume that all nodes have identical and fixed transmission range r in the ad-hoc network, and that if the distance between two nodes $\|AB\| \leq r$, they are *in-range nodes* (or *neighboring nodes*) that are able to communicate directly with a *single-hop* wireless link, denoted by $\overline{AB} = 1$, otherwise they are *out-of-range nodes* with $\overline{AB} = 0$. If there exists a multi-hop wireless communication path between A and B interconnected by in-range wireless links, we claim that A and B is mutually *reachable*.

We first define the term *Adjacently Grouped Pair* (AGP) of nodes.

Definition 1. *Nodes A and B form an **Adjacently Grouped Pair** (AGP), denoted by $A \overset{0}{\sim} B$, if $\|AB\|$ obeys **normal distribution** with a mean $\mu < r$, and a standard deviation $\sigma < \sigma_{max}$, where $\|AB\|$ denotes the distance between A and B.*

In practice, rather than the absolute value of σ_{max}, one is often interested in the ratio of the standard deviation to the mean of a distribution, commonly referred to as *coefficient of variation* $CV_{max}, (0 < CV_{max} < 1)$, where $\sigma_{max} = r * CV_{max}$.

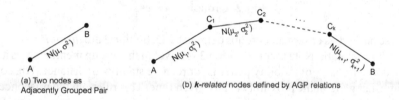

(a) Two nodes as Adjacently Grouped Pair

(b) *k-related* nodes defined by AGP relations

Fig. 1. Using Adjacently Grouped Pairs to Form a Group.

Figure 1(a) shows such a pair of nodes. Intuitively, this definition captures the fact that if two adjacent nodes are in the same group over a period of time, the distance between them stabilizes around a mean value μ with small variations, while $\mu < r$ so that they can communicate wirelessly. Although it is possible that they may be out of range from each other ($\|AB\| > r$) intermittently, the probability is low based on the density function of normal distribution. In addition, σ represents the degree of variations. The mobility patterns of nodes are more similar to each other with a smaller σ.

We now define the term k-*related* with Adjacently Grouped Pairs.

Definition 2. *Nodes A and B are k-related, denoted by $A \overset{k}{\sim} B$ ($k \geq 1$), if there exist intermediate nodes $C_1, C_2, \ldots, C_k$, such that $A \overset{0}{\sim} C_1, C_1 \overset{0}{\sim} C_2, \ldots, C_i \overset{0}{\sim} C_{i+1}, \ldots, C_k \overset{0}{\sim} B$.*

Figure 1(b) illustrates such definition. We further define the nodes A and B as *related*, denoted by $A \sim B$, if either $A \overset{0}{\sim} B$ or there exists $k \geq 1$, such that $A \overset{k}{\sim} B$. Note that even if $A \sim B$, $\|AB\|$ does not necessarily obey normal distribution. In addition, it may be straightforwardly derived that the relation $A \sim B$ is both *commutative* (in that if $A \sim B$, then $B \sim A$), and *transitive* (in that if $A \sim B$ and $B \sim C$, then $A \sim C$).

We now formally define the term **group** in our group mobility model.

Definition 3. *Nodes* $A_1, A_2, \ldots, A_n$ *are in one **group** G, denoted by* $A \in G$, *if* $\forall i, j, 1 \leq i, j \leq n, A_i \sim A_j$.

It may be proved[2] that for nodes A and B and groups G, G_1, G_2,

- if $A \in G$ and $A \sim B$, then $B \in G$.
- if $A \in G$ and $\neg(A \sim B)$, then $B \notin G$.
- if $A \in G_1, B \in G_2$ and $A \sim B$, then $G_1 = G_2$.
- if $A \in G_1, B \in G_2$ and $\neg(A \sim B)$, then $G_1 \neq G_2$.
- if $A \in G_1$ and $A \in G_2$, then $G_1 = G_2$.

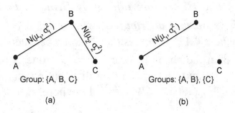

Fig. 2. Grouping Nodes.

These properties ensure that groups defined by Definition 3 are *disjoint sets* of nodes in an ad-hoc network. Note that Definition 3 is novel in that group memberships are determined by similarity of mobility patterns (or relative stability of distances) discovered over time, not geographic proximity at any given time. This rules out the misconception that as long as A and B are *neighboring* nodes, they belong to the same group. Figure 2 gives an example. Figure 2(a) shows that $A \overset{0}{\sim} B$ and $B \overset{0}{\sim} C$, hence $A \sim B \sim C$, which forms one group $\{A, B, C\}$. In comparison, Figure 2(b) shows only $A \overset{0}{\sim} B$. In this case, although A and C (or B and C) are neighboring nodes, $A \sim C$ (or $B \sim C$) does not hold. We thus have two disjoint groups $\{A, B\}$ and $\{C\}$. This scenario may arise when two groups are briefly merged geographically but separated again, due to different directions of travel.

3   Theoretical Analysis

The motivation of proposing the group mobility model is to accurately identify groups of nodes that show similar mobility pattern and maintain a stable structure over time. Therefore, it is with high probability that nodes within the same group tend to be mutually reachable. For an adaptive service that includes multiple identical service instances running on individual nodes, this is particularly beneficial to the goal of improving service accessibility with minimum resources. Intuitively, the ideal case is that, should we have an algorithm to capture grouping information with perfect accuracy at any given time, we would have placed *one* service instance in each of the groups, and trivially achieved the best service accessibility with minimum resource overheads.

---

[2] Proof omitted for space limitations.

However, in reality there are two difficulties that prevent us to achieve the ideal scenario. First, groups are detected on the fly with a distributed algorithm based on local states, and thus may not be able to be identified with perfect accuracy. Second, with dynamic group membership, service instances may need to be created and terminated even with a perfect grouping algorithm. To address these problems, a realistic approach is to first quantitatively define a QoS parameter as the optimization goal with regards to the adaptive service, then theoretically derive the upper and lower bounds of such a QoS parameter, and finally design a best possible algorithm in realistic scenarios.

3.1 Service Efficiency

We first define two parameters to quantitatively analyze different aspects of QoS in service provisioning. At any given time t, let N be the total number of nodes in the network, $N_s(t)$ be the number of service instances, and $N_a(t)$ be the number of nodes that are reachable from at least one node that runs a *service instance*, thus having access to the adaptive service. We then define *service coverage* S_{cover} and *service cost* S_{cost} as

$$S_{cover}(t) = \frac{N_a(t)}{N}, \text{ and } S_{cost}(t) = \frac{N_s(t)}{N} \tag{1}$$

The objective is obviously to have the maximum service coverage while incurring the lowest possible service cost. This objective is characterized by the definition of a new QoS parameter, *service efficiency* S, defined as

$$S(t) = \frac{S_{cover}(t)}{S_{cost}(t)} = \frac{N_a(t)}{N_s(t)} \tag{2}$$

There is one additional detail related to the definition $S(t)$. Our primary goal for the adaptive service is to reach as many nodes as possible, while reducing the service cost is only secondary. However, (2) treats $S_{cover}(t)$ and $S_{cost}(t)$ with equal weights, which may not yield desired results. For example, assume that we have N nodes and two groups with a split of $2N/3$ and $N/3$, all nodes in each of the groups are reachable from each other. To maximize $S(t)$ in (2), we only need to place one service instance in the larger group and enjoy a service efficiency of $2N/3$, rather than placing two service instances in both groups, having a service efficiency of only $N/2$. Therefore, assuming we have $K(t)$ groups at time t, we need to rectify the definition of $S(t)$ as:

$$S(t) = \frac{S_{cover}(t)}{S_{cost}(t)} = \frac{N_a(t)}{N_s(t)}, \text{ while satisfying } N_s(t) \geq K(t) \tag{3}$$

We may then proceed with the optimization objective of maximizing the service efficiency $S(t)$.

3.2 Theoretical Analysis

In this section, we derive the lower and upper bounds for $S(t)$. Ideally, in an ad-hoc network with N nodes and $K(t)$ groups at time t, if there exists a perfect grouping algorithm to accurately group all nodes, we may trivially select one representative node in each group to host an instance of the adaptive service. We thus have

$$S_{cost}(t) = K(t), \text{ and } S(t) = \frac{N_a(t)}{K(t)} \quad (4)$$

If it happens that for all groups, at time t_0, all nodes in each of the group are able to access the representative node, we have $N_a(t_0) = N$, and $S(t_0) = N/K(t_0)$, which is its optimal value. However, this may not be the case since there exists a low probability that a small subset of nodes in the same group is not reachable from the service instance. For the purpose of deriving global the lower and upper bounds for $S(t)$, we start from examining a group consisting of m nodes. In such a group, the service efficiency is equivalent to the number of nodes that have access to the service instance. We show upper and lower bounds of the average service efficiency in such a group, and then extend our results to the ad-hoc network.

Lemma 1. *If $A \overset{0}{\sim} B$, i.e., $\|AB\|$ obeys $N(\mu, \sigma^2)$ based on Definition 1, then $Pr(\overline{AB} = 0) = 1 - \Phi(\frac{r-\mu}{\sigma})$, where $\Phi(x)$ is defined as $\frac{1}{\sqrt{2\pi}} \int_{-\infty}^{x} e^{-y^2/2} dy$.*

Proof. The probability that A and B are out-of-range nodes is $Pr(\|AB\| > r)$. Hence, $Pr(\overline{AB} = 0) = Pr(\|AB\|) > r) = 1 - Pr(d \leq r) = 1 - \Phi(\frac{r-\mu}{\sigma})$. □

When the transmission range r is definite, $Pr(\overline{AB} = 0)$ increases monotonically as μ increases to approach r and as σ increases.

Lemma 2. *Assume that (1) $A \overset{0}{\sim} B$, i.e., $\|AB\|$ obeys $N(\mu, \sigma^2)$ based on Definition 1; (2) μ obeys uniform distribution in the interval $[0, r]$; and (3) σ also obeys uniform distribution in $[0, \sigma_{max}]$. The **average probability** $p = Pr(\|AB\|) > r) = 1 - \frac{\int_0^r \int_0^{\sigma_{max}} \Phi(\frac{r-\mu}{\sigma}) d\mu d\sigma}{r * \sigma_{max}}$.*

Proof.

$$p = \frac{\int_0^r \int_0^{\sigma_{max}} 1 - \Phi(\frac{r-\mu}{\sigma}) d\mu d\sigma}{\int_0^r \int_0^{\sigma_{max}} 1 d\mu d\sigma}$$

$$= \frac{\int_0^r \int_0^{\sigma_{max}} 1 d\mu d\sigma - \int_0^r \int_0^{\sigma_{max}} \Phi(\frac{r-\mu}{\sigma}) d\mu d\sigma}{\int_0^r \int_0^{\sigma_{max}} 1 d\mu d\sigma}$$

$$= 1 - \frac{\int_0^r \int_0^{\sigma_{max}} \Phi(\frac{r-\mu}{\sigma}) d\mu d\sigma}{r * \sigma_{max}} \qquad □$$

We may then derive the upper and lower bounds for the average number of nodes that are reachable from the service instance, in a group G with m nodes ($m > 0$).

Lemma 3. *If node A_1 hosts the only service instance in G and $A_1 \overset{0}{\sim} A_i, i = 2, \ldots, m$ (Fig. 3), then the average number of nodes that are reachable from A_1 is $m - (m-1)p$.*

Proof. Consider X, the number of nodes that are **not** reachable from the service instance A_1 in group G. Its distribution is a **binomial distribution** $B(m-1, p)$, i.e.,

$$Pr(X = k) = \binom{m-1}{k} p^k (1-p)^{m-1-k}$$

X has a mean of $(m-1)p$, and the average number of nodes that are reachable from A_1 is $m - (m-1)p$. □

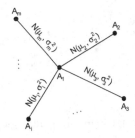

Fig. 3. Star-Grouped Nodes.

Theorem 1. *The upper bound of* $S(t)$ *in* G *is* $m - (m-1)p$.

Proof. We claim that the average number of nodes that are reachable from A_1 is maximized when group G is formed as a star structure as in Fig. 3. This may be proved as follows. Assume that a node A_i is only reachable from A_1 via an intermediate node A_j. The average probability of this reachability is $Pr(\|A_i A_j\| \le r)Pr(\|A_j A_1\| \le r) = (1-p)^2$, obviously it is smaller than $1-p$, which is the average probability of reachability via only a single-hop link. The star structure in a group ensures that all nodes $A_i, i = 2, \ldots, m$ enjoy single-hop reachability to A_1 without depending on intermediate nodes.   □

Theorem 2. *Given an ad-hoc network with average group size* m*, the upper bound of* $S(t)$ *is* $m - (m-1)p$.

Proof. Assume that there are K groups $G_1, G_2, \ldots, G_K$, with sizes $m_1, m_2, \ldots, m_K$. The total number of nodes in this network is $N = \sum_{i=1}^{K} m_i$, and the average group size $m = N/K = (\sum_{i=1}^{K} m_i)/K$. To achieve the global upper bound of $S(t)$ in the network, upper bounds of $S_i(t)$ should be achieved locally within each of the groups. According to Theorem 1, each group should be formed as a star structure, which achieves an ***upper bound*** of $S_i(t) = m_i - (m_i - 1)p, 1 \le i \le K$. Therefore, the upper bound of $S(t)$ is:

$$\frac{\sum_{i=1}^{K}[m_i - (m_i - 1)p]}{K} = \frac{\sum_{i=1}^{K} m_i + Kp - \sum_{i=1}^{K} m_i p}{K} = \frac{N + Kp - Np}{K}$$
$$= m - (m-1)p \quad \square$$

Lemma 4. *In group* G *with* m *nodes* $(m > 0)$*, if node* A_1 *hosts the only service instance and* $A_i \overset{0}{\sim} A_{i+1}$*,* $i = 1, 2, \ldots, m-1$ *(Fig. 4), then the average number of nodes that are reachable from* A_1 *is* $\frac{1-(1-p)^m}{p}$.

Proof. Consider X, the number of nodes that are **not** reachable from the service instance A_1 in group G. The reachability of A_i from A_1 depends on all the links $\overline{A_j A_{j+1}}$, $j = 1, 2, \ldots, i-1$, i.e., if A_{i-1} is not reachable, $A_j, j \ge i$ are not reachable as a result. This leads to

$Pr(X = k) = Pr$(nodes $A_i, i = m, m-1, \ldots, m-k+1$ are not reachable, all other nodes $A_j, j = 1, 2, \ldots, m-k$ are reachable) $= Pr(\overline{A_{m-k} A_{m-k+1}} = 0, \overline{A_j A_{j+1}} = 1,$ $j = 1, 2, \ldots, m-k-1) = p(1-p)^{m-k-1}$

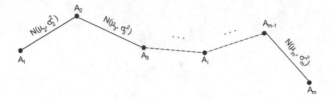

Fig. 4. Chain-Grouped Nodes.

Here, if $\overline{A_{i-1}A_i} = 0$, nodes A_j are not reachable for all $j \geq i$, independent from $\overline{A_jA_{j+1}} = 0$ or $\overline{A_jA_{j+1}} = 1$. Observing this, the average number of nodes that are not reachable is $p \sum_{k=1}^{m-1} k(1-p)^{m-k-1}$, denoted by A_{m-1}.

Let $B_{m-1} = \sum_{k=1}^{m-1} k(1-p)^{m-k-1}$, multiply by $1-p$ on both sides, we have

$$(1-p)B_{m-1} = (1-p)\sum_{k=1}^{m-1} k(1-p)^{m-k-1} = \sum_{k=1}^{m-1} k(1-p)^{m-k}$$

$$= \sum_{j=0}^{m-2} j(1-p)^{m-j-1} + \frac{(1-p)-(1-p)^m}{p}$$

$$= B_{m-1} - (m-1) + \frac{(1-p)-(1-p)^m}{p}$$

Subtract B_{m-1} on both sides, we have

$$pB_{m-1} = (m-1) - \frac{(1-p)-(1-p)^m}{p} \tag{5}$$

Therefore, the average number of nodes that are **not** reachable from A_1 is $A_{m-1} = pB_{m-1} = m - \frac{1-(1-p)^m}{p}$, and the average number of nodes that are reachable from A_1 is $m - A_{m-1} = \frac{1-(1-p)^m}{p}$. $\qquad\square$

Theorem 3. *The lower bound of $S(t)$ in G is $\frac{1-(1-p)^m}{p}$.*

Proof. Based on the proof of Theorem 1, for any node A_i, the more intermediate nodes required from the service instance A_1, the less probable that it is reachable from A_1 at time t. Obviously, the worst case is reached when all nodes in the group form a chain structure as in Fig. 4. $\qquad\square$

Theorem 4. *Given an ad-hoc network with minimum group size m_{min}, the lower bound of $S(t)$ is $\frac{1-(1-p)^{m_{min}}}{p}$.*

Proof. Assume there are K groups $G_1, G_2, \ldots, G_k$, with sizes $m_1, m_2, \ldots, m_K$. The smallest group size is $m_{min} = \min\{m_1, m_2, \ldots, m_K\}$. To achieve the global lower bound of $S(t)$ in the network, lower bounds of $S_i(t)$ should be achieved locally within each of the groups. According to Theorem 3, each group should be formed as a chain structure, achieving a ***lower bound*** of $S_i(t) = \frac{1-(1-p)^{m_i}}{p}, 1 \leq i \leq K$. Hence, for the entire network, we have

$$S(t) = \frac{\sum_{i=1}^{K}[1-(1-p)^{m_i}]/p}{K} = \frac{K - \sum_{i=1}^{K}(1-p)^{m_i}}{Kp}$$

$$\geq \frac{K - \sum_{i=1}^{K}(1-p)^{m_{min}}}{Kp} = \frac{1 - (1-p)^{m_{min}}}{p}$$

Therefore, the lower bound of $S(t)$ is $\frac{1-(1-p)^{m_{min}}}{p}$. $\square$

4 Adaptive Service Provisioning Algorithm

Taking the definition of group mobility model (Sect. 2) and its analytical properties, we propose a fully distributed algorithm, referred to as the *adaptive service provisioning algorithm*, that enables dynamic service instance creation and termination in each of the nodes. For this purpose, the algorithm first identifies group memberships of nodes by leveraging the definition of the group mobility model, then selects representative nodes that require creating and terminating service instances. The objective is to maximize *service efficiency* in the network, so that it converges to the upper bound derived in our theoretical analysis. From the proofs of previous theorems, we believe that the upper bound is achieved by having exactly one service instance for each group, if the group mobility model can be utilized to accurately identify the groups at any given time.

In order to address the group division problem in the network, we start by determining if, at time t_0, two neighboring nodes form an *Adjacently Grouped Pair* (AGP). For this purpose, the distance between two neighboring nodes is measured and recorded for a fixed number of rounds l, where l is a pre-determined size of the sampling buffer. The average distance $\bar{d}$ and the standard deviation s may thus be derived from these l samples, which are used to approximate the mean value μ and standard deviation σ in the normal distribution. If the approximated μ and σ complies with **Definition 1** in Sect. 2, the two nodes are identified as an AGP. The advantage of this measurement-based approach is that *Adjacently Grouped Pairs* can be identified at run-time by only relying on local states of each node, e.g., its distances to all neighboring nodes. This conforms with our design objective of minimizing local states and message exchange overheads.

In this algorithm, we assume that each node $A_i(i = 1, \ldots, N)$ has an unique physical ID $id(A_i)$, and at the initial time t_0, there are K_s nodes ($1 \leq K_s << N$) in the network that host service instances of the adaptive service. Our goal is to converge to the upper bound of service efficiency by dynamically initiating new service instances or terminating existing ones, based on identification of groups. On each node, the following local states are maintained:

- **Service Instance ID** [$sid(A_i)$]: the physical ID of the node that hosts the service instance that is currently reachable from A_i.
- **Profile of Measurements** [$P(A_i)$]: a two-dimensional profile in which each row represents one of the neighboring nodes, and each column represents distances to all neighboring nodes obtained from one round of measurements. After l measurements, l samples of distances to A_i are obtained for each neighboring node, denoted by $d_i^{(k)}, k = 1, \ldots, l$.
- **Neighboring nodes in the same group as** A_i [$G_n(A_i)$]: the subset of neighboring nodes that has been identified as in the same group as A_i itself. Note that rather than

---

At time t**:**

out-list := $\emptyset$;

for each node C_i in $G_n(A), 1 \le i \le |G_n(A)|$ **do**

    if A and C_i are out-of-range nodes **then** out-list := out-list $+ \{C_i\}$;

Current list of neighboring nodes of A is $\{B_1, B_2, \dots, B_k\}$;

At time $t + (i - 1) * \Delta t, 1 \le i \le l$**:**

for each neighboring node $B_j, 1 \le j \le k$ **do**

    Record the distance $d_j^{(i)}$ between A and B_j;

    if A and B_j are out-of-range **then** $d_j^{(i)} := +\infty$;

At time $t + (l - 1) * \Delta t$**:**

if $d_j^{(2)} == +\infty$ **then** $d_j^{(2)} := $ r;

for $i = 3, \dots, l$ **do if** $d_j^{(i)} == +\infty$ **then** $d_j^{(i)} := \mathbf{max}(r, d_j^{(i-1)} + |d_j^{(i-1)} - d_j^{(i-2)}|)$;

for each neighboring node $B_j, 1 \le j \le k$ **do**

    calculate the average distance $\bar{d}_j := \sum_{i=1}^{l} d_j^{(i)}/l$;

    calculate the sample estimate of a standard deviation $s_j := \sqrt{\frac{\sum_{i=1}^{l}(d_j^{(i)} - \bar{d}_j)}{l-1}}$;

    if $\bar{d}_j < r$ and $s_j < \sigma_{max}$ **then**

        $G_n(A) := G_n(A) + \{B_j\}$;

        if $A \notin G_n(B_j)$ **then** $G_n(B_j) := G_n(B_j) + \{A\}$;

    else if $B_j \in G_n(A)$ **then** $G_n(A) := G_n(A) - B_j$;

for each node C_i in out-list, $1 \le i \le |$out-list$|$ **do**

    if A and C_i are out-of-range **then** $G_n(A) := G_n(A) - \{C_i\}$;

if A hosts a service instance **then**

    $sid(A) := \mathbf{max}\{id(A), sid(A), sid(A_j)$ while $A_j \in G_n(A)\}$;

    if $sid(A) \ne id(A)$ **then** terminate the service instance on A;

else

    $sid(A) := \mathbf{max}\{sid(A), sid(A_j)$ while $A_j \in G_n(A)\}$;

if $sid(A) == -1$ **then**

    if there exists a neighboring node $B_j \notin G_n(A), sid(B_j) \ne -1$

    and $sid(B_j)$ is reachable from A **then**

        A sends a *service replication request* to the group of (B_j);

        A starts to execute a new service instance;

        $sid(A) = id(A)$;

Fig. 5. The Adaptive Service Provisioning Algorithm.

maintaining all nodes in the same group as A_i, this set only contains *neighboring nodes* that are in the same group.

The algorithm to be executed on a specific node A is given in Fig. 5. Its highlights are illustrated as follows. Initially, at time t_0 when the algorithm starts, all nodes are assigned initial states $G_n(A_i) = \emptyset, i = 1, \dots, N$, and $sid(A_i) = i$ if A_i hosts a service instance, otherwise $sid(A_i) = -1$. The algorithm then starts to be executed periodically in each of the nodes, updating local states $sid(A_i)$, $P(A_i)$ and $G_n(A_i)$. For a node A, the algorithm may be divided into four phases.

- **Preparation Phase.** As A starts to run the algorithm, it first examines if any nodes in $G_n(A)$ is currently out of range. If so, this node may be previously added to $G_n(A)$ by mistake[3], we thus temporarily add it into a locally maintained *out-list*.
- **Measurement Phase.** For each neighboring node, A measures the distance for l times between itself and its neighboring node[4].
- $G_n(A)$ **Calculation Phase.** According to the profile, if A finds that a neighboring node, e.g., A_k, is in its group, A and A_k will add each other in their respective $G_n(A_i)$. On the other hand, existing nodes in $G_n(A)$ may be removed if measurements do not show AGP properties.
- $sid(A)$ **Update Phase.** If A is hosting a service instance, the service instance ID of A is updated as:

$$sid(A) = \max\{id(A), sid(A), sid(A_j) \text{ while } A_j \in G_n(A)\} \tag{6}$$

If the updated service instance ID is not $id(A)$, which means another node in the same group is currently hosting a service instance, A will then terminate its own instance. On the other hand, if another node A_i does not host any service instances, and it can not find any service instances in its $G_n(A_i)$, it will probe its non-AGP neighboring nodes and examine if they have access to any service instances. If so, it creates an identical replication of the service instance. Otherwise, the group that A_i is in will continue to be out of reach from any service instances, and they will regularly poll their new non-AGP neighboring nodes to examine if a service instance may be replicated. Once it is replicated, the changes of service instance IDs will be propagated to the entire group.

5   Performance of Adaptive Service Provisioning Algorithm

We conduct simulation experiments to evaluate the performance of the adaptive service provisioning algorithm. The performance metrics that are measured include (1) number of identified groups with different CV_{max} values; (2) service coverage and service cost; (3) service efficiency; (4) *service turnovers*, i.e., migration of service instances due to creations and terminations. This gauges the probability of having stable service instances remain on the same nodes.

The simulated mobile ad-hoc network consisted of 100 mobile hosts roaming in a square region of $800 * 800$ meters, with all boundaries connected, i.e., nodes reaching one edge of the region will emerge on the opposite edge and continue to move on in its previous direction. The transmission range r is set to be $60m$.

We assume that there exists group mobility behavior in the network. When approximating such group movements, we divide the nodes into 10 disjoint sets, each set has a

---

[3] For example, the two nodes may happen to be close to each other when the algorithm was executed in a previous round.

[4] If the samples can not be obtained momentarily because of node mobility, the distance is estimated assuming constant velocity.

randomly generated size and has independent group-wise mobility pattern. The movement of a particular node consists of a motion vector following group mobility, and another motion vector showing its own random movement. Please note that nodes in the same set are not necessarily identified as in one **group** based on our adaptive service provisioning algorithm, since the algorithm is designed to detect groups strictly based on local states in the nodes themselves.

Initially each node A_i is assigned a unique ID $id(A_i)$ and it is in a group of its own. Other parameters in the simulation include (1) $CV_{max} = 0.25$ (except for Fig. 6(a), where we investigate the impact of CV_{max} on the number of groups); (2) The initial number of service instances is 5. (3) Distance sampling size l is 20; (4) Each node runs the algorithm every 100 time units. The simulation runs for 1000 time units.

Figure 6(a) shows that the algorithm is effective and efficient in classifying nodes into groups. The number of groups converges rapidly to a stable value with a small degree of fluctuations. In addition, we have observed that the parameter CV_{max} may affect both the convergence rate and stable values. Such observations are as expected, since larger CV_{max} represents more relaxed criteria for identifying groups. However, we have observed that the effects of CV_{max} on the stable number of groups are insignificant.

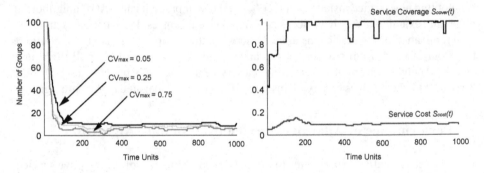

(a) Number of Groups with Different CV_{max} Values.

(b) Service Coverage and Service Cost.

Fig. 6. Experimental Results: Part I.

Figure 6(b) shows the service coverage $S_{cover}(t)$ and service cost $S_{cost}(t)$. We have observed that after the initial stage of convergence, $S_{cover}(t)$ is generally stable. There are some brief time periods that $S_{cover}(t)$ decreases, due to the fact that a subset of nodes roam away from a larger group and are thus temporarily out of service. However, the adaptive service resumes after this subset of nodes creates a new service instance by replicating from another passing-by group. With respect to the service cost $S_{cost}(t)$, Figure 6(b) has shown that it remains near a constant and low level.

Figure 7(a) compares the service coverage achieved with and without executing the adaptive service provisioning algorithm. Since the average number of service instances is approximately 8 to 12 when the algorithm is executed on all nodes, we assign 10 service instances in the simulation in which the algorithm is not used. Since the initial number of service instances is only 5 for the case with the algorithm, it is normal that initially the service coverage with the algorithm is less, compared to that without the algorithm executing. However, after a stabilizing period, the service coverage with the algorithm is shown to be better and much more stable than that without the algorithm.

During the simulation, we record the list of service nodes every 10 time units, which is compared to its counterpart in previous time instants. *Service turnovers* are characterized as follows. When a particular node begins to host a new service instance, or when an existing node terminates its service instance, we increment the measurement by 1. Shown in Fig. 7(b), the measured values essentially indicates the frequency of service migration from one node to another. It is only during the starting stage of the simulation that service turnovers are as large as 3, due to initial service replication. Afterwards, service turnovers remain around 0, except for very few time periods when the service instances are rearranged to adapt to behavioral changes in the network.

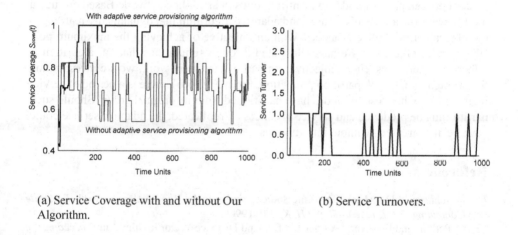

(a) Service Coverage with and without Our Algorithm.

(b) Service Turnovers.

Fig. 7. Experimental Results: Part II.

Figure 8 shows the service efficiency $S(t)$ with its upper and lower bounds, theoretically derived in Sect. 3. Recall that the upper and lower bounds depend on the m and m_{min}, which is the average and smallest group size, respectively. Since m and m_{min} varies over time, the upper and lower bounds are not constants and vary accordingly. After the initial stabilizing period, $S(t)$ generally remains between the derived upper and lower bounds, except for very rare cases where the observed $S(t)$ is slightly over the upper bound. The reason is as follows. When the upper bound is derived and proved, a node is considered to be out of service if it is not able to access the service in its group; however, in our simulations, there are rare cases in which a particular node can not access any service instances in its own group, but is able to occasionally eavesdrop within its neighboring group. Finally, we may also observe from Fig. 8 that our fully distributed algorithm is able to achieve a service efficiency that effectively converges to long-term stable values of its derived upper bound.

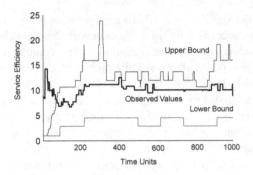

Fig. 8. Service Efficiency $S(t)$ and Its Derived Upper and Lower Bounds.

6 Conclusions and Future Work

In this paper, we have presented a novel group mobility model that depends only on distances between pairs of nodes to identify groups in an ad-hoc network. Based on such a model, we show a fully distributed and adaptive algorithm that dynamically rearranges the placement of service instances, with an objective of achieving the maximum possible service efficiency. We have illustrated through simulations that our algorithm is effective to achieve such an objective. As part of the future work, we are investigating the problem of network partition prediction. From Fig. 6(b), there is a period of service interruptions when a set of nodes have partitioned from its original group. Should such partitioning be predicted and service instances be replicated, the adaptive service could have been guaranteed without interruptions.

References

[1] D. Johnson and D. Maltz, "Dynamic Source Routing in Ad Hoc Wireless Networks," *Mobile Computing (ed. T. Imielinski and H. Korth)*, 1996.

[2] C. Perkins and E. Royer, "Ad-hoc On-Demand Distance Vector Routing," in *Proceedings of 2nd IEEE Workshop on Mobile Computing Systems and Applications (WMCSA '99)*, 1999.

[3] J. Chang and L. Tassiulas, "Energy Conserving Routing in Wireless Ad-hoc Networks," *IEEE INFOCOM 2000*, 2000.

[4] S. Chakrabarti and A. Mishra, "QoS Issues in Ad Hoc Wireless Networks," *IEEE Communications Magazine*, pp. 142–148, February 2001.

[5] G. Karumanchi, S. Muralidharan, and R. Prakash, "Information Dissemination in Partitionable Mobile Ad Hoc Networks," in *Proceedings of the 18th IEEE Symposium on Reliable Distributed Systems*, 1999, pp. 4–13.

[6] X. Hong, M. Gerla, G. Pei, and C. Chiang, "A Group Mobility Model for Ad Hoc Wireless Networks," in *Proceedings of the 2nd ACM International Workshop on Modeling and Simulation of Wireless and Mobile Systems*, 1999.

[7] G. Pei, M. Gerla, and X. Hong, "LANMAR: Landmark Routing for Large Scale Wireless Ad Hoc Networks with Group Mobility," in *Proceedings of IEEE/ACM MobiHOC 2000*, 2000.

[8] C. W. Reynolds, "Flocks, Herds, and Schools: A Distributed Behavioral Model," in *Proceedings of ACM ACM SIGGRAPH 87*, July 1987.

Experimental Extensions to RSVP — Remote Client and One-Pass Signalling

Martin Karsten

Industrial Process and System Communications, Darmstadt University of Technology
Merckstr. 25, D-64283 Darmstadt, Germany
Martin.Karsten@KOM.tu-darmstadt.de

Abstract. We present and evaluate two experimental extensions to RSVP in terms of protocol specification and implementation. These extensions are targeted at apparent shortcomings of RSVP to carry out lightweight signalling for end systems. Instead of specifying new protocols, our approach in principle aims at developing an integrated protocol suite, initially in the framework set by RSVP. This work is based on our experience on implementing and evaluating the basic RSVP specification. The extensions will be incorporated in the next public release of our open source software.

1 Introduction

There have been numerous proposals for QoS signalling protocols, which exhibit differences along certain, partially interdependent, characteristics, for example with respect to participating entities, interaction mode, flexibility, generality, supported services, among others. The *Resource ReSerVation Protocol* (RSVP) [1] provides a rich set of functionality and has been chosen for standardization by the IETF [2]. However, the basic specification of RSVP considerably has shortcomings in a variety of contexts, in which the specific set of RSVP's features are either over- or under-dimensioned.

- Embedded end systems often have strict limitations with regard to their processing power and memory equipment. Therefore, it is imperative to keep the respective requirements of any signalling protocol as low as possible. Running a full RSVP daemon on such an end system might not be the appropriate configuration.
- A number of valid service models exist, in which the performance can be described as transmission rate over certain time intervals. In this case, RSVP's ability to collect path characteristics might not be needed. Furthermore, RSVP is designed to support multi-point to multi-point communication. This design requirement imposes a receiver-oriented reservation model and thus, a two-way session setup, which might not be needed for simple unicast communication. Therefore, a sender-oriented one-way reservation setup can be a sensible extension to RSVP.

The eventual goal of our work is to design an integrated protocol suite, which can be broken down to a few well-defined subsets for specific scenarios. Our current work is based on RSVP, because it seems to be a good candidate to start this investigation. We expect to either be able to actually design such a protocol suite within the framework set by RSVP, or alternatively, to gain important insight to design such a protocol suite from scratch, if RSVP turns out not to be an appropriate basis.

L. Wolf, D. Hutchinson, and R. Steinmetz (Eds.): IWQoS, LNCS 2092, pp. 269-274, 2001.

The rest of this paper is structured as follows. In Section 2, we present two extensions to RSVP. A performance-related evaluation is presented in Section 3 and the paper is wrapped up with a conclusion and an outlook to future work in Section 4.

2   RSVP Extensions

As discussed in Section 1, there are several circumstances under which the current RSVP specification is improvable to accommodate specific requirements. In this section, we present according protocol extensions for RSVP.

2.1   Remote Clients

RSVP defines two alternative methods to transmit messages between RSVP-capable nodes. RSVP messages are either transmitted as raw IP packets or using UDP encapsulation [2]. When using UDP encapsulation, packets are addressed to well-defined ports. If multiple clients run on a single end system, this addressing scheme requires a central manager entity (usually the RSVP daemon) to receive and dispatch incoming messages. For outgoing messages, it seems to be possible to use the same port numbers by multiple application processes, but this might not be supported on all platforms. For embedded devices, the effort of running a dedicated RSVP daemon might be prohibitively expensive, even if this daemon does not need the full functionality. An elegant solution is to define additional protocol mechanisms which allow an RSVP daemon running on the first RSVP-capable hop to administer and communicate remotely with a number of clients. These clients in turn only need to implement RSVP stubs and except for the special addressing scheme, participate in the full RSVP signalling procedure. This interaction is shown as *remote API* in Figure 1.

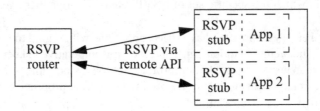

Fig. 1. Remote Client Extension.

The remote client extension can be realized through a new message type, *InitAPI*, and reusing the LIH field of the RSVP_HOP object. In the notation of [2], the InitAPI message is defined as follows.

```
<InitAPI Message> ::=    <Common Header> [ <INTEGRITY> ]
                         <SESSION> <RSVP_HOP>
```

An additional flag in the SESSION object distinguishes whether a message is used to register or de-register a client. Of course, the detailed representation of protocol elements could be chosen differently, if necessary for any purpose. Both registration and de-registration messages carry the local IP address of the client system as part of the RSVP_HOP object. The LIH field of this object is used to carry the local UDP port, which is chosen arbitrarily by the clients. Clients communicate to the remote RSVP daemon through a well-known port. In general, from the point of view of the

RSVP daemon, a client operates similar to a regular RSVP hop, distinguished only by the registration process and UDP communication. Client registration is done using soft state, i.e. clients have to regularly refresh their registration, otherwise all respective state is timed out at the RSVP daemon. The periodic refresh is triggered by the RSVP daemon and other protocol messages are not refreshed between the daemon and the client in order to avoid complicated timer management at the client side. The application using the client API can optionally initiate retransmission of requests, if desired. In order to enable end-to-end consensus about established reservations, confirmation messages do not terminate at the daemon as in [2], but are forwarded to the client. Of course, the first-hop RSVP node must be in the path between the client and the other end system. Additionally, the client system is responsible for exerting traffic control on incoming reservation requests and allocating resources. This is identical to regular RSVP processing and even mostly independent of the signalling protocol at all, but rather on the actual link technology and its dimensioning.

2.2   One-Pass Reservations

In its basic form, RSVP uses a bidirectional message exchange to set up an end-to-end simplex reservation. This procedure is called *one-pass with advertising* (OPWA) [2] and used for the following purposes. In order to support heterogeneous requests from multiple receivers within a multicast group, reservations are requested and established from the receiver to the sender. The advertising phase is needed to route reservation requests along the reverse data path to the sender. Furthermore, to flexibly support a variety of service classes and to enable precise calculation of reservation parameters for delay-bounded services, appropriate data are collected during the advertisement phase and delivered to the receiver.

As discussed in Section 1, there are a number of scenarios in which both features are not needed. In such cases, the original OPWA procedure represents an unnecessary signalling overhead for both end systems and intermediate nodes. Additionally, there might be situations where an initial (potentially duplex) reservation establishment by the initiator is desirable as fast as possible, which can later optionally be overridden by appropriate signalling requests from the responder and in turn the initiator. We have designed a true one-pass service establishment mechanism, which allows to handle such situations. It fully interacts with traditional RSVP signalling, such that it is possible to optionally override an initial one-pass reservation with later requests. The operation of a one-pass reservation as duplex request is shown in Figure 2. The figure shows the situation for a responder overriding a reservation installed by the initiator. Below, we specify the protocol elements for this extension.

A new message type, *PathResv*, is defined to indicate that reservations based on the transmitted *TSpec* shall be established through the transmission of this message. Other than the message type, the syntax is exactly the same as for a *Path* message. In order to request a duplex reservation, the following object can optionally be added to a PathResv message

```
DUPLEX_Object ::=   <SenderReceivePort><ReceiverSendPort>
```

The DUPLEX object carries the reverse port information, assuming that the same transport protocol is used in both directions. Again, this specification can easily be changed or extended, if necessary for any purpose. The duplex extension is only sen-

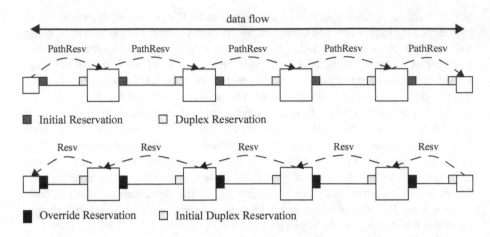

Fig. 2. One-Pass Duplex Request with Subsequent Override.

sible, when symmetric paths can be assumed between two end systems and further-more, only for unicast communication. Consequently, duplex requests for multicast sessions must be ignored at intermediate nodes.

The advantages of such an extension are quite obvious. First, it reduces signalling complexity for end systems, by offering a one-pass request model without active in-volvement of the responder. Optionally, a confirmation message could be send back to the initiator, in order to assure the end-to-end service establishment, but we have not implemented that, yet. By reducing the overall signalling effort to a single pass, intermediate nodes are relieved from processing effort, as well, because of fewer total messages. Thereby, this mechanism enables lightweight signalling in the framework of RSVP. These advantages are increased even further when one-pass duplex signal-ling is employed. Optionally, one-pass session establishment can be overridden by lat-er requests from both initiator and responder. In this case, any state that has been in-directly created through one-pass mechanisms is replaced by regular state. While this usage scenario eventually leads to the same overall signalling costs as using tradition-al RSVP, it allows for a faster initial session establishment, because only one half of the round-trip is needed. As a side effect, the remote API extension also allows to bet-ter integrate legacy and new RSVP-incapable end-systems, because no interaction with low-level system services is needed to port it to such platforms.

3   Evaluation

The extensions presented in Section 2 have been implemented in our RSVP engine [3]. In this section, we present and discuss the consequences of the proposed RSVP extensions. This investigation is focused on performance-related aspects.

3.1   Remote Clients

In order to evaluate the remote client extension, there is not much virtue in running large scale performance experiments, because in reality, a first-hop RSVP node is less likely to be challenged by requests from a lot of clients. In general, the number of ses-sions that can be handled with this implementation can be estimated to be in the same order of magnitude than what can be sustained at a regular router. It is more interesting to study the effects of the remote client extensions on actual client applications. We look at two interesting numbers, which give an indication that the usage of the remote client API probably does not constitute a severe difficulty, even on small embedded systems. We have taken a very simple rate-based UDP sender and compiled it with and without using the remote RSVP API. The library has been statically linked and we report the size of executables as well as the size of memory allocation for various platforms.

Table 1. Size of Client's Executables and Memory Allocation in Bytes.

Platform	Linux 2.2 (Intel)		FreeBSD 3.4 (Intel)		Solaris 2.6 (Sparc)	
Memory Type	Footprint	Data	Footprint	Data	Footprint	Data
with RSVP	96532	980K	171912	1180K	268736	1832K
without RSVP	12488	904K	83368	1016K	141736	1648K
Delta (RSVP)	84044	76K	88544	164K	127000	184K

These results listed in Table 1 remain to be interpreted in the context of real embedded systems, but bearing in mind that the example client is a very simple program consisting of less than 300 lines of code, it can be concluded from these numbers that the increase in executable size and memory allocation due to enabling RSVP capability does not seem prohibitively expensive.

3.2   One-Pass RSVP Signalling

In this section, we report a series of experiments comparing the performance of traditional RSVP signalling with one-pass signalling. Because our RSVP implementation is continually worked on and improved, we report new numbers for traditional signalling, instead of taking them from [4]. All experiments are carried out in the same environment as reported in [4], namely a topology of 450MHz standard Pentium III based PCs running FreeBSD 3.4. For all experiments, we generate a number of sessions and then periodically create and delete sessions in order to simulate an average lifetime of 4 minutes. In all experiments, we report the worst-case CPU processing load and memory allocation at intermediate nodes. Each experiment has run for several minutes and the CPU load number has always stabilized around a value smaller than the peak load. There are no memory leaks in our software, such that the memory allocation remains stable for a given number of flows, as well.

The performance figures for traditional RSVP signalling can be found in Table 2. Although there are slight differences to the earlier numbers reported in [4], it can be concluded that the results are quite similar in their essence. The main difference is given by a decreased variable memory allocation per flow of approximately 1450 bytes, compared to approximately 1850 bytes reported in [4]. In order to evaluate the one-pass reservation mechanism, the same experiment has been run, but employing the one-pass reservation scheme. The results are given in Table 2, as well.

Table 2. Performance of Traditional and One-Pass Signalling.

Experiment Settings		Traditional Signalling		One-Pass Signalling	
Number of Flows	Average Lifetime	Load (% CPU)	Memory (in KB)	Load (% CPU)	Memory (in KB)
0	--	0.00	2932	0.00	3004
20000	240 sec	24.56	31628	16.70	27288
40000	240 sec	49.56	60340	34.18	51588
60000	240 sec	74.56	89060	52.25	75888
80000	240 sec	--	--	70.17	100188

Although the implementation has not been optimized for one-pass reservations, at all, a significant improvement of the overall performance is visible. This can be explained mainly by the lower amount of messages that are transmitted. The performance of one-pass signalling is linear to the number flows, as expected, and the memory usage is decreased by more than 200 bytes per flow, compared to traditional signalling. This result is definitely promising with respect to further consideration and potential optimization of this mechanism.

4   Conclusions and Future Work

In this paper, we have evaluated two experimental extensions to RSVP. These extensions are targeted at different scenarios, in which the current specification of RSVP does not provide an adequate set of functionality. The extensions have been implemented and tested to investigate their effect on RSVP's implementation and processing effort. It turns out that the extensions can be realized and used with acceptable effort.

Since the eventual goal of this work is to investigate and design a flexible QoS signalling suite, much additional work remains to be carried out. There are plenty of other potential protocol mechanisms, for example in the field of reservation aggregation. By experimental combination of such mechanisms in a common framework set by our initial RSVP implementation, we hope to gain further insight towards the goal of designing a flexible and modular signalling protocol suite.

References

[1]   L. Zhang, S. Deering, D. Estrin, S. Shenker, and D. Zappala. RSVP: A New Resource ReSerVation Protocol. *IEEE Network Magazine*, 7(5):8–18, September 1993.
[2]   R. Braden, L. Zhang, S. Berson, S. Herzog, and S. Jamin. RFC 2205 - Resource ReSerVation Protocol (RSVP) – Version 1 Functional Specification. Standards Track RFC, September 1997.
[3]   M. Karsten. KOM RSVP Engine, 2001. http://www.kom.e-technik.tu-darmstadt.de/rsvp.
[4]   M. Karsten, J. Schmitt, and R. Steinmetz. Implementation and Evaluation of the KOM RSVP Engine. In *Proceedings of the 20th Annual Joint Conference of the IEEE Computer and Communications Societies (INFOCOM'2001)*. IEEE, April 2001.

Extended Quality-of-Service for Mobile Networks

Jukka Manner and Kimmo Raatikainen

University of Helsinki, Dep. of Computer Science
P.O. Box 26, FIN-00014, Finland

Abstract. Guaranteed QoS for multimedia applications is based on reserved resources in each intermediate node on the whole end-to-end path.
This can be achieved more effectively for stationary nodes than for mobile
nodes. Many multimedia applications become useless if the continuity is
disturbed due to end-to-end or slow re-reservations of resources each
time a mobile node moves so that its point-of-presence in the IP network
changes. Additionally, due to lack of QoS support from the correspondent
node, mobile nodes would need a way to reserve at least local resources,
especially wireless link resources. This paper proposes small modifications to the standard Internet resource reservation protocol, RSVP, so
that initial resource reservations and re-reservations due to terminal mobility can often be done locally in an access network. This is clearly a
significant improvement to the current RSVP.

1 Introduction

Future mobile networks will be based on IP technology. This implies that, on
the network layer, all traffic, both traditional data and streamed data like audio or video, is transmitted as independent packets. At the same time different
multimedia applications are becoming increasingly popular. These applications
require better than best-effort service from the connecting network. They require strict Quality of Service (QoS) with guaranteed minimum bandwidth and
maximum delay. Other applications, such as telnet-type applications, would also
benefit from a differentiated treatment.

The Internet Engineering Task Force (IETF) has proposed two main models for providing differentiated treatment of packets in routers. The Integrated
Services (IntServ) model [3] together with the Resource Reservation Protocol
(RSVP) [4][11] provides per flow guaranteed end-to-end transmission service.
The Differentiated Services (DiffServ) framework [2] provides non-signaled flow
differentiation that usually provides but does not guarantee proper transmission
service. The problems with these architectures are that RSVP requires support
from both communication end points and from the intermediate nodes. DiffServ
requires support from the underlying network. The Internet Architecture Board
has outlined additional issues related to these two architectures [7].

Let's consider a scenario, where a fixed network correspondent node (CN)
would be sending a data stream to a mobile node behind a wireless link. If the

L. Wolf, D. Hutchison, and R. Steinmetz (Eds.): IWQoS 2001, LNCS 2092, pp. 275–280, 2001.

correspondent node does not support RSVP it cannot signal its exact traffic characteristics to the network and request specific forwarding services. Likewise, with DiffServ, the multimedia stream will get a relative prioritized service which may result in changing visual and audio quality in the receiving application; even if the connecting wired network is over-provisioned, the wireless link bandwidth is limited.

In the absence of end-to-end QoS, a mobile node could still inform the access network of its resource requests for both outgoing and incoming traffic. This would be very beneficial, even if the QoS concepts are only in the access network. Furthermore, the same solution would allow, for example, the mobile node to reserve wireless resources independently from end-to-end resources. Therefore, we will consider, for illustration purposes, a scenario with a mobile access network and mobile nodes that are aware of the proposed signaling mechanism (Figure 1). Reserving local resources is especially important in wireless access networks, where the bottleneck resource is most probably the wireless link.

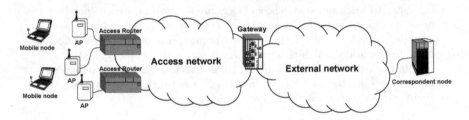

Fig. 1. Reference Architecture with a Mobile Access Network.

We propose a signaling mechanism based on a slightly modified RSVP. Access network specific reservations would be distinguished from the end-to-end reservations. The mobile node does not need to know the access network topology or the nodes that will reserve the local resources. The reservation message itself identifies the intention and the IP routing will find the network node to respond to the reservation.

We propose to use one of the reserved bits in the RSVP common header to identify the local messages. The proposed mechanism is mobile node centric in that the mobile needs to request the treatment for all its flows, both outgoing and incoming. The mechanism would operate alongside the end-to-end mechanisms and complement the range of services offered by the access network. However, the scheme is not tied to only mobile networks but can be used in any network that needs flexible local resource management.

2   Overview of the Solution

Currently we can identify two primary ways to signal QoS requirements to an access network: DiffServ Code Points (DSCP) and RSVP with IntServ. In the

DiffServ-based solution the mobile node can mark the upstream packets but must know the proper DSCP values. For the downstream the gateway on the edge of the access network must be able to mark the incoming packets with proper code points. This can be accomplished with default values for different micro flows in the Service Level Agreement negotiated between the client and the ISP.

Another way to find the code point values could be to use a Bandwidth Broker [9][12][5] that dynamically returns the proper code point for each flow: when the first packet of a flow arrives, the gateway requests the proper code point from the Bandwidth Broker. The gateway maintains a mapping from micro flow identification to the code point (soft state) so that future packets can be directly identified and labelled. A third way would be to define a protocol that the mobile node could use to dynamically adjust the mapping information stored in the gateway.

RSVP can provide the signalling mechanism for QoS requirements to the access network. For upstream reservations, the mobile node would send the PATH message to the gateway, which would return the RESV message and set up the reservations. The gateway would act as an RSVP proxy [6]. However, the reservation in the downlink direction is not as straightforward since the downlink reservation needs to be initiated by the RSVP proxy. We would need a way to trigger the proxy to initiate the RSVP signalling for a downlink flow.

These mechanisms do not seem to solve the problem entirely. The DiffServ mechanisms cannot provide explicit resource reservations and are less flexible for giving specific treatment to different flows. The problem with the RSVP proxy approach is that the proxy cannot automatically distinguish reservations that would be answered by the correspondent node and reservations that would require interception. Additionally, the RSVP proxy needs a way to know when to allocate resources for incoming flows. Mobile access networks also add to the problems, since mobile nodes can frequently change their point of presence in the network and resource allocations need to be re-arranged.

Our proposed solution is based on the RSVP proxy and the RSVP local repair mechanism. We suggest using the left most bit of the four flag bits in the RSVP common header to differentiate reservations that are internal to the access network. We call the flag the *RSVP Proxy* flag (RP). The enhanced RSVP proxy is called the *Correspondent RSVP Proxy* server. We also add a new message type called *Proxy PATH* message. Alternatively, the flag could be replaced by a full object similar to the CAP object [10], but this would introduce 8 new bytes for the transfer of a single bit over the wireless link.

When a mobile node wants to reserve resources in the local network, it uses the RP flag to indicate a local reservation. The structure of the RSVP messages follow the standard, even the intended receiver is set to be the host that the mobile node is communicating with. The correspondent RSVP proxy that intercepts the RSVP message will notice that the flag was set, does not forward the message further and responds according to the following description.

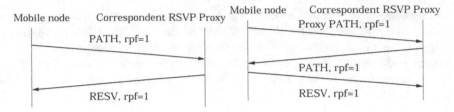

Fig. 2. Upstream Reservation.   **Fig. 3.** Downstream Reservation.

Upstream Transfers

Setting upstream reservations is straightforward and follows the RSVP Proxy functionality (Figure 2). The mobile node sends the PATH message to the correspondent node with the RP flag set. When the proxy receives the PATH message, it notes that the reservation is meant to stay within the access network and responds with a RESV message. It will not forward the PATH message further to the next hop.

Downstream Transfers

For downstream flows we need a way to signal to the correspondent RSVP proxy to initiate the RSVP reservation setup on behalf of the correspondent node. To do this, the mobile node sends a Proxy PATH message to the correspondent node with the RP flag set (Figure 3). The Proxy PATH message is identical to a standard PATH message apart from the message type field. When the correspondent RSVP proxy intercepts this message, it notes that the message is meant to stay within the access network. The message type indicates that the proxy must initiate an RSVP reservation for a downstream flow and use the information in the arrived message to fill the field in the new PATH message it must send back to the mobile node. Thus, the proxy copies the information from the Proxy PATH message to the PATH message, sets the RP flag, and sends the message to the mobile node. It will also store the state to enable further refresh messages to be sent. The mobile node receives this message and responds with an RESV message with the RP flag set. This reserves the resources within the access network for the downstream.

All the other RSVP features operate in the standard way including the local repair mechanism and reservation tear-down. All related messages must have the RP flag set in order to keep the signaling within the access network. Intermediate RSVP routers between the mobile node and correspondent RSVP proxy must forward the Proxy PATH message as an ordinary IP packet.

The mechanism also allows RSVP to be used to signal DiffServ Code Points in a DiffServ access network using the RSVP DCLASS object [1]. The DCLASS object is used to represent and carry DiffServ code points within RSVP messages.

The mobile node can use the DCLASS object to instruct the correspondent RSVP proxy to mark incoming traffic with certain DiffServ code points to trigger different forwarding behavior within the access network. However, the mobile node needs to be aware of the different code point values and the related services.

Alternatively, the proposed mechanism could be used to signal required QoS information to the correspondent RSVP proxy to enable the selection of the proper code point. Thus, the mechanism can be used to signal relative priority to specific flows, without explicit resource reservations. The mechanism would work well in a network that supports both IntServ and DiffServ, as in [8].

Fast Local Repair

The Proxy PATH message could potentially be used in mobile networks to initiate a local repair for incoming flows when the mobile node is changing its point of presence in the network. In standard RSVP, when the mobile node has moved, it will need to wait until a PATH message is sent downstream, refreshing the reservation states on the new route. If mobility and the refresh messages are not coupled in any way, there might be a time interval during which the mobile node will not get the requested service.

When the mobile node moves, it should send the Proxy PATH message immediately after the handover. The message is forwarded through the intermediate RSVP routers until it finds the cross-over RSVP router that has the reservation for the mobile node stored on a different interface. The message instructs the cross-over router to initiate a local repair by sending the needed PATH message.

If the reservation was set for the local network, the RP flag must be set. This will prevent the Proxy PATH message to be routed out of the local network if the cross-over router would be located after the correspondent RSVP proxy. Thus, with local reservations, the closest of the correspondent RSVP proxy and cross-over router will respond to the routing change.

In some access networks, the access network gateways could also act as correspondent RSVP proxies. If the movement of the mobile node results in packets flowing through a new gateway (and new proxy), the Proxy PATH message would re-reserve the local resources for the new path; when the mobile node moved, it sent a Proxy PATH message and re-reserved the resources on the new leg. However, asymmetric upstream and downstream routing can create problems.

3   Concluding Remarks

The proposed enhancement is simple to implement in an RSVP router. The most significant change is that the enhanced RSVP client needs to trigger the correspondent RSVP proxy to carry out an RSVP reservation. The enhancement, as we see it, would make RSVP more interesting as a QoS signaling protocol in future 4th generation mobile networks. Furthermore, the proposed scheme does not tie the location of the correspondent RSVP proxy servers; one scenario would be to use the signaling to allocate only wireless link resources, whereas some access network could allow local reservations between the network gateway and the mobile nodes.

The main problems of the enhancement are how to distinguish between end-to-end and local reservations at the mobile node, and the effect of asymmetric routing on downstream reservations: a Proxy PATH sent upstream may not find the right correspondent Proxy server through which the actual stream will arrive. A refinement to the scheme might be to use multicast to send the Proxy PATH and thus set reservations on all proxies; unused reservations could be left to timeout. Future work on the proposed local signaling include an implementation and performance study.

Acknowledgement

This work is partially based on our work in the Information Societies Technology project IST-1999-10050 BRAIN, which is partly funded by the European Union. The authors would like to acknowledge their colleagues in the project, particularly Markku Kojo from the University of Helsinki and Tapio Suihko from VTT Information Technology, for fruitful discussions.

References

1. Y. Bernet. Format of the rsvp dclass object. RFC 2996, IETF, November 2000.
2. S. Blake, D. Black, M. Carlson, E. Davies, Z. Wang, and W. Weiss. An architecture for differentiated services. RFC 2475, IETF, December 1998.
3. R. Braden, D. Clark, and S. Shenker. Integrated services in the internet architecture: an overview. RFC 1633, IETF, June 1994.
4. R. Braden, L. Zhang, S. Berson, S. Herzog, and S. Jamin. Resource reservation protocol (rsvp), version 1 functional specification. RFC 2205, IETF, September 1997.
5. P. Chimento and B. Teitelbaum. Qbone bandwidth broker architecture. Technical report, June 2000. http://qbone.inetrnet2.edu/bb/bboutline2.html, checked 11.4.2001.
6. S. Gai, D. G. Dutt, N. Elfassy, and Y. Bernet. Rsvp proxy. Internet draft (work in progress), IETF, december 2000. (draft-ietf-rsvp-proxy-01.txt).
7. G. Huston. Next steps for the ip qos architecture. RFC 2990, IETF, November 2000.
8. J. Manner, M. Kojo, A. Laukkanen, and K. Raatikainen. A qos architecture framework for mobile networks. In *Proceedings of the IEEE International Conference on Third Generation Wireless and Beyond (3Gwireless'01), San Francisco, USA,* May 2001.
9. K. Nichols, V. Jacobson, and L. Zhang. A two-bit differentiated services architecture for the internet. RFC 2638, IETF, July 1999.
10. H. Sued. Capability negotiation: The rsvp cap object. Internet draft, IETF, February 2001. (draft-ietf-issll-rsvp-cap-02.txt).
11. J. Wroclawski. The use of rsvp with ietf integrated services. RFC 2210, IETF, September 1997.
12. R. Yavatkar, D. Hoffman, Y. Bernet, F. Baker, and M. Speer. Subnet bandwidth manager. RFC 2814, IETF, May 2000.

Quality of Service Schemes for IEEE 802.11
A Simulation Study

Anders Lindgren, Andreas Almquist, and Olov Schelén

Division of Computer Science and Communications
Department of Computer Science and Electrical Engineering
Luleå University of Technology, SE - 971 97 Luleå, Sweden
{Anders.Lindgren,Andreas.Almquist,Olov.Schelen}@sm.luth.se

Abstract. This paper analyzes and compares four different mechanisms for providing QoS in IEEE 802.11 wireless LANs. We have evaluated the IEEE 802.11 mode for service differentiation (PCF), Distributed Fair Scheduling, Blackburst, and a scheme proposed by Deng *et al.* using the *ns-2* simulator. The evaluation covers medium utilization, access delay, and the ability to support a large number of high priority mobile stations. Our simulations show that PCF performs badly, and that Blackburst has the best performance with regard to the above metrics. An advantage with the Deng scheme and Distributed Fair Scheduling is that they are less constrained, with regard to the characteristics of high priority traffic, than Blackburst is.

1   Introduction

As usage and deployment of wireless Local Area Networks (WLANs) increases, it is reasonable to expect that the demands to be able to run real-time applications will be the same as on wired networks. Given the relatively low bandwidth in these networks, the introduction of Quality of Service is indispensable.

The IEEE 802.11 standard [6] for WLANs is the most widely used WLAN standard today. It contains a mode for service differentiation, but that has been shown to perform badly and give poor link utilization [8]. We study and evaluate four schemes for providing QoS over IEEE 802.11 wireless LANs, the PCF mode of the IEEE 802.11 standard [6], Distributed Fair Scheduling [7], Blackburst [4], and a scheme proposed by Deng *et al.* [1].

1.1   IEEE 802.11

IEEE 802.11 has two different access methods, the mandatory Distributed Coordinator Function (DCF) and the optional Point Coordinator Function (PCF). The latter aims at supporting real-time traffic.

Distributed Coordinator Function. The Distributed Coordinator Function is the basic access mechanism of IEEE 802.11. It uses a Carrier Sense Multiple Access with Collision Avoidance (CSMA/CA) algorithm to mediate access to the shared medium [6].

Before sending a frame, the medium is sensed, and if it is idle for at least a DCF interframe space (DIFS), the frame is transmitted. Otherwise, a backoff time B (measured in time slots) is chosen randomly in the interval $[0, CW)$, where CW is the Contention Window. Whenever the medium has been idle for at least a DIFS, the backoff timer is

L. Wolf, D. Hutchison, and R. Steinmetz (Eds.): IWQoS 2001, LNCS 2092, pp. 281–287, 2001.

decremented with one each time slot the medium remains idle. When the backoff timer reaches zero, the frame is transmitted. If a collision is detected (which is done by the use of a positive acknowledgment scheme), the contention window is doubled and a new backoff time is chosen. The backoff mechanism is also used after a successful transmission before sending the next frame. After a successful transmission, the contention window is reset to its start value, CW_{min}.

Point Coordinator Function. PCF is a centralized, polling-based access mechanism which requires the presence of a base station that acts as Point Coordinator (PC). If PCF is to be used, time is divided into superframes where each superframe consists of a contention period where DCF is used, and a contention-free period (CFP) where PCF is used. The CFP is started by a beacon frame sent by the base station, using the ordinary DCF access method. Therefore, the CFP may be shortened since the base station has to contend for the medium.

During the CFP, the PC polls each station in its polling list (the high priority stations), when they are clear to access the medium. To ensure that no DCF stations are able to interrupt this mode of operation, the interframe space (IFS) between PCF data frames is shorter than the usual IFS (DIFS). This time is called a PCF interframe space (PIFS). To prevent starvation of stations that are not allowed to send during the CFP, there must always be room for at least one maximum length frame to be sent during the contention period.

1.2 DENG

Deng and Chang proposes a method (which we call the DENG scheme) for service differentiation with minimal modifications of the IEEE 802.11 standard [1]. It uses two properties of IEEE 802.11 to provide differentiation: the interframe space (IFS) used between data frames, and the backoff mechanism. If two stations use different IFS, a station with shorter IFS will get higher priority than a station with a longer IFS. To further extend the number of available classes, different backoff algorithms are used depending on the priority class. Table 1 shows the four defined priority classes [1].

Table 1. DENG Priority Classes. Combining nackoff algorithms and IFS gives priorities 0-3. ρ is a random variable in the interval $(0, 1)$, and i means the ith backoff procedure for this frame.

Priority	IFS	Backoff algorithm
0	DIFS	$B = \frac{2^{2+i}}{2} + \left\lfloor \rho \times \frac{2^{2+i}}{2} \right\rfloor$
1	DIFS	$B = \left\lfloor \rho \times \frac{2^{2+i}}{2} \right\rfloor$
2	PIFS	$B = \frac{2^{2+i}}{2} + \left\lfloor \rho \times \frac{2^{2+i}}{2} \right\rfloor$
3	PIFS	$B = \left\lfloor \rho \times \frac{2^{2+i}}{2} \right\rfloor$

1.3 Distributed Fair Scheduling

In [7] an access scheme called Distributed Fair Scheduling (DFS) which utilizes the ideas behind fair[1] queuing [2] in the wireless domain is presented. It uses the backoff mechanism of IEEE 802.11 to determine which station should send first. Before transmitting a frame, the backoff process is always initiated. The backoff interval calculated is proportional to the size of the packet to send and inversely proportional to the weight of the flow. This causes stations with low weights to generate longer backoff intervals than those with high weights, thus getting lower priority. Fairness is achieved by including the packet size in the calculation of the backoff interval, causing flows with smaller packets to get to send more often. This gives flows with equal weights the same bandwidth regardless of the packet sizes used. If a collision occurs, a new backoff interval is calculated using the backoff algorithm of the IEEE 802.11 standard.

1.4 Blackburst

The main goal of Blackburst [4,5] is to minimize the delay for real-time traffic. Unlike the other schemes it imposes certain requirements on the high priority stations. Blackburst requires: 1) all high priority stations try to access the medium with equal, constant intervals, t_{sch}; and 2) the ability to jam the medium for a period of time.

When a high priority station wants to send a frame, it senses the medium to see if it has been idle for a PIFS and then sends its frame. If the medium is busy, the station waits for the medium to be idle for a PIFS and then enters a black burst contention period. The station now sends a so called black burst to jam the channel. The length of the black burst is determined by the time the station has waited to access the medium, and is calculated as a number of *black slots*. After transmitting the black burst, the station listens to the medium for a short period of time (less than a black slot) to see if some other station is sending a longer black burst which would imply that the other station has waited longer and thus should access the medium first. If the medium is idle, the station will send its frame, otherwise it will wait until the medium becomes idle again and enter another black burst contention period. By using slotted time, and imposing a minimum frame size on real time frames, it can be guaranteed that each black burst contention period will yield a unique winner [4].

After the successful transmission of a frame, the station schedules the next transmission attempt t_{sch} seconds in the future. This has the nice effect that real-time flows will synchronize, and share the medium in a TDM fashion [4]. This means that unless some low priority traffic comes and disturbs the order, very little blackbursting will have to be done once the stations have synchronized.

Low priority stations use the ordinary CSMA/CA access method of IEEE 802.11.

2 Evaluation

2.1 Simulation Setup

To evaluate the above described methods described, we used the network simulator *ns-2* [3] which already has IEEE 802.11 DCF functionality. We extended the simulator with implementations of IEEE 802.11 PCF and the other schemes, and ran the simulation scenarios described below to measure three different metrics: throughput, access delay and maximum number of high priority stations.

---

[1] Fair in the sense that each flow is allocated bandwidth proportional to some *weight*.

Scenarios. Our simulation topology consisted of several wireless stations and one base station (connected to a wired node which serves as a sink for the flows from the wireless domain) in the wireless LAN.

The traffic in our simulations was generated by each station generating constant bit rate flows to the sink. We always used 230 byte frames (including IP and UDP headers), but varied the inter-frame interval between the simulations to vary the offered load, calculated as shown in (1).

$$load = \frac{n_{stations} \cdot \frac{size_{pkt}}{interval_{pkt}}}{c_{bitrate}} \tag{1}$$

Each point in our plots is an average over ten simulation runs, and the error bars indicate the 95% confidence interval. In the delay and throughput comparison simulations, we had 20 wireless stations and varied the fraction of high priority stations. When we investigated the maximum number of high priority stations we used a variable number of stations.

Metrics. The metrics we have used are *throughput*, *access delay*, and *maximum number of high priority stations*. To be able to see the differentiation and medium utilization of the schemes, we have looked at both the average throughput for the stations at each priority level, and the total throughput for all stations together. To compare the graphs from different levels of load, we plot a normalized throughput on the y axis which is calculated as the fraction of the offered data actually delivered to the destination.

To determine to what extent the schemes were able to provide good service to high priority traffic, we ran simulations where the stations sent 65.7 kbit/s streams. We fixed the low priority traffic load at certain levels, and gradually increased the number of high priority stations to see how many simultaneous high priority stations that could get good service. We used two definitions of good service. The first considers throughput, and requires that 95% of the offered data is delivered, while the second requires access delay to be below 20 *ms*.

Method Specific Details. Table 2 shows the parameter values used in our simulations. For further explanation and description of the parameters, we refer to [6, 4, 1, 7].

Table 2. Parameter Values Used in Uur Simulations.

Parameter	Value	Parameter	Value	Parameter	Value	Parameter	Value
DIFS	50 μs	Time slot	20 μs	Deng high prio	3	DFS high weight	0.075
PIFS	30 μs	$c_{bitrate}$	2 Mbit/s	Deng low prio	1	DFS low weight	0.025
Superframe	110 TU^2	$size_{pkt}$	230 bytes	Deng DIFS	100 μs	DFS Scaling_Factor	0.02
Max CFP	108.85 TU	CW_{min}	31	Black slot	20 μs		

When using PCF, during a CFP the Point Coordinator (the base station) polls the stations in its polling list in a round robin fashion. If all stations have been polled once, the CFP will be ended prematurely. If there is not enough time to poll all stations the next station in the list will be polled first in the next CFP. To enhance the performance of DFS when there is much low priority traffic, we decided to use exponential mapping [7] of the backoff intervals.

---

[2] 1 $TU = 1024 \ \mu s$.

2.2   Results

Our initial simulations compared the performance of the different schemes with regard to *throughput*. The simulations show that even at low loads, PCF gives low priority flows significantly lower throughput than the other schemes do. The PCF high priority stations perform acceptable at this low load, but the performance for these starts to deteriorate when the amount of high priority traffic increases. Fig. 1 shows how well the different schemes provides service differentiation with regard to throughput, and Fig. 2 shows the total throughput, which indicates how well the different schemes utilizes the medium. We have run simulations with several levels of load but because of space limitations we only present the most interesting graphs here.

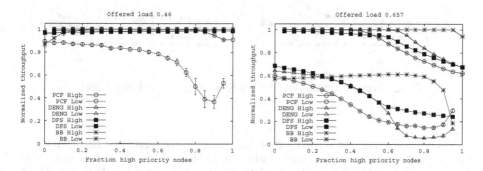

Fig. 1. Average throughput for a Station at the Given Priority Level.

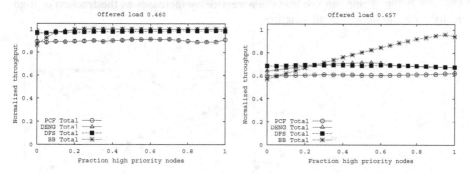

Fig. 2. Total throughput for the QoS Schemes.

As we increased the load in our simulations (see Fig. 1, the right graph), none of the schemes were capable of delivering all data of the high priority stations when there are only high priority stations in the system. Blackburst gives the best performance both for high and low priority traffic. This also implies that Blackburst has the best medium utilization, verified in Fig. 2.

An interesting observation is that the throughput for low priority traffic cases increases slightly for PCF and DENG when there is only one low priority station. Our hypothesis about this is that all high priority stations will send their frames in what appears to the low priority stations as a big "chunk" (not letting any low priority traffic get in between their frames). After that, all high priority stations will start decrementing

their backoff timers, not contending for the medium. During this time, low priority stations can access the medium. When there is only one low priority station, it will get to
send, without contending with some other low priority station. A similar phenomenon
occurs for Blackburst.

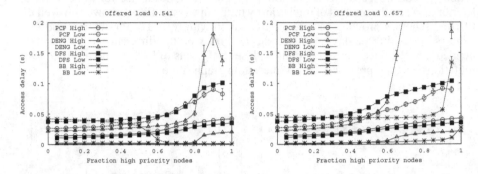

Fig. 3. Average Access Delay for a Station at each Priority Level.

When investigating the second metric, *access delay*, we found that Blackburst and
Deng performs well for high priority traffic. Blackburst also gives low access delay
to low priority traffic as long as the load is relatively low, but it should be noted that
when the network becomes heavily loaded Blackburst totally starves low priority traffic.
As shown in Fig. 3 one can see that the access delay increases as the fraction of high
priority traffic increases for all schemes.

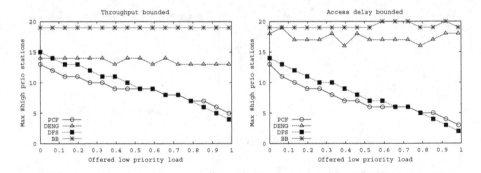

Fig. 4. Maximum Number of High Priority Stations with Good Performance.

The investigation of the third metric, *maximum number of high priority stations*,
indicates that Blackburst is the scheme capable of supporting the largest number of
prioritized stations both with regard to throughput and access delay. As Fig. 4 shows,
both Blackburst and DENG are able to give the high priority stations good service,
regardless of the amount of low priority traffic. The reason why DFS doesn't perform

that well is due to the fact that DFS tries to distribute the bandwidth fairly among the stations according to their weights instead of trying to give perfect service to high priority traffic. For PCF this is both because it has poor medium utilization, and because there must always be room for a low priority frame during the contention period.

In a real life scenario, it is not likely that all traffic is CBR and therefore we also ran simulations with burstier traffic[3] which did not affect the high priority traffic in any significant way. Thus we feel that the results presented here would be valid even with other characteristics of low priority traffic.

3 Conclusions

From our simulations we can conclude that the PCF mode of the IEEE 802.11 standard performs poorly in the metrics studied compared to the other schemes evaluated. Blackburst gives the best performance to high priority traffic both with regard to throughput and access delay. A drawback with Blackburst is the requirements it imposes on the high priority traffic. If these can not be met, DENG might be a suitable alternative since it can serve quite many high priority stations, while giving them very low access delay. An major advantage of DFS is that it will try to achieve fairness, and will not starve low priority traffic, which in many cases is a desirable property of a scheme. Further, our simulations show that Blackburst is the scheme among those studied here that gives the best medium utilization, which is important, given the scarcity of bandwidth in wireless networks.

Finally, we conclude with the observation that there might not be one scheme that is the best to choose in all situations, but the choice of QoS scheme should instead depend on the expectations of the traffic, and other circumstances. Before deciding on what QoS scheme to use in a network, an analysis of what the network should be used for, and what kind of services that is needed should be done.

References

1. D-J. Deng and R-S. Chang. A priority scheme for IEEE 802.11 DCF access method. *IEICE Transactions on Communications*, E82-B(1), January 1999.
2. S. J. Golestani. A self-clocked fair queueing scheme for broadband applications. In *Proceedings of IEEE INFOCOM*, 1994.
3. S. McCanne and S. Floyd. ns network simulator version 2.1b6, August 2000.
4. J. L. Sobrinho and A. S. Krishnakumar. Real-time traffic over the IEEE 802.11 medium access control layer. *Bell Labs Technical Journal*, pages 172–187, Autumn 1996.
5. J. L. Sobrinho and A. S. Krishnakumar. Quality-of-Service in ad hoc carrier sense multiple access networks. *IEEE Journal on Selected Areas in Communications*, 17(8):1353–1368, August 1999.
6. The Institute of Electrical and Electronics Engineers, Inc. *IEEE Std 802.11 - Wireless LAN Medium Access Control (MAC) and Physical Layer (PHY) specifications*, 1999 edition.
7. N. H. Vaidya, P. Bahl, and S. Gupta. Distributed fair scheduling in a wireless LAN. In *Sixth Annual International Conference on Mobile Computing and Networking*, Boston, August 2000.
8. M. A. Visser and M. El Zarki. Voice and data transmission over an 802.11 wireless network. In *Proceedings of PIMRC'95, Toronto, Canada*, pages 648–652, September 1995.

---

[3] ON-OFF sources with ON and OFF periods from a Pareto distribution.

Differentiated Services over Shared Media

Pascal Anelli and Gwendal Le Grand

Laboratoire d'Informatique de Paris 6, University of Paris 6
8 rue du Capitaine Scott, 75015 Paris, France
{Pascal.Anelli,Gwendal.Le-Grand}@lip6.fr

Abstract. The growing use of multimedia communication applications with specific bandwidth and real time delivery requirements has created the need for a new Internet in which traditional best effort datagram delivery can coexist with additional enhanced *Quality of Service* (QoS) transfers. There are many aspects in QoS control. In this article, we address the problem of the support of Expedited Forwarding over shared media. Shared media can be found in broadcast networks operating in packet mode. One problem in this environment is unsteady bandwidth. On these networks, the total bandwidth which is used depends on the offered load. In case of excess load, the total bandwidth decreases when it should be reaching its maximal value. Therefore, it is difficult to manage the bandwidth since it does not remain at the same level. In this article, we propose a distributed algorithm to manage the bandwidth efficiently and which enables QoS for a DiffServ environment.

1   Introduction

The current Internet consists of a multitude of networks built from various link layer technologies which rely on the *Internet Protocol* (IP) to interoperate. IP makes no assumption about the underlying protocol stacks and offers an unreliable, connectionless network layer service which is subject to packet loss and delay, all of which increase with the network load. Because of the lack of guarantees, the IP delivery model is referred to as best-effort. However, some applications may require a better service than the simple best effort service. It is the case for many multimedia applications which may require a fixed bandwidth, a low delay and little jitter. There are various aspects in *Quality of Service* (QoS) management. In this article, we address the problem of bandwidth allocation and guaranteed bandwidth over shared media also known as broadcast networks (e.g. an Ethernet network or a wireless LAN). The push for inclusion of wireless capabilities in laptop computers becomes unstoppable. However, bandwidth in wireless networks is still limited and it is thus necessary to manage it in order to provide a good level of QoS to the users. The work which is presented in this paper can be applied to any shared medium, but it is rather clear that wired environments do not have the same requirements since bandwidth is not limited in the same way as it is on a wireless link.

The article is organized as follows : we present related work in section 2. Then, section 3 describes and evaluates a new scheme for bandwidth management for

L. Wolf, D. Hutchison, and R. Steinmetz (Eds.) : IWQoS 2001, LNCS 2092, pp. 288-293, 2001.

Differentiated Services over shared media. Finally, we conclude and describe future work in section 4.

2   Related Work

2.1   QoS on Shared Media (Intserv)

Asynchronous shared media (like Ethernet or IEEE 802.11 for example) do not guarantee any type of quality of service. When the network load increases, the total bandwidth decreases when it should be reaching its maximal value. Moreover, it is impossible for a specific flow to have a fixed throughput since Ethernet guarantees some kind of fairness.

In the Intserv context, many works have been carried out to handle these limitations. Yavatkar, Hoffman and Bernet [9] propose a centralized architecture for subnet bandwidth management using a centralized algorithm. Their approach uses a dedicated manager per LAN and depends highly on *Resource Reservation Protocol* (RSVP) [3]. Moreover, [9] does not deal explicitly with best effort traffic related issues.

The *Controlled Load Ethernet Protocol* (CLEP) described in [2] is an implementation of the Controlled Load service over Ethernet defined by Wroclawski in [8]. It provides the client data flow with a quality of service approximating the quality of service this flow would receive on an unloaded network. This service is obtained by incorporating an access controller on the outgoing interfaces of the nodes. This allows to control the load of the broadcast network. As in *Medium Access Protocol* (MAC) layer, CLEP's bandwidth management is distributed. The mains principles used are: (1) flow control of all the streams, (2) protocol to exchange states between access controllers. Access controller is built around token bucket filters. Specific packets can be provided with a guaranteed quality of service. These packets are organized in different privileged flows. Traffic without guaranteed QoS is handled as best effort traffic. All the flows (best effort and privileged flows) use the controlled-load service to control packets admission in the Ethernet network. Packets are admitted only if there is enough bandwidth for them. CLEP provides the shared medium with the following properties:
− a steady bandwidth in overload condition,
− a guaranteed bandwidth for streams which have a reservation,
− a fair share of bandwidth for best effort streams (requires no QoS),
− an isolation of the streams which have QoS requirements
However, this solution requires some QoS signalling before transmitting any data. In some cases, bandwidth usage can be low.

2.2   DiffServ

Recently, DiffServ was proposed by the Internet community to support various services [1]. The key aspects of differentiated services concern scaling [7]:

- traffic streams are reduced to a small number of traffic aggregations. Each aggregation is identified by a single *Per-Hop Behaviour* (PHB) on the routers,
- signalling and all the inherent costs are eliminated.

Differentiated services paradigm is made by an architecture which separates clearly forwarding from control. Control is executed at the edges of the DiffServ network. Control actions can be policing, shaping, marking and depend of the *Traffic Conditioning Agreement* (TCA). Apart from *Default* (DE) which handles the traffic in a best effort manner, two forwarding behaviours are defined : *Expedited Forwarding* (EF) [4] and *Assured Forwarding* (AF) [5]. The first PHB is dedicated to support a service with a strong QoS requirement about delay. The second PHB allows to use a service for which the average throughput is "guaranteed". Routers of a differentiated services network handle IP datagrams in different traffic streams and forward them using different PHBs. The PHB to be applied is based on mechanisms which process either drop or temporal priorities. These mechanisms are located on the output interfaces.

The DiffServ architecture relies on a centralized processing for each PHB. This means the total amount of bandwidth is dedicated to a single output interface. This is the case of point to point links in switched networks. But in broadcast networks, the link is multipoint i.e. access to the link is distributed. For these types of network, it raises some difficulties to apply coherent PHBs:

- the bandwidth is shared between the output interfaces of all the nodes connected to the link. Moreover, access is fair like in Ethernet networks. It is hard to assign any level of bandwidth to a particular source.
- the bandwidth is unsteady; it depends on the offered load. This is true for Ethernet networks where traffic in overload condition is inversely proportional to the offered load (cf. Figure 2)
- Distributed access prevents from scheduling all the packets like it can be done in single access.

Finally, a source on a broadcast network cannot have any QoS guarantee for any of its streams. As QoS is an end to end concept, the QoS provided to the destination node is that of the network the least efficient on the path from the source to the destination. In corporate environments, broadcast networks are mainly used in the access network to the internet. As for switched networks, it is important for this type of network to support the DiffServ architecture. This paper presents a solution to activate the deployment of DiffServ over broadcast networks. It describes a system to manage bandwidth in order to support different PHBs.

3   Bandwidth Management for DiffServ over Shared Media

3.1   Principle

In the following, we propose a system to control bandwidth in order to support EF on a shared LAN. This system called DS-CLEP (Differentiated Services CLEP) is derived from works on CLEP but it changes by the stream management and the lack of dynamic reservation. The mains objectives are:

- manage QoS streams by aggregation at the network level,
- no dynamic signalling to avoid to manage it by applications (e.g. RSVP),
- get statistical gain in keeping the isolation between streams.

Using this scheme, a network which only has a best effort service can be integrated in a DS domain with two PHBs: DE and EF. EF means that traffic is limited. Excess traffic is dropped and no drop priorities are used. Statistical gain can be accomplished at two levels:

- locally, on each node. The extra bandwidth allocated to EF streams is used to send best effort traffic. However, the gain depends on the best effort load of the node. If it is small, the gain is small. In this article, we propose a solution which provides the nodes with a local statistical gain.
- globally, for a LAN. The extra EF bandwidth is used the best effort traffic of all the nodes. This may however have a negative impact on the EF traffic since the extra bandwidth may need a long time to be recovered when the stream needs it.

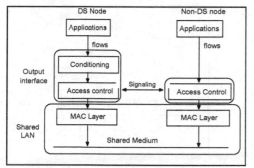

Fig. 1. Functional Elements of a DiffServ Network on a Shared Medium.

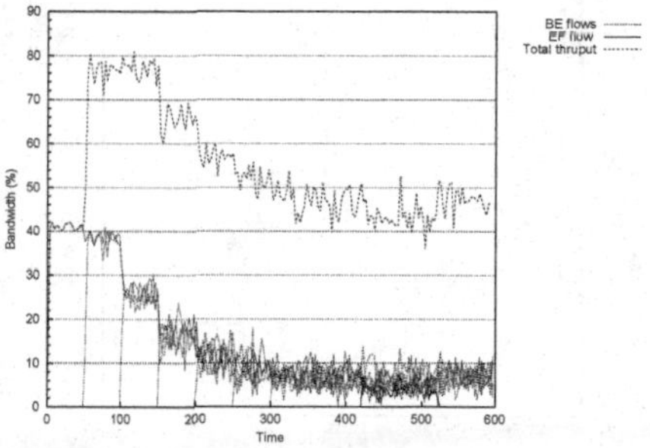

Fig. 2. Classical Behaviour of an Ethernet Under Heavy Load.

As shown on Figure 1, we add an access control module and a conditioning module (for DiffServ nodes only) to the traditional architecture of a node (i.e. in an architecture which does not support QoS). The conditioning module aims at marking and limiting the EF traffic as specified within a TCA. A TCA is set at the nodes by the network administrator. Without conditioning, the node cannot send any EF traffic. Actions between access controllers are synchronized with a signalling protocol which exchanges states for internal purposes and are not seen by the upper layers (e.g. IP).

3.2   Evaluation

The comparison between the different solutions are made by simulation, using NS-2 [6]. All the evaluations involve the same topology and the same scenario. The topology comprises 8 nodes out of which 7 are traffic sources and one is a traffic sink for all the flows. One of the traffic sources has two flows (a DE and an EF flow) whereas all the other sources only have a DE flow.

Each traffic source produces a DE flow at a constant bit rate of 410 kbit/s with a packet size of 512 bytes. The starting time for these flows is laid with a step of 50s.

Node 1 also benefits from a high level of QoS. A high priority flow starts at t= 420s and lasts 100s. The rate of this flow is set to 200kbit/s and the packet size is 512 bytes, as for the other best effort flows.

The network model is studied under heavy load condition. Thus, we set the link bandwidth to 1 Mbit/s. This value is very low compared to the actual bandwidth usage in the networks. The motivation is to demonstrate the algorithm behaviour and a model for a high speed network changes nothing to the algorithm. In truth, the latter requires more simulation time to process the huge quantity of events produced.

The maximum flow capacity of the network is around 75% of the link bandwidth. The difference is consumed by the MAC layer like collisions resolution, interframe gap, etc. This value has been kept to indicate the available bandwidth with our management system. This scenario is played on 3 different simulations models. We measure the throughput received by the destination. Throughput is expressed as a percentage of link bandwidth.

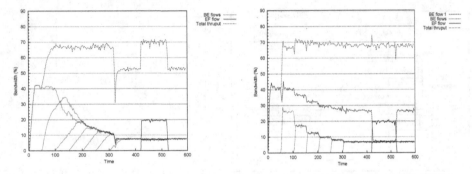

Fig. 3. Bandwidth Share with CLEP.   **Fig. 4.** Bandwidth Share with DS-CLEP.

The first model involves a classical Ethernet without any bandwidth management. Figure 2 shows the well-known result. In this case, no flow can have any guarantee of throughput. We can see strong variations and an overall throughput decreasing as load increases. The total used bandwidth depends on the offered load. In case of excess load, the total bandwidth decreases when it should be reaching its maximal value. In the second model, Ethernet is extended with the bandwidth management system CLEP presented in [2]. In Figure 3, the overall load increases as a more sources start using the link, but the total bandwidth usage is steady regardless of the load. However, at t=320s the total bandwidth usage decreases because an explicit reservation at 200 kbit/s is set. Until the flow starts at t= 420s, the bandwidth usage is smaller (200 kbit/s less) until the flow with required QoS starts. In a sense, the system is not work conservating, the link can be in the idle state when there are packets awaiting transmission. The second thing to see is that DE flows converge to the fair share. In the last model, we use our proposal of bandwidth management system to enable DiffServ on shared media. Recall, this system wants to be work conservating by searching to assign locally unused bandwidth of EF flows. In the model, node 1 contains a source in DE (noted flow 1 in Figure 4) and a TCA for an EF flow at 200kbit/s. In Figure 4, while the EF flow has not yet started, flow 1 gets its fair share and all the bandwidth unused by the EF flow. The total bandwidth stays nearly the same. At t=320s, the EF flow starts and flow 1 on the same node releases some bandwidth for the EF flow. When the EF flow stops, bandwidth is retrieved by the flow DE on the same node. In this system, a throughput guarantee is given to particular flows without managing any signalling overhead. Unused bandwidth reserved for a PHB is collected by DE flows on the same node.

The latter solution allows the network administrator to distribute TCAs for EF traffic between different nodes of a broadcast network without decreasing DE traffic. A source can transmit a QoS flow at any time without generating any signalling. This behaviour is highly desirable in order to support a DiffServ environment.

4   Conclusion

In this proposal, we have shown a control of bandwidth of shared media can be done in a distributed manner in order to activate a deployment of DiffServ paradigm on this type of network. However, our study is made with throughput parameters. Delay and packet losses are the other important parameters which characterize QoS. This study must be extended with the analysis of these parameters.

Moreover, the algorithm can be improved by a global recovery (involving all nodes) of unused EF bandwidth. Our solution must yet be extended with the support of AF PHB i.e. permit to transmit out of profile traffic when the network is in low load condition. But the first step presented here is cheerful for the future. Although this work applies to wired networks, the scope is mainly concerning wireless networks with more limited bandwidth, which seem to have a brilliant future.

References

1.   S. Blake, D. Black and M. Carlson, RFC 2475, An Architecture for Differentiated Services, December 1998.
2.   Bouyer, E. Horlait, Bandwidth Management and Reservation over Shared Media, SFBSID'97, Fortaleza, Brasil, November 1997
3.   Braden, L. Zhang, S. Berson, S. Herzog, S. Jamin, Resource Reservation Protocol (RSVP) – Verison 1 Functional Specification, RFC 2205, September 1997
4.   Jacobson, K. Nichols and K. Poduri, RFC 2598, An Expedited Forwarding PHB, June 1999
5.   Heinanen, F. Baker, W. Weiss and J. Wroclawski, RFC 2597, Assured Forwarding PHB Group, June 1999.
6.   /, The Network Simulator - ns-2.
7.   W. Weiss, Bell Labs Technical Journal, Vol. 3, N, 4,   , QoS with Differentiated Services, October 1998.
8.   J. Wroclawski, Specification of the Controlled-Load Network Element Service, RFC 2211, September 1997.
9.   R. Yavatkar, D.Hoffman, Y. Bernet, F. Baker, M. Speer, RFC 2814, SBM (Subnet Bandwidth Manager): A Protocol for RSVP-based Admission Control over IEEE 802-Style Networks, May 2000.

End-to-Edge QoS System Integration: Integrated Resource Reservation Framework for Mobile Internet

Yasunori Yasuda[1], Nobuhiko Nishio[2], and Hideyuki Tokuda[3]

[1] NTT Information Sharing Platform Labs.
1-1 Hikarinooka Yokosuka-Shi Kanagawa 239-0847 Japan
yasunori@isl.ntt.co.jp
[2] Graduate School of Media and Governance, Keio University
5322 Endo, Fujisawa Kanagawa 252-8520 Japan
vino@sfc.keio.ac.jp
[3] Faculty of Environmental Information, Keio University
5322 Endo, Fujisawa Kanagawa 252-8520 Japan
hxt@ht.sfc.keio.ac.jp

Abstract. In order to support QoS-aware applications in a mobile Internet environment, it is essential to achieve effective end-to-end QoS control so that these applications can dynamically adapt to various changes in the network resources or environment. In this paper, we describe the design and implementation of our integrated "End-to-Edge QoS framework" which consists of mechanism for resource reservation, the QoS translation, and the QoS arbitration.

1 Introduction

Recently, lots of efforts have gone into ensuring quality of multimedia communication in a mobile Internet environment. Such effort includes the introduction of RSVP, RTP and RTCP that are new protocols designed to facilitate real-time communication and the investigations conducted by dedicate working groups such as intserv and diffserv. However, most research entities have overlooked an important element: end-host. In order to realize guaranteed End-to-End communication, especially in the mobile internet, first/last-one-hop are the most crucial segments that affect the network performance. Therefore, one of our core objectives is to scrutinize the inner workings of the end-hosts in a QoS system.

Lately, computing resources and appliances have proliferated and network appliances have gotten a wide variety in its structure and performance, such as wireless and broadband, which presents us various characteristics in their networking performance. In such incredibly increasing digital computing environment, users tend to form a relatively small computer network rather than merely putting a stand-alone PC on the desktop. Such type of small networks appear both in offices and homes, some of which are capable of offering remote access functionality connecting among distributed areas. Because each of such networks is a unit of administration, it is important to manage it in terms of an

L. Wolf, D. Hutchison, and R. Steinmetz (Eds.): IWQoS 2001, LNCS 2092, pp. 294–299, 2001.

appropriate policy in order to offer a comfortable computing and communication environment for common users.

In this paper, the term "edge router" means a router which functions as a gateway to the outer ISP (Internet Service Providers) at the frontier of such small-size networks. The "End-to-End" path is divided into three segments: two "End-to-Edge" segments and a "Edge-to-Edge" segments. Since lots of research efforts have gone into the field of "Edge-to-Edge" communication, our research interest lies in the edge router "End-to-Edge". This paper describes how we apply resource reservation and coordination mechanism for the End-to-Edge segments. In the following, we give an explanation of our integrated resource-centric communication mechanism named "End-to-Edge QoS framework". This system consists of three parts: '*i*Reserve' (*i*ntegrated resource Reservation) which resides in the end-host and gives rigid and integrated resource reservation function, 'S-MAX' (System for Mobility, Adaptability and eXtensibility) controls network traffic on the end-host in response to the characteristics of current available NIC (Network Interface Card) and realizes dynamic exchange of NIC devices, and 'QSTAR' (QoS Specification, Translation, Arbitration and Registration Framework) which resides in the edge router, coordinates multiple network resource requirements submitted from applications on the end-hosts.

In the next section, we give a general overview and structure of the End-to-Edge QoS framework. Following that, we report the experimental result of the End-to-Edge system. In the last section, we discuss a number of directions for the future research and conclude this paper.

2 End-to-Edge QoS Framework

This section gives a design overview of our End-to-Edge QoS framework.

2.1 Overview

In most mobile network environments, the links between mobile hosts and edge routers are the bottleneck. Therefore, this is the determining factor of maximum end-to-end QoS that can be achieved. Our main contribution is to control QoS between mobile hosts and an edge router in order to achieve end-to-end QoS efficiently.

A typical target environment of our framework is shown in Figure 1. Suppose some applications on both MH on home network and CH on foreign network are communicating with each other and MH will dynamically move between home network and foreign network using Mobile IP, End-to-Edge QoS framework must control CPU capacity and bandwidth of every application on same mobile host, and control bandwidth of traffic between mobile hosts and an edge router.

Our integrated resource reservation framework provides CPU capacity and network bandwidth reservation mechanism, QoS translation mechanism and QoS arbitration mechanism to the adaptive applications on mobile hosts that make use of IETF Mobile IP[4]. This framework combines three frameworks '*i*Reserve'[1], 'S-MAX'[2] and 'QSTAR' into one.

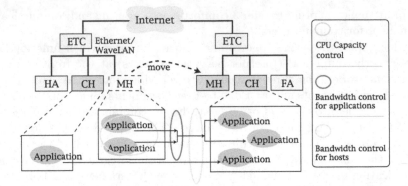

Fig. 1. The target environment of our integrated resource reservation framework. MH stands for Mobile Host, HA and FA stand for Home Agent and Foreign Agent of IETF Mobile IP, CH stands for Correspondent Host, and ETC stands for Edge Router.

*i*Reserve. The '*i*Reserve' framework provides integrated CPU capacity profiling and reservation mechanism. Currently we have developed processor and memory resource reservation mechanisms [5] into Real-Time Mach microkernel, then QoS ticket model and Q-Thread library on top of the microkernel extend the programming environment[6]. A middleware for continuous media processing which can manage the three-layer QoS representation and *static* translation is presented [1]. The *i*Reserve architecture is introduced for an integrity and re-organization of our resource management middleware. One of the most significant features is to be capable of profiling and calibrating actual resource usage (such as CPU cycles) requirement which are platform-dependent.

S-MAX. The 'S-MAX' framework provides packet scheduling mechanism that can cope with IETF Mobile IP and dynamic network interface switching for adaptive applications. The key insight of this framework is using notification of changes to trigger adaptation in applications and resource enforcing mechanism for bandwidth control. *S-MAX* is a unique framework since it has both the resource enforcement mechanism and the change notification mechanism in a mobile multimedia environment. There has been a significant number of proposals for QoS framework that support adaptive applications[7][8]. Due to the lack of resource enforcement mechanism with these frameworks, applications must control resource usage completely on their own; therefore application programmers have to write more complicated codes than in *S-MAX*.

QSTAR. The *QSTAR* framework, which is based on the extensible object model for QoS specification in adaptive QoS system[3], provides QoS Translation and Arbitration mechanism. The extensible object model herein lets users and application programs specify list of their QoS preference, each of which specifies an objective QoS level against a subjective utility value. With the *QSTAR* framework, the user can specify not only the desired QoS, but also a QoS range to minimize the quality degradation resulting from resource shortage. QoS translation between the layers of a system for a set of QoS parameters is discussed in [9] [10], but mapping of the user's preferences are not considered. To minimize the effect of resource shortage, we must realize controlled fallback by taking into account the user's preference.

Therefore, our End-to-Edge QoS framework not only provides sufficient functionalities to support adaptive applications on the mobile hosts, but also offers an easy way to implement adaptive applications.

2.2 Structure of the End-to-Edge QoS framework

The structure of the End-to-Edge QoS framework is shown in Figure 2.

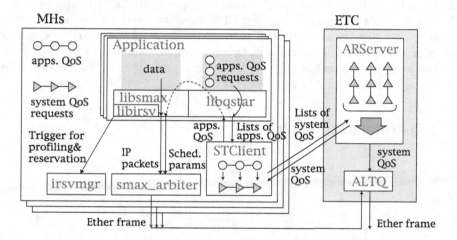

Fig. 2. Implementation of End-To-Edge QoS Framework on RT-Mach.

This End-to-Edge QoS framework consists of several components of each framework, namely irsvmgr and libirsv from *i*Reserve, smax_arbier and libsmax from *S-MAX*, libqstar, STClient, ARServer from *QSTAR*.

Our *i*Reserve implementation consists of server module (irsvmr) and library module(libirsv). The server mainly works as resource coordination and the library module takes care of QoS profiling.

The *S-MAX* has three major components: the smax_arbiter, the smax_notifier and the libsmax. The smax_arbiter is a packet scheduler in order to control bandwidth and delay of application traffic, and the smax_notifier handles NIC changes or Mobile IP state changes and notifies to the smax_arbiter to adapt available network resources. The libsmax is a library, which provides APIs for using smax_arbiter.

The *QSTAR* consists of two facilities, namely the 'STClient' to translate between application QoS parameters and system QoS parameters, and the 'AR-Server' to choose appropriate system QoS parameters for each applications among requested list of system QoS parameters.

At first, an application on a mobile host sends lists of application QoS parameters to the STClient using the libqstar. The STClient translates the lists of application QoS parameters to the lists of system QoS parameters(i.e. packet scheduling parameters), and sends them to the ARServer on an ETC(edge router). The ARServer arbitrates the QoS requests from the STClient, sets the arbitrated system QoS parameters to the ALTQ[11], and returns the arbitrated system QoS parameters to the STClient. When the STClient receives arbitrated system QoS parameters, STClient is translates the system QoS parameter to the application QoS parameter and then returns it to the application. Once application receives arbitrated QoS parameter, application can settle the packet scheduling parameter to the smax_arbiter using API the libsmax. The API for setting a packet scheduling parameter invokes the API of the irsvmg for trigger of CPU usage profiling. The irsvmgr starts profiling the application's CPU usage and then reserves profiled CPU capacity for the application. After finishing all processes as mentioned above, application can send IP packets to the smax_arbiter. Data stream sent by every application on same mobile host is controlled by smax_arbiter, and data stream sent by every hosts on same network is controlled by ALTQ on the ETC.

3   Experimental Result

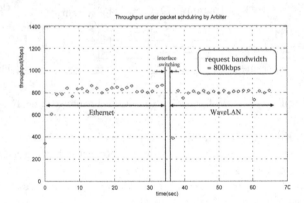

Fig. 3. UDP Throughput in Adaptation.

We evaluated throughput for adaptation handling. Figure 3 shows throughputs of the adaptive application when network interface changes from 10Mbps ethernet to 2Mbps WaveLAN. The test adaptive application requests to allocate 800 kbps bandwidth, whenever NIC changes between ethernet and WaveLAN. We can see these throughputs are almost 800kbps both on Ethernet and WaveLAN using our End-to-edge QoS Framework. The difference between requested and measured throughputs on both Ethernet and WaveLAN is in the ±8.5% at the worst.

4   Conclusion

This paper proposes an End-to-Edge QoS framework, which consists of three main modules, 'iReserve,' 'S-MAX' and 'QSTAR'. End-to-Edge part of the network includes last/first-one-hop segment which can affect the overall performance of the network communication. In addition, resource management and policy enforcement are relatively easy to control in this segment. We take account of three crucial issues on mobile multimedia communication: (i)QoS profiling and resource reservation in end-host, (ii)QoS adaptation for packet scheduling in end-host and (iii)QoS translation and QoS arbitration in edge router. By introducing a QoS translation mechanism, we allow applications or user-level QoS to be usable for applications so we can realize a rigid resource arbitration enforcement in the system-level. In the future, we plan to integrate our End-to-Edge QoS system to other "Edge-to-Edge" QoS systems and evaluate the overall system performance.

Acknowledgments

We express our appreciation of the implementing QSTAR, assistance and advice provided by Yasunori Matsui who was with NTT labs.

References

1. N. Nishio and H. Tokuda , Simplified Method for Session Coordination Using Multi-Level QOS Specification and Translation, Proceedings of 5th IWQoS, 1997.
2. Yasunori Yasuda and Hiroshi Inamura, S-MAX: A Framework for Adaptive Applications using Change Notification and Resource Enforcement., Proceedings of MoMuc'98, 1998, Oct.
3. Yasunori Matsui et al. , An Extensible Object Model for QoS Specification in Adaptive QoS Systems, Proceedings of ISORC'99, 1999, May.
4. C. Perkins ,IP Mobility Support, RFC 2002, 1996.
5. C.W. Mercer and S. Savage. and H. Tokuda , Processor Capacity Reserves: An Abstraction for Managing Processor Usage, IEEE Multimedia94, 1994.
6. K. Kawachiya et al., A New Execution Model for Dynamic QOS Control of Continuous-Media Processing, Proceedings of NOSSDAV'96, 1996.
7. Jon Inouye, Shanwei Cen, Calton Pu and Jonathan Walpole , System Support for Mobile Multimedia Applications, Proceedings of NOSSDAV'97, 1997.
8. Brian D. Noble, M. Satyanarayanan, et al., Agile Application-Aware Adaptation for Mobility, Proceedings of SOSP-16, 1997.
9. Gopalakrishna, R., Parulkar, G.M., Efficient Quality of Service Support in Multimedia Computer Operating Systems, Washington University Technical Report WUCS-TM-94-04, 1994.
10. Nahrstedt, K., Smith, J., The QoS Broker, IEEE Multimedia, 1995, Spring.
11. Kenjiro Cho, Alternate Queueing (ALTQ) for FreeBSD including CBQ, RED, WFQ, http://www.csl.sony.co.jp/person/kjc/programs.html, 1997.

Problems of Elastic Traffic Admission Control in an HTTP Scenario

Joachim Charzinski

Siemens Information and Communication Networks
Hofmannstr. 51, D-81359 Munich, Germany
j.charzinski@ieee.org

Abstract. Admission control for elastic traffic has been advocated in order to maintain performance (i.e. ensure a minimum bandwidth) for each admitted flow and to avoid unnecessary traffic in the network due to retransmissions of packets or even whole transfers after a temporary overload situation. This paper aims at indicating problems of admission control for elastic traffic on a per-TCP-connection basis is problematic in the context of Web traffic: (i) A TCP connection is not equivalent to a transfer. (ii) It is the variance in connection volumes rather than the connection arrival rate that causes most overload situations. (iii) From an application point of view, the target of maintaining performance for admitted flows under high offered load is not met.

Keywords: Admission Control; HTTP; Elastic Traffic; Application Level.

1 Introduction

A major part of the traffic transported in today's Internet is elastic traffic [1], using the rate sharing principle designed into the Internet's Transmission Control Protocol. Under ideal circumstances, this rate sharing can be modeled by Processor Sharing models to describe the amount of time it takes to transfer a given amount of data [2, 3]. The relative offered load $\rho = \lambda\theta/C$ to a link is a function of the arrival rate λ of new transfers, the mean transfer size θ and the capacity C of the link under consideration. In order to maintain a stable operation of the network and to avoid unnecessary retransmissions at packet or file level, ρ must be less than 1 on each link – a limit that has far less restrictions than the applicability of the processor sharing model itself. In order to ensure this condition in a real network, admission control has been proposed for elastic traffic [4, 5, 6, 7, 8] on a per-TCP-connection basis. This can be done without changing end-system protocol stacks by either intercepting (dropping) TCP connection set-up (SYN) packets in the network [6, 7] or by sending artificial TCP connection reset (RST) packets to the end systems [9] where the latter approach has the disadvantage of potentially faster application level retries.

As the major part of the elastic traffic transported in the Internet currently is Web traffic [1], an admission control method for elastic traffic should be able to achieve its goal for this application, not only at the level of IP packets, but also at an application level. This paper discusses three issues arising if a simple per-connection admission control is employed for Web traffic: (i) The relevant unit of transfers is not necessarily a TCP connection, (ii) temporarily increased offered load is caused by the variance in

L. Wolf, D. Hutchison, and R. Steinmetz (Eds.): IWQoS 2001, LNCS 2092, pp. 300–304, 2001.
© Springer-Verlag Berlin Heidelberg 2001

transfer sizes in addition to arrival rates and (iii) a simple per-connection admission control fails to meet its goals of maintaining quality of service for admitted traffic and avoiding unnecessary transfers if regarded from a user perspective.

2   The Unit of Transfer

A Web page typically consists of multiple elements that need to be downloaded when a page is requested. These elements are loaded using separate HTTP GET requests, serialized in one or multiple parallel TCP connections to the corresponding server(s). Many elements are small enough to be transmitted in just one IP packet [10]. In addition, there can be a significant delay between multiple transfers in one connection [11]. Both facts prevent the TCP flow control from acting as idealized in Processor Sharing models. Fig. 1 confirms this by showing that the second and following items transfered in a TCP connection are received with significantly higher data rates than the first item, which is due to an increased average TCP window size.

The mean transfer rates in Fig. 1 represent the downstream bit rates, neglecting the fact that after the considered transfer, there will be a phase of no downstream traffic until the last acknowledgment is received by the sender. This rate is equal to the link's line rate if only one packet is transmitted – an effect that causes the reduction in average rates with increasing item size for small item sizes.

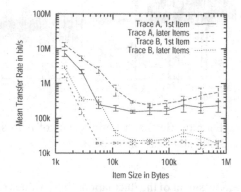

Fig. 1. Mean download speed of single items as a function of their size with 95 % confidence intervals. Parameter: position in HTTP/TCP connection. Traces are described in [10].

An admission control algorithm accepting a TCP connection for a link will not be aware of the structure within that connection. Assuming that an open TCP connection will always take its fair share is too pessimistic as there is a significant number of idle connections [12] and at the same time too optimistic as a server can start a new transfer in an open connection with a large window size even during a congestion period if the same TCP connection has been previously used for transfers. On the other hand, excluding idle TCP connections from the set of flows bandwidth is allocated to will cause problems as those flows can start transmitting data without being admitted by access control.

3 Causes for Overload

Classical approaches for stream admission control in multiservice networks assume that the fluctuation of offered load is basically due to a fluctuation of arrival rates. Due to the heavy-tailed distribution of requested item sizes [13] and consequently also of traffic volumes transported in HTTP/TCP connections, this is not the case with HTTP traffic. Here the fluctuations in requested item sizes have a larger share in the short-term workloads described by $\lambda\theta$ than the fluctuations in arrival rates. This can be seen in Fig. 2 where for the busiest hours (21:00–23:00) the number of HTTP/TCP connection set-ups and the mean of the downstream volumes to be transfered in those connections have been plotted for each one minute interval over the five weeks of measurement of Trace B from [10], in order to give short-term estimates for the two load components λ and θ. Only connections that transfer at least one item were considered.

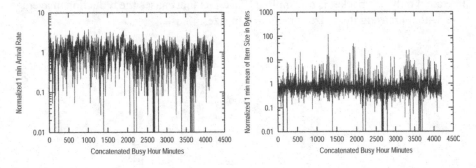

Fig. 2. Trace of 1 min GET rates (left) and 1 min means of item sizes in concatenated 1min intervals between 21:00 and 23:00 in Trace B.

In order to ease the assessment of the fluctuations, both values have been normalized to their means during those hours. The means were $\bar{\lambda}_{1min} = 36.7 \pm 0.7$ per minute and $\bar{\theta}_{1min} = 11.2 \pm 0.9$ kByte with 2σ intervals for 95 % confidence, i.e. the one minute estimates of the arrival rate varied between zero and around 145 TCP connection set-ups per minute whereas the one minute estimates of the mean requested file size varied between zero and more than 1 MB. The same behavior was also observed by evaluating traces from a higher loaded link serving around 1000 HTTP/TCP connection set-ups per minute [14].

The duration of averaging intervals of 1 minute has been chosen to roughly reflect the time scale of admission control. A comparison of the left and right plots in Fig. 2 shows that the variation in arrival rates is less than the variation in mean connection volumes. The same situation can be found if the 15 min mean values or per-GET-request numbers are considered (data not shown). An investigation of the corresponding distributions confirms that the distribution of *1min mean* connection volumes is heavy-tailed whereas the distribution of the number of arrivals per minute is not (although it is more variant than the Poisson distribution suggested by [15] and used in most Processor Sharing analyses).

Note that of course the worst overload situations will occur when a peak in arrival rate coincides with a peak in mean requested file sizes. However, performing admission control on a per-TCP connection arrival basis will only reduce the peak in connection arrivals to the network and not the second and more variant load component, the mean volume transfered in a connection. Whereas this may be optimal from a goodput (*successful* transmissions) point of view, it can constitute a problem in terms of Grade of Service as determined by the blocking probability. The overall number of blocked requests might decrease if connections with larger θ were blocked with a higher probability than short-lived connections [16].

4   Application View

Admission control is usually employed to serve two purposes: (i) to make sure the performance of already admitted flows does not suffer from a new flow, i.e. to guarantee a minimum throughput and (ii) to maintain a high goodput in the network.

From a Web user's point of view, the first purpose translates into *"Either a click should be rejected by admission control or the page should load completely at a minimum transfer rate"*. Admission control solutions that simply drop TCP connection set-up packets under high load have the effect of causing a repeated set-up attempt after the default TCP timeout of 3 seconds[1]. Correspondingly, loading the page will just take longer but the loading process will not be blocked. A user will not recognize this as "maintained performance" but rather as a performance degradation comparable to not performing admission control at all.

In transaction oriented scenarios like home banking, electronic commerce or electronic business applications, the situation is even worse: Even if user activities were admitted on a per-click basis instead of a per-TCP-connection basis, blocking a part of a longer transaction due to overload is not what a user will consider as acceptable performance. From the user's point of view, admission control in this case should allow or block a whole transaction.

5   Conclusions

If admission control is employed for elastic traffic, it should not only ensure network goodput but also be compatible with today's most important elastic application, i.e., Web access and its use in e-business scenarios. The simple solution of performing admission control on a per-TCP-connection basis is difficult for several reasons presented above. Other ideas more suited to the Web should be investigated.

---

[1] In practice, this value varies between 0.7 and 6 s for the initial SYN packet and between 0.2 and 1.4 s for data packets [7].

One option to maintain reactiveness to short flows is to preempt longer transfers. This could be done by relying on recovery mechanisms to resume a download at a checkpoint [17, Sec. 14.36]. The question is, however, if short flows should be admitted into a highly loaded network. If the congestion is caused by long flows (high θ) and buffers are long enough, admitting additional short flows reduces blocking. On the other hand, short flows effectively do not participate in the TCP flow control, so if the congestion is mainly caused by a high connection arrival rate (high λ), short flows should not be admitted when a link is overloaded.

From a user's point of view, the "ideal" admission control would work on a per-click basis (browsing) or on a per-transaction basis (e-commerce), which is both very hard to implement due to the distributed nature of the Internet as not all target hosts and routes needed to load all elements are known when a user requests a new page. In addition, the rate requirements of transactions depend on user reaction times, so that any admission control algorithm is forced to either waste bandwidth or to assign very low bandwidth to the transaction.

References

[1] McCreary, S., Claffy, K.: Trends in Wide Area IP Traffic Patterns: A View from the Ames Internet eXchange. In *Proc. ITC Spec. Sem. on IP Traffic*. Monterey, CA, USA (2000)
[2] Kleinrock, L.: *Queueing Systems Vol. 2*. Wiley, New York, NY, USA (1976)
[3] Heyman, D., Lakshman, T., Neidhardt, A.: A New Method for Analysing Feedback-Based Protocols with Applications to Engineering Web Traffic over the Internet. *ACM Perf. Eval. Review* **25** (1997) 24–38
[4] Roberts, J., Massoulié, L.: Bandwidth sharing and admission control for elastic traffic. In *Proc. ITC Specialist Seminar*. Yokohama, Japan (1998)
[5] Massoulié, L., Roberts, J.: Arguments in favour of admission control for TCP flows. In *Proc. ITC 16*. Edinburgh, UK (1999)
[6] Roberts, J., Oueslati-Boulahia, S.: Quality of Service by Flow Aware Networking. *Phil. Trans. Royal Soc. of London, Series A* **358** (2000)
[7] Mortier, R., Pratt, I., Clark, C., Crosby, S.: Implicit Admission Control. *IEEE JSAC* **18** (2000) 2629–2639
[8] Roberts, J.: Traffic Theory and the Internet. *IEEE Communications Mag.* **39** (2001) 94–99
[9] Kumar, A., Hegde, M., Anand, S. V. R., Bindu, B. N., Thirumurthy, D., Kherani, A. A.: Nonintrusive TCP Connection Admission Control for Bandwidth Management of an Internet Access Link. *IEEE Comm. Mag.* **38** (2000)(5) 160–167
[10] Charzinski, J.: HTTP/TCP Connection and Flow Characteristics. *Perf. Eval.* **42** (2000) 149–162
[11] Charzinski, J.: Web Performance in Practice – Why We are Waiting. *AEÜ Intl. J. Elec. Comm.* **55** (2001) 37–45
[12] Charzinski, J.: Measured HTTP Performance and Fun Factors. submitted (2001)
[13] Crovella, M., Bestavros, A.: Self-Similarity in World Wide Web Traffic: Evidence and Possible Causes. *IEEE/ACM Trans. Networking* **5** (1997) 835–846
[14] WAND Research Group: Auckland-II Trace.
http://wand.cs.waikato.ac.nz/wand/wits/auck/2/
[15] Nabe, M., Murata, M., Miyahara, H.: Analysis and modeling of World Wide Web traffic for capacity dimensioning of Internet access lines. *Perf. Eval.* **34** (1999) 249–271
[16] Lindberger, K.: Dimensioning and Design Methods for Integrated ATM Networks. In *Proc. ITC 14*. Antibes, France (1994)
[17] Fielding, R., et al: Hypertext Transfer Protocol – HTTP/1.1. RFC 2068 (1997)

Aggregation and Scalable QoS: A Performance Study

Huirong Fu and Edward W. Knightly

Department of Electrical and Computer Engineering
Rice University
{hrfu,knightly}@ece.rice.edu
http://www.ece.rice.edu/networks

Abstract. The IETF's Integrated Services (IntServ) architecture together with reservation *aggregation* provide a mechanism to support the quality-of-service demands of real-time flows in a scalable way, i.e., without requiring that each router be signaled with the arrival or departure of each new flow for which it will forward data. However, reserving resources in "bulk" implies that the reservation will not precisely match the true demand. Consequently, if the flows' demanded bandwidth varies rapidly and dramatically, aggregation can incur significant performance penalties of under-utilization and unnecessarily rejected flows. On the other hand, if demand varies moderately and at slower time scales, aggregation can provide an accurate and scalable approximation to IntServ. In this paper, we develop a simple analytical model and perform extensive trace-driven simulations to explore the efficacy of aggregation under a broad class of factors. Example findings include (1) a simple single-time-scale model with random noise can capture the essential behavior of surprisingly complex scenarios; (2) with a two-order-of-magnitude separation between the dominant time scale of demand and the time scale of signaling and moderate levels of secondary noise, aggregation achieves performance that closely approximates that of IntServ.

1 Introduction

Flow-based resource reservation schemes as embodied by the IETF's Integrated Services protocol (IntServ) [6] provide a means to guarantee each flow's quality-of-service requirements. However, since processing reservation requests on a per-flow basis may not be feasible in high speed core routers, *aggregation* has been proposed as a mechanism to significantly reduce the signaling demands placed on core routers (e.g., [2]).

With aggregation, the per-flow guarantees of IntServ can be achieved without per-flow signaling of core routers. In particular, edge routers can maintain a long-time-scale aggregate reservation between a pair of ingress-egress routers. With this existing reservation, individual flows need only signal the ingress node which locally accounts for resources along the path and independently accepts or rejects new flows. Occasionally, when the aggregate reservation is determined to be too large or too small as compared to the actual demand, it can be readjusted via a "bulk" reservation adjustment in the core. Thus, core nodes are infrequently signaled to achieve scalability, yet without sacrificing the service model of per-flow guarantees and ideally, with minimal sacrifice in network

L. Wolf, D. Hutchison, and R. Steinmetz (Eds.): IWQoS 2001, LNCS 2092, pp. 307–324, 2001.

utilization. Thus, aggregation has the potential to simultaneously achieve scalability, per-flow quality-of-service, and high utilization. [1]

However, the performance of aggregation depends on a number of factors, the most important of which is the traffic characteristics of the underlying flows. For example, in one extreme in which a class' aggregate traffic is relatively constant over time, the core reservation can be nearly static and reserved-resource utilization will be high given the close match between the reservation and the actual traffic. At the other extreme, if a class' aggregate demanded bandwidth oscillates quickly and with high variance, aggregation would have relatively poor performance. In this case, the choice would be to either rapidly re-adjust the core reservation to track the demand (thereby frequently signaling and losing the advantage of scalability), or incur inaccuracies between the demand and the reservation (thereby suffering from under-utilization).

In this paper, we explore the fundamental roles of the timescales and variance of traffic demand and the timescales of aggregate control on the performance of an aggregate reservation scheme. Using a combination of modeling, analysis, and trace-driven simulations, we provide conditions under which aggregation is an accurate and high-performance approximation to the baseline IntServ. Our contributions are as follows.

First, we devise a simple model for aggregate traffic consisting of a sinusoid with random phase and additive white uniform noise. While clearly omitting many facets of realistic workloads, the model serves to isolate the effects of a single demand time-scale as well as the effects of additional variance. Second, we develop a theoretical model which, under the above traffic demands, provides a closed-form expression for the system's key performance measures such as overload probability. Third, we perform a set of simulation and numerical investigations into the performance of the basic model, and consider the impact of a number of simulated extensions to the basic model, such as correlated, rather than white additive noise. Finally, we perform a set of trace-driven simulations. This study provides practical insights into a number of factors not included in the theoretical model such as the role of network topology, correlated demand phases, and aggregating the traffic aggregates. Moreover, we study the accuracy of the simplified demand model as well as via the theoretical results.

Example findings are as follows. First, we find that the basic demand model and theoretical result are able to predict the performance of complex and trace-driven scenarios. For example, in experiments with QBone traces, we found that the model is able to predict the overload probability to within 11% accuracy, reserved resource utilization to within 1% accuracy and the available bandwidth to within 19% accuracy when the ratio of control to demand time scales is 1/36. Second, we find via trace- and model-driven simulations as well as the theoretical model, that if the control and demand time scales are separated by two orders of magnitude and additional variance is moderate, then aggregation provides performance quite similar to that of IntServ. For example, we find that if the control and demand time scales are separated by a factor of 72 and the range of the additive noise is 0.42 times the range of primary demand, then aggregation achieves a utilization of 97% of the utilization achieved by IntServ. However,

---

[1] In this way, the combination of IntServ and aggregation differs from *DiffServ* [3], as the latter cannot provide (per-flow) guaranteed service without additional mechanisms such as those described above.

for more highly variable NLANR traces in which the additive noise dominates the sinusoidal demand with a range nearly twice as large, both the model and trace driven simulations show that with the control and demand time scales separated by two orders of magnitude, aggregation achieves a utilization of 44% of that of IntServ.

Previous research on aggregation addresses both the protocols (i.e., mechanisms and architectures) and algorithms (i.e., policies) required for aggregate reservation. For example, an architecture for RSVP aggregation describing how to create and remove aggregate reservations is described in [2]. Furthermore, mechanisms have been devised for aggregation over label switched paths [1], multiple domains [9], and via RSVP tunnels [14] as well as via reservation agents [10]. Aggregation *policies* address issues such as how to accurately characterize an aggregate flow [11] and how to predictively make efficient bulk allocations including considerations of hysteresis [13]. In contrast, our work presents the first performance study to explore the role of traffic characteristics in the efficacy of aggregation, that is, to determine the regime under which aggregation is a high-performance mechanism. Finally, alternate architectures (than aggregation) have been proposed to provide scalable per-flow quality of service. Examples include end-point control via probing [4], combined end-point and router control [7], and "dynamic packet state" [12]. However, discussion of the relative merits of such architectures is beyond scope of this work.

The remainder of this paper is organized as follows. In Section 2, we define the system and demand models, describe the problem formulation, and develop an analytical method to characterize the impacts of control time scale, demand time scale, and mean and variance of demand on the performance tradeoffs of aggregate reservations. Next, in Section 3 we use model-driven simulation and numerical examples to study the performance impacts of periodic primary demand and additive secondary demand. In Section 4, we present a set of trace-driven simulation experiments to further evaluate the performance tradeoffs of aggregation under a broader set of scenarios not treated by the basic model. Finally, in Section 5 we conclude.

2 System and Demand Models and Analysis

As described in the introduction, aggregation provides a mechanism to reserve network resources on behalf of multiple traffic flows. In this section, we develop a simplified model to capture the key elements of the performance of aggregation, namely, we introduce a single-time-scale demand model in which the aggregate reservation is characterized by a sinusoid with random phase and additive random noise. We describe a baseline scenario in which such aggregate flows are multiplexed onto a backbone link and describe three relevant performance measures: the overload probability, the reserved resource utilization, and the normalized available bandwidth. Finally, we derive an expression for overload probability for the basic scenario.

2.1 System Model

We first consider a network model as shown in Figure 1(a). In this model, a number of flows (indexed by j) are multiplexed onto a class or link (indexed by i), and flow j

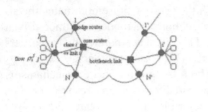

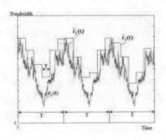

(a) Simplified Network Model.   (b) Aggregate Demand $r_i(t)$, Request $\hat{r}_i(t)$, and Reservation $\tilde{r}_i(t)$.

Fig. 1. System Model.

of link i has bandwidth requirement ρ_i^j. The network has a single bottleneck link with capacity C and all other links have infinite capacity.

Ignoring delay requirements and considering only bandwidth, IntServ's guaranteed service can admit any set of flows such that $\sum_i \sum_j \rho_i^j < C$, whereas flows are rejected when the total reserved rate would exceed C.

With aggregate resource reservation, individual flows do not signal the core routers. Instead, a flow signals its ingress router which makes "bulk" or aggregate resource reservations in the core, and accepts or rejects incoming flow requests according to whether there is sufficient available capacity in the bulk reservation. The ingress node will then periodically adjust the reservation in the core node according to its current demand.

The aggregate *demand* of link-i is simply $\sum_j \rho_i^j$ which we define by $r_i(t)$, a time varying function since the number of flows and their rates change over time due to flow arrivals and departures. Similarly, we denote the aggregate *reservation* at time t by $\tilde{r}_i(t)$. Consequently, when a new flow with rate ρ_i^* requests admission, if the ingress node has a current aggregate reservation such that $\sum_j \rho_i^j + \rho_i^* < \tilde{r}_i(t)$, then the flow is admitted. Otherwise, when $\tilde{r}_i(t)$ is insufficient, the ingress node will signal the core node for aggregate *request* $\hat{r}_i$, at time t denoted by $\hat{r}_i(t)$. Typically, the requested increment $(\hat{r}_i(t) - \tilde{r}_i(t))$, often referred to as the bulk reservation, is substantially larger than ρ_i^* to avoid rapid subsequent requests to core routers. Then, if $\sum_{l \neq i} \tilde{r}_l(t) + \hat{r}_i(t) < C$, the core node will grant the request $\hat{r}_i(t)$ and the new aggregate reservation level $\tilde{r}_i(t) = \hat{r}_i(t)$ will be established and the new flow will be accepted; if $\sum_{l \neq i} \tilde{r}_l(t) + \hat{r}_i(t) \geq C$ but

$$C - \sum_l \tilde{r}_l(t) > \rho_i^*, \text{ then the new aggregate reservation level } \tilde{r}_i(t) = C - \sum_{l \neq i} \tilde{r}_l(t)$$

will be established; otherwise, the current reservation level is maintained and the flow is rejected.

Likewise, if the ingress node determines that the current demand $r_i(t)$ is significantly less than the current aggregate reservation $\tilde{r}_i(t)$, then a *decrease* in reserved bandwidth will be requested in order to more efficiently utilize network resources.

Figure 1(b) illustrates the temporal behavior of aggregation. From a trace described in Section 4, the figure depicts the aggregate *demand* of a single ingress node $r_i(t)$

as well as the sequence of aggregate *requests* denoted by $\hat{r}_i(t)$ and the sequence of aggregate *reservations* denoted by $\tilde{r}_i(t)$.

2.2 Demand and Aggregation Model

Aggregation introduces a tradeoff. If the aggregate reservation $\tilde{r}_i(t)$ is infrequently adjusted, the signaling overhead in the core network is minimal. However, if the demand $r_i(t)$ varies rapidly, it will diverge from $\tilde{r}_i(t)$ and cause either under-utilization of the reservation or unnecessarily blocked flows. On the other hand, if the aggregate reservation is rapidly adjusted to match the current demand level, the system will achieve high utilization, yet the requirements of the signaling system are increased and in the limit (adjusting the reservation level for each flow), are identical to IntServ.

Here, we introduce a simple model to study the relationship between system performance, control (or signaling) and demand time scales, and demand variance. In particular, we consider as our "basic model", an aggregate demand of class (or link) i characterized by

$$r_i(t) = m_i + a_i \cos\left(\frac{2\pi}{T}t + \theta_i\right) + Z_i(t), \tag{1}$$

where m_i is the mean rate and a_i is the amplitude of a sinusoid with period T. The random nature of the demand is further modeled by additive white noise $Z_i(t)$ (i.e., $E Z_i(t) Z_i(t+s) = 0$ for $s \neq 0$) that has uniform distribution, that is, $Z_i(t) \sim U[-b_i, b_i]$. Finally, the sinusoids have random phase θ_i which is also uniformly distributed with $\theta_i \sim U[0, 2\pi]$. We denote $p_i(t) = m_i + a_i \cos\left(\frac{2\pi}{T}t + \theta_i\right)$ as the *primary* demand and $Z_i(t)$ as the *secondary* demand such that $r_i(t) = p_i(t) + Z_i(t)$.

While the model clearly omits properties of realistic traffic, it serves to isolate the performance impact of two key factors: demand time scale and demand variance (via T, a_i and b_i). Moreover, despite its simplicity, the model exhibits coarse resemblance to some traces of traffic aggregates. For example, considering the trace of Figure 1(b), the traffic exhibits a near-deterministic periodic long-term trend with additional variability.

To characterize the aggregate *reservation* $\tilde{r}_i(t)$, we consider periodic reservation adjustments at exactly intervals of τ seconds. Moreover, we assume that the requested reservation level for a bulk reservation at time t, $(k_i - 1)\tau \leq t < k_i\tau$, is given by $\hat{r}_i(t) = \max_{(k_i-1)\tau \leq s < k_i\tau} r_i(s)$, where $k_i = 1, \cdots, \frac{T}{\tau}$. To avoid triviality, we assume that $\frac{T}{\tau}$ is an integer. In other words, the aggregate bandwidth reservation is adjusted every τ seconds with a requested rate sufficient for the future interval (i.e., "perfect prediction" of the future demanded rate). While in practice, the adjustment interval might be made adaptive and perfect prediction is impossible, the model serves to also isolate the control time scale τ.

Thus, under the above scenario, we study the relative impact of demand and control time scales as well as demand variance on system performance, using the performance measures defined next. Moreover, we show experimentally in Sections 3.3 and 4.1, that conclusions derived from the above "basic model" can generalize to significantly more complex scenarios.

2.3 Performance Analysis

To evaluate the effectiveness of aggregate-based resource reservation, we consider three performance metrics that we describe as follows. First, the *overload probability*, denoted by P_{ol}, is the ratio of the overloaded traffic (which cannot be admitted) to the total demand, i.e.,

$$P_{ol} = \frac{E(\sum_{i=1}^{N} (r_i - \tilde{r}_i)^+)}{E(\sum_{i=1}^{N} r_i)}, \tag{2}$$

where r_i denotes a random variable with the steady state distribution of $r_i(t)$.

Second, *reserved resource utilization*, denoted by U_r, refers to the fraction of an aggregate reservation that has been utilized by the underlying traffic, i.e.,

$$U_r = \frac{(1 - P_{ol}) \cdot E(\sum_{i=1}^{N} r_i)}{E(\sum_{i=1}^{N} \tilde{r}_i)}. \tag{3}$$

Finally, the *normalized available bandwidth*, denoted by b_A, also reflects the efficiency of aggregation by describing the fraction of bandwidth available after accounting for all aggregate reservations, i.e.,

$$b_A = \frac{C - E(\sum_{i=1}^{N} \tilde{r}_i)}{C}. \tag{4}$$

Under the basic model of aggregate demand and the above performance measures, we compute the overload probability of aggregate resource reservation as follows.

Aggregation Performance. Consider N aggregate demands sharing a single bottleneck link with capacity C as described in the basic model. If the aggregate demand of class i is $r_i(t) = m_i + a_i \cos\left(\frac{2\pi}{T}t + \theta_i\right) + Z_i(t)$, $i = 1, 2, \cdots, N$, where $Z_i(t)$ is white uniform noise with $Z_i(t) \sim U[-b_i, b_i]$, then the overload probability is approximately

$$P_{ol} \approx \frac{(\frac{\tau}{T})^N \cdot \sum_{k_1=1}^{\frac{T}{\tau}} \cdots \sum_{k_N=1}^{\frac{T}{\tau}} [\sum_{i=1}^{N} f_{i,k_i} - C]^+}{\sum_{i=1}^{N} (m_i + \frac{2\tau}{T} \cdot a_i + b_i)}, \tag{5}$$

where $f_{i,k_i} = \max_{(k_i-1)\tau \le s < k_i\tau} [m_i + a_i * \cos\left(\frac{2\pi}{T}s\right)]$.

A "sketch" derivation of the result is as follows. To simplify the analysis, we first consider the phases θ_i to be *discretely* uniform in $[0, \tau, 2\tau, \cdots T]$. In other words, aggregate reservation requests from different classes occur at identical epochs. Second, we decouple the impact of the primary and secondary demands and observe that over a window τ, the secondary demand satisfies $P(\max_{0 \le s \le \tau} Z_i(s) = b_i) \approx 1$ such that, to ensure sufficient bandwidth is available over the entire window τ, an additional bandwidth b_i must be reserved.[2] Next, we exploit the odd symmetric characteristics of the cosine wave at points $\pi/2$ and $3\pi/2$ to compute the mean discrete primary demand i as

$$\frac{\tau}{T} \sum_{k_i=1}^{\frac{T}{\tau}} f_{i,k_i} = \frac{\tau}{T} \left(\frac{T}{\tau} \cdot m_i + 2a_i \right),\tag{6}$$

which simplifies to $m_i + \frac{2\tau}{T} a_i$. Finally, we compute Equation (2) by conditioning on the relative phases of the different aggregates, and after some manipulation, Equation (5) follows.

Due to space limitations, the detailed derivation of all three performance measures is presented in [8]. However, we do consider analytical results for P_{ol}, U_r, and b_A in the numerical and simulation studies that follow.

3   Experiments with the Basic Model

The theoretical model described above characterizes the relationship among the time scale of demand, the demand variance, the control time scale, and the performance of aggregation. In this section, we present numerical and simulation investigations into these issues. In particular, using the basic demand model described in Section 2, we quantify the role of the demand time scale and demand variance for the basic model. Moreover, we show that alternate models of primary demand having different periodic functions, and alternate models of secondary demand having temporal correlation, have little impact on system performance.

3.1   Control and Demand Time Scales

Here, we isolate the roles of control and demand time scales by exploring the performance of the basic model under the special case of $Z_i(t) = 0$, i.e., no secondary demand. With this scenario, one can ask what frequency of reservation $(1/\tau)$ is required for aggregate-based resource reservation to achieve performance similar to IntServ's flow-based resource reservation? Similarly, if the control time scale τ is limited by scalability constraints (e.g., routers have a known upper limit on the frequency for which they can be signaled) what is the performance "cost" of aggregating demand? We first consider a simple scenario with $N = 2$ classes, a bottleneck link capacity of $C = 3$, and demand of both classes given by $m_i = 1$, and $a_i = 1$.

We begin by illustrating the performance tradeoffs of aggregate-based resource reservation as the control time scale τ varies from 0 to T. The results are depicted in Figure 2(a)-(c) for a fixed demand time scale of $T = 2\pi \approx 6.28$ (for discussion, we refer to the units of T as hours). We make the following observations about the figures.

First, regarding the extreme cases of $\tau = 0$ and $\tau = T$, observe that $\tau = 0$ corresponds to the case of *no* aggregation, or IntServ, that is, the core's requested reservation corresponds precisely to the flows' total demanded bandwidth (or equivalently, the

---

[2] This argument can be made rigorous by discretizing the interval τ and taking limits of the maximum noise in the window.

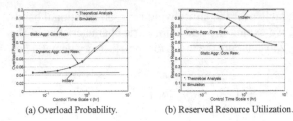 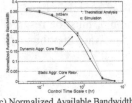

(a) Overload Probability.   (b) Reserved Resource Utilization.   (c) Normalized Available Bandwidth.

Fig. 2. Impact of Control Time Scale τ.

aggregate reservation is continuously adjusted). This provides an upper bound to the efficacy of aggregation, which under the given workload is given by an overload probability of 4.4%, a reserved resource utilization of 100% and an available bandwidth of 36%. At the other extreme, when $\tau = T$, the aggregate reservation is *static*, and corresponds to the maximum total flow demand over the entire period. In this case, the overload probability is 15.9%, the utilization is 56.1% and the available bandwidth is 0. This scenario provides a lower bound for the performance of aggregation.

Second, observe that as compared to a static aggregate reservation, system performance rapidly improves as the control time scale τ is decreased from the extreme of T.[3] Furthermore, most of this improvement is incurred with moderate values of τ indicating little further performance improvements for extremely small values of τ and rapid signaling. Two interpretations of this behavior are as follows. First, the curves describe the signaling frequency required to achieve a certain level of performance. For example, the figure shows that when τ is less than 1% of T, aggregation achieves near ideal performance. In other words, if the control and demand time scales are separated by two orders of magnitude, the performance of aggregation is nearly indistinguishable from that of IntServ. Second, the curves can be viewed in terms of "bulk size", i.e., the required increase or decrease in reserved bandwidth in order to achieve a certain performance level. Observe that the mean bulk size is simply given by $\dfrac{\sum_{i=1}^{N} \sum_{k_i=1}^{T/\tau-1} |f_{i,k_i+1} - f_{i,k_i}|}{N \times T/\tau}$ so that conclusions regarding time scales of control can be converted to conclusions regarding the magnitude of the reservation updates.

Figure 3 depicts the reserved resource utilization as a function of the demand time scale T for a fixed control time scale τ of 5.9 minutes. This figure characterizes a scenario in which performance limitations of core routers dictate a maximum signaling frequency of once per 5.9 minutes (per class, the *total* number of signaling messages increases with the number of classes). The curve then quantifies the performance penalty for performing aggregation rather than IntServ as a function of the demand time scale. Observe that for aggregation to achieve performance within 10% of IntServ, the system period must be no smaller than 1.57 hr when the control time scale τ is 5.9 minutes.

---

[3] Curve fitting yields a near precise match between the P_{ol} vs. τ curve of Figure 2 (a) and the function $0.162 - 0.12e^{-0.4\tau}$. However, we have not yet been able to establish this exponential relationship analytically.

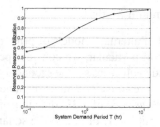

Fig. 3. Impact of the Demand Time Scale T.

3.2 Variance of the Secondary Demand

Here, we explore the role of additional variation in the demand on the performance of aggregation. Namely, we consider secondary demand given by $Z_i(t) \sim U[-b_i, b_i]$, $b_i \leq m_i - a_i$, as in the basic model described in Section 2. We consider one bottleneck link with $C = 8$ and two traffic aggregates with $m_i = 2$ and $a_i = 1$, and variance of the secondary demand given by $\sigma_i^2 = b_i^2/3$.

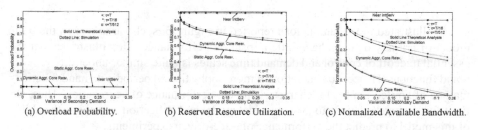

(a) Overload Probability.   (b) Reserved Resource Utilization.   (c) Normalized Available Bandwidth.

Fig. 4. Impact of Secondary Demand.

From Figure 4 it is clear that variance in secondary demand hinders the efficacy of aggregation. For a static aggregate reservation ($\tau = T$), the impact is quite severe as reserved resource utilization decreases from 100% to 53% when the variance of the secondary demand is 0.2. For aggregation with an adjustment time scale of $\tau = T/16$, the effects are mitigated, e.g., reserved resource utilization decreases to 70% under the same variance. Regardless, sufficient "noise" in the demand can degrade the performance of aggregation to levels comparable to a static reservation. Alternatively, if the noise is moderate, performance similar to IntServ can still be achieved. For example, to achieve a reserved resource utilization within 20% of IntServ with aggregation and $\tau = T/16$, the variance of the secondary demand must be limited to 0.05. This corresponds to a range of noise 0.39 times the range of the primary demand (i.e., $b_i = 0.39a_i$). Of course, the detrimental effects of such variance can be alleviated with faster signaling (and reduced τ).

3.3 Alternate Primary Demand Models

Here, we consider the impact of alternate models of primary demand in addition to the sinusoid with random phase. In particular, we consider periodic sawtooth and square waves with random phase, and in all cases set the secondary noise $Z_i(t)$ to 0.[4] For these three primary demand models, we consider a mean demand m_i of 2, variance 1.33, and period $T = 2\pi$. To achieve a variance of 1.33, the sinusoid has amplitude a_i of 1.63, whereas the sawtooth has amplitude a_i of 2, and the square wave has amplitude a_i of 1.15. Let $C = 6$ and $N = 2$.

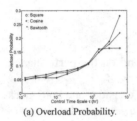
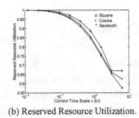
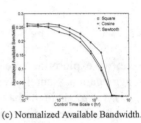

(a) Overload Probability.   (b) Reserved Resource Utilization.   (c) Normalized Available Bandwidth.

Fig. 5. Alternate Primary Demand Models.

As illustrated in the simulations reported in Figure 5, such variations on the basic model of primary demand have little impact on performance. This illustrates that the essential tradeoff of control and demand time scales is quite similar under different demand functions. Hence, consideration of more sophisticated periodic demand functions may be of limited impact for characterizing the performance of aggregation. Thus, we limit further investigations to the sinusoidal model and in Section 4 evaluate the ability of this model to predict the performance of trace-driven experiments.

3.4 Alternate Secondary Demand Models

In this section, we use simulations to consider the performance impact of an alternate secondary demand model as compared to the uniform white noise considered in the basic model. Specifically, we consider a $Z_i(t)$ to be given by a sawtooth wave with random phase. In the experiments below, we consider a sawtooth with mean 0, variance 0.33, maximum 1, minimum -1, and period equal to $T/4$, and compare the performance with white noise with the same mean, variance, and range.

Figure 6 illustrates the impact of temporal correlation in secondary demand $Z_i(t)$ on overload probability for $C = 6$, $N = 2$, $m_i = 2$, $a_i = 1$ and $b_i = 1$. The figure shows that for small control time scales τ, correlated secondary noise improves performance whereas for larger τ it degrades performance. Regardless, the difference is minimal, as

---

[4] For example, if the phase θ_i is 0, and $0 \le t \le T/2$, the sawtooth's demand is given by $(m_i - a_i) + \frac{4a_i}{T} \cdot t$, whereas the square wave's demand is given by $(m_i + a_i)$. In the range $[T/2, T]$, the sawtooth's demand is $(m_i - a_i) - \frac{4a_i}{T} \cdot (t - T)$, whereas the square wave is $(m_i - a_i)$.

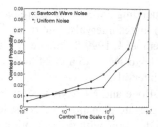

Fig. 6. Alternate Secondary Demand Model.

the figure depicts a worst-case scenario in which the period of the secondary demand sawtooth wave is $1/4^{th}$ that of the primary demand period T, and $b_i = a_i$. For smaller periods of temporally correlated secondary demand and $b_i < a_i$, the difference is even smaller.

4   Trace Driven Simulations

In this section, we broaden our experimental investigation to consider more realistic scenarios and trace-driven simulations. In particular, we study issues such as the ability of the basic model to predict the performance obtained in trace-driven scenarios, as well as the impact of network and protocol characteristics in aggregation's performance.

4.1   Simulation Source and Scenarios

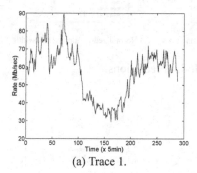

(a) Trace 1.

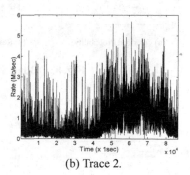

(b) Trace 2.

Fig. 7. Traces.

The trace depicted in Figure 7(a) depicts aggregate measurements obtained from the QBone "PSC" ingress node on November 16, 2000.[5] The mean, variance and demand period T of the aggregate traffic are 56.8 Mb/sec, 191 and 24 hours, respectively.

---

[5] Available at http://tombstone.oar.net/sitemap.html.

Figure 7(a) shows the corresponding trace of the aggregate traffic. Measurements were reported as averages over 5 minutes intervals. The trace depicted in Figure 7(b) is obtained from NLANR on December 1, 1999.[6] The mean, variance and demand period T of the aggregate traffic is 0.74 Mb/sec, 0.45 and 24 hours, respectively. Measurements were reported over 1 second intervals.

In simulations, we use QBone and NLANR traces to represent the aggregate demand $r_i(t)$. For multiple aggregate demands, we consider collections of traces each with random phase over their duration, with the exception of one experiment where we study the effects of synchronized phase (identical θ_i). We consider a number of network topologies ranging from the single bottleneck of the baseline scenario to more complex meshes obtained using the topology generator of [5]. Moreover, we consider perfect prediction of future demand such that a core reservation request at time t of aggregate i is for maximum bandwidth required over the next τ second interval, i.e., $\max_{t \le s < t+\tau} r_i(s)$. Finally, for each scenario, we conduct 100 independent simulation runs to empirically obtain the average of the performance parameters. For each run, we simulate four demand periods, and discard results from the first cycle as transient. Further details of each scenario, including link capacities, the number of aggregate demands and their spatial distributions are described in the corresponding subsection.

4.2   Validation of the Basic Model

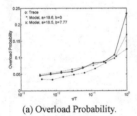 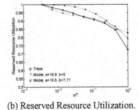 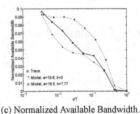

(a) Overload Probability.   (b) Reserved Resource Utilization.   (c) Normalized Available Bandwidth.

Fig. 8. QBone Simulations and Model Predictions.

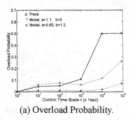 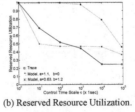 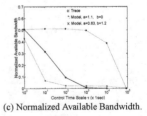

(a) Overload Probability.   (b) Reserved Resource Utilization.   (c) Normalized Available Bandwidth.

Fig. 9. NLANR Simulations and Model Predictions.

---

[6] Available at http://moat.nlanr.net/Traces/Kiwitraces/auck2.html.

Here, we consider a single bottleneck link with capacity 120 Mb/sec for trace 1 and 3 Mb/sec for trace 2 and compare the performance of the trace-driven simulations with that predicted by the basic model. To compute the parameters of the basic model, we compute the mean, variance, and demand timescale of both traces. For trace 1, considering only the primary demand, the basic model yields parameters $r_i(t) = 56.8 + 19.6 \cdot \cos(2\pi t/T + \theta_i)$, whereas considering both primary demand and secondary demand, we have $r_i(t) = 56.8 + 18.5 \cdot \cos(2\pi t/T + \theta_i) + Z_i(t)$, $Z_i(t) \sim U[-7.77, 7.77]$. For trace 2, considering only the primary demand, the basic model yields parameters $r_i(t) = \max\{0, 0.65 + 1.1 \cdot \cos(2\pi t/T + \theta_i)\}$, whereas considering both primary demand and secondary demand, we have $r_i(t) = \max\{0, 0.63 + 0.63 \cdot \cos(2\pi t/T + \theta_i) + Z_i(t)\}$, $Z_i(t) \sim U[-1.2, 1.2]$, with the "max" required to ensure that even with the high variance of secondary demand, the aggregate rate is non-negative.

Figures 8-9 show the performance comparison between the trace driven simulations and the model predictions for the three performance measures described in Section 2. For trace 1, we observe that the sinusoidal model with random phase, while highly simplifying the details of the true trace, is able to capture the basic behavior of the system. For example, for $\tau/T = 1/36$, the predictions of overload probability, reserved resource utilization, and normalized available bandwidth are with 11%, 1%, and 19% of the simulated values. Furthermore, characterizing variance via additive random noise $Z_i(t)$ rather than purely through the sinusoid with random phase further improves the prediction, i.e., consideration of primary and secondary demand in general outperforms consideration of only primary demand. Finally, we observe that as predicted by the model, if demand and control time scales are separated by two orders of magnitude and the secondary demand is moderate, aggregation attains performance nearly identical to IntServ. For example, under $\tau = 10$ minutes, $T = 144\tau = 24$ hours, the overload probability is 4.56% for IntServ and 5% for aggregation.

For the NLANR experiments, we observe that considering only the primary sinusoidal demand and ignoring secondary demand introduces large prediction errors. However, characterizing demand variance via additive random noise $Z_i(t)$ rather than purely through the sinusoid with random phase is still able to capture the basic performance characteristics of the system. Finally, we observe that, as predicted by the model, variance in secondary demand hinders the efficacy of aggregation. For example, if the demand and control time scales are separated by two orders of magnitude, since the range of the additive noise is nearly twice (1.2/0.63 = 1.9 times) the range of primary demand, aggregation achieves a utilization of only 44.2% of that achieved by IntServ.

4.3   Network Topology

We next study the impact of different network topologies, including dumbbell, star, tree, mesh and freeway with on-ramps. Figure 10 shows the corresponding network topologies, traffic distribution (arrow lines) and link capacities (Mb/sec). For the dumbbell, star, and mesh, all links are potential bottlenecks and result in overload whereas only some of the links are bottlenecked for freeway with on-ramps and the tree (bottleneck links are represented by the bold lines in Figure 10).

Figure 11 shows the overload probability, reserved resource utilization and available bandwidth versus the control time scale for different network topologies. We depict the

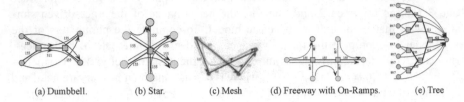

(a) Dumbbell.   (b) Star.   (c) Mesh   (d) Freeway with On-Ramps.   (e) Tree

Fig. 10. Simulation Topologies.

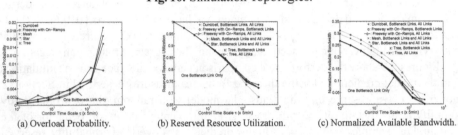

(a) Overload Probability.   (b) Reserved Resource Utilization.   (c) Normalized Available Bandwidth.

Fig. 11. Impact of Network Topology.

performance of both bottleneck links (solid lines) and all links (dotted lines), and as a benchmark, also depict the performance of one bottleneck link with capacity 155 Mb/sec shared by two aggregate demands.

We make three observations about the experiments. First, considering all links, the freeway topology has slightly lower utilization but higher available bandwidth than other topologies due to resource contention among aggregate demands in both the freeway and cross traffic on-ramps. Second, considering only bottleneck links, the performance difference for different network topologies is very small when all links are bottlenecked and the traffic is balanced (with the exception of freeway). Similarly, there is little performance impact between a single and multiple bottleneck links in the different topologies. In other words, from the perspective of bottleneck links with QBone-like demand, aggregate reservation incurs nearly the same performance tradeoffs as in the single-bottleneck scenario.

4.4   Number of Aggregate Demands

In this section, we study the role of the number of aggregate demands on aggregate resource reservation by considering 2, 4, and 8 aggregate demands sharing a bottleneck link with capacity scaled to 155, 311, and 622 Mb/sec respectively.

Because of statistical multiplexing among aggregate demands, one may expect that like flow-based resource reservation, an increased number of aggregate demands (with a proportional increase in capacity) will reduce the overload probability and improve resource utilization for aggregate reservation. However, we find that this is not always the case.

Figures 12(b) and (c) indicate that under aggregate reservation, an increased number of aggregate demands always reduces the reserved resource utilization and available

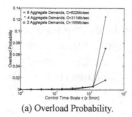 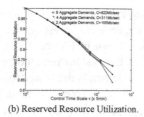 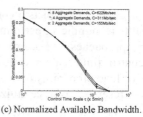

(a) Overload Probability.   (b) Reserved Resource Utilization.   (c) Normalized Available Bandwidth.

Fig. 12. Impact of the Number of Traffic Aggregates.

bandwidth. In addition, as shown in Figure 12(a), when the control time scale τ is smaller than $T/2$, the system can achieve slightly lower overload probability under aggregate reservation (as with flow-based resource reservation). However, when the control time scale τ is greater than $T/2$, an increased number of aggregate demands cause significantly higher overload probability under aggregation reservation, unlike the behavior of flow-based reservation. For example, when the control time scale is T, the overload probability for 2 aggregate demands sharing one bottleneck link of 155 Mb/sec is less than 2% but more than 12% for 8 aggregate demands sharing one bottleneck link with capacity 622 Mb/sec. This is because under a large control time scale, the negative effect of quantization error from bulk reservation is cumulative. However, we observe that the performance impact of the number of aggregate demands is quite limited for faster control and $\tau \le T/2$.

4.5 Merging Aggregate Demands

In the above experiments, each aggregate demand reserves its bandwidth independently, and as described in Section 2, a new reservation is admissible only if the total rate of all aggregate demands is less than the link capacity. An alternate possibility is to *merge* multiple aggregate demands into a single reservation rather than to reserve resources for each aggregate's demanded bandwidth independently (which we refer to as *isolation*). We consider 2, 4, and 8 aggregate demands sharing a bottleneck link with capacity scaled to 155, 311, and 622 Mb/sec respectively.

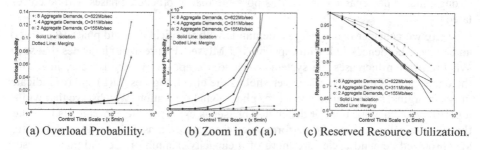

(a) Overload Probability.   (b) Zoom in of (a).   (c) Reserved Resource Utilization.

Fig. 13. Impact of Merging.

Figure 13 depicts a comparison of these two scenarios. Observe that merging results in significant performance improvements, especially when the control time scale τ approaches T. For example, as shown in Figure 13(a) and Figure 13(c), when the control time scale τ is as large as the system demand period T, merging makes the overload probability of 8 aggregate demands decrease from 13% (isolation) to 0, while the reserved resource utilization increases from 64% to more than 79%.

As an alternate viewpoint, the experiments illustrate that to achieve the same performance as isolation, merging can allow an increase in the control time scale τ. For example, as shown in Figure 13(b), to keep the overload probability to zero for 4 aggregate demands, isolation requires $\tau \leq T/128$ while merging requires only that $\tau \leq T/32$.

Finally, Figure 13(c) illustrates that such gains increase with the number of aggregate demands. For example, when the control time scale is $\tau = T/4$, 8 aggregate demands can achieve a 10% gain, whereas 4 aggregate demands achieve a 5% gain. Thus, exploiting the effects of statistical multiplexing for aggregate demands themselves can have an important effect, especially under larger control time scales.

4.6   Demand Phases

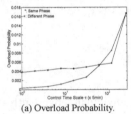 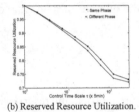 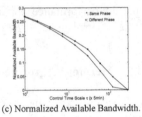

(a) Overload Probability.   (b) Reserved Resource Utilization.   (c) Normalized Available Bandwidth.

Fig. 14. Impact of Demand Phase.

For the final experiments, we consider the case of synchronized demands. That is, both the theoretical model and the simulations are based on each aggregate demand having a uniformly independent phase. However, as many traces' behavior indicates strong-time-of-day characteristics, it is possible that in practice, phases will be correlated.

Here we consider two aggregate demands with identical demand phase ($\theta_1 = \theta_2$) and a single bottleneck link with capacity 155 Mb/sec. Figure 14(a) indicates that such synchronization increases the system overload, except under very large τ, in which the coarseness of the reservation overwhelms the effect. Figures 14(b) and (c) indicate only a marginal performance impact for phase synchronization. We observe that while a performance degradation for dependent phases is expected, the experiments indicate that they equally degrade the performance of aggregation as well as IntServ. Hence, synchronized demand cycles are more of a capacity planning issue and play a lesser role in the efficacy of aggregation itself.

5   Conclusions

In this paper, we studied the problem of aggregate resource reservation and investigated the conditions under which aggregation can simultaneously achieve high utilization and scalability. We presented a simple single-time-scale model with random noise and provided a derivation of overload probability for such aggregates. Moreover, we used numerical and simulation experiments to explore the design space outside the scope of the basic model and found that aggregate reservation is largely insensitive to the particular shape of the (periodic) primary demand, as well as to temporal correlation in the secondary demand. However, both the model and simulations indicate that the performance of aggregation is strongly related to the relationship between the demand and control time scales as well as to the variance of the secondary demand. Finally, trace-driven simulations corroborated the conclusions obtained with the theoretical model and moreover showed that the model is able to characterize the performance of aggregation even under quite complex scenarios. Example findings include that a separation of time scales of two orders of magnitude between demand and control (i.e., between the dominant traffic time scale and the time scale for adjustment of the aggregate reservation) ensure excellent performance of aggregation, provided that additional "noise" (random secondary demand in addition to the primary periodic demand) is moderate. We found that in one trace the noise was sufficiently moderate whereas in a second trace the noise dominated the primary demand and aggregation incurred a 44% utilization penalty. While neither trace is an ideal representation of aggregate real-time traffic as both traces are dominated by TCP flows, our results regardless provide both an insight into the basic performance tradeoffs of aggregation as well as a simple model-based technique for performance prediction.

References

1. D. Awduche et al. Extensions to RSVP for LSP Tunnels. Internet Draft, draft-ietf-mpls-rsvp-lsp-tunnel-07.txt, August 2000.
2. F. Baker, C. Iturralde, F. Le Faucheur, and B. Davie. Aggregation of RSVP for IP4 and IP6 Reservations. Internet Draft, draft-ietf-issll-rsvp-aggr-02.txt, March, 2000.
3. D. Black, S. Blake, M. Carlson, E. Davies, Z. Wang, and W. Weiss. An Architecture for Differentiated Services, 1998. Internet RFC 2475.
4. L. Breslau, E. Knightly, S. Shenker, I. Stoica, and H. Zhang. Endpoint admission control: Architectural issues and performance. In *Proceedings of ACM SIGCOMM 2000*, Stockholm, Sweden, August 2000.
5. K. Calvert, M. Doar, and E. Zegura. Modeling Internet topology. *IEEE Communications Magazine*, pages 160–163, June 1997.
6. D. Clark, S. Shenker, and L. Zhang. Supporting real-time applications in an integrated services packet network: Architecture and mechanism. In *Proceedings of ACM SIGCOMM '92*, Baltimore, Maryland, August 1992.
7. T. Ferrari, W. Almesberger, and J. Le Boudec. SRP: a Scalable Resource Reservation Protocol for the Internet. In *Proceedings of IWQoS '98*, Napa, CA, May 1998.
8. H. Fu and E. Knightly. Aggregation and scalable QoS: A performance study. Rice University ECE Technical Report 00-07, February 2001.

9. P. Pan, E. Hahne, and H. Schulzrinne. BGRP: A tree-based aggregation protocol for inter-domain reservations. *Journal of Communications and Networks*, 2(2):157–167, June 2000.
10. O. Schelen and S. Pink. Aggregating resource reservations over multiple routing domains. In *Proceedings of IWQoS '98*, Napa, CA, May 1998.
11. J. Schmitt, M. Karsten, L. Wolf, and R. Steinmetz. Aggregation of guaranteed service flows. In *IWQoS '99*, London, UK, May 1999.
12. I. Stoica and H. Zhang. Providing guaranteed services without per flow management. In *Proceedings of ACM SIGCOMM '99*, Cambridge, MA, August 1999.
13. A. Terzis, L. Wang, J. Ogawa, and L. Zhang. A two-tier resource management model for the Internet. In *Proceedings of Global Internet Symposium '99*, Rio de Janeiro, Brazil, December 1999.
14. A. Terzis, L. Zhang, and E. Hahne. Making reservations for aggregate flows: Experiences from an RSVP tunnels implementation. In *Proceedings of IWQoS '98*, Napa, CA, May 1998.

Customizable Cooperative Metering for Multi-ingress Service Level Agreements in Differentiated Network Services

Syed Umair Ahmed Shah and Peter Steenkiste

School of Computer Science, Carnegie Mellon University
5000 Forbes Avenue, Pittsburgh, PA 15213, USA
{umair,prs}@cs.cmu.edu

Abstract. In the Differentiated Service architecture an interesting class of SLAs provides guarantees for the traffic flowing from multiple ingress nodes into a single or group of egress nodes. For many distributed services, like a VPN service, where multiple ingress points are sending to the same egress, customers would like to control the bandwidth distribution across the ingress routers, based on their specific requirements. In this paper we present a simple strategy called CCDM (Customized Coordinated Dynamic Metering) that supports a customizable distribution of bandwidth across ingress routers. The bandwidth distribution policy is specified by customers in the form of router extensions (code modules) that execute in the control plane of the ingress routers. We describe and motivate the CCDM design. We also describe an implementation of CCDM in the context of the Darwin system and present measurement results for three different customized policies, demonstrating the customizability aspect of our solution.

1 Introduction and Problem Motivation

The Differentiated Services (DiffServ) framework [2] supports network quality of service using a simple network core that treats packets belonging to one of a small number of service classes "the same way". Traffic is policed at the entry points to the network according to *service level agreements* (SLAs). The SLA between the service provider and the customer (end-user or another service provider) defines the traffic contract and the guarantees that the customer should receive from the network based on the customer's needs and the provider's policies.

Simple SLAs require only static enforcement of the traffic contract. An example is an SLA for an ordinary dialup customer where the traffic enforcement need only be done at the ingress router. The more interesting case of managing an SLA contract is when it involves multiple ingress nodes (Fig. 1). For such an SLA, an important issue is how the egress link bandwidth is shared among the ingress nodes. For example, an organization that is connected to the service provider's network through multiple ingress points can use such an SLA. In such a setup, the organization would like to efficiently control how the different

L. Wolf, D. Hutchison, and R. Steinmetz (Eds.): IWQoS 2001, LNCS 2092, pp. 325–341, 2001.
© Springer-Verlag Berlin Heidelberg 2001

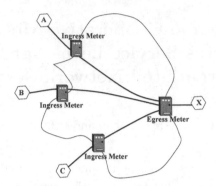

Fig. 1. Multiple ingress domains *(A, B, C)* sending to a single egress X.

ingress nodes share this egress bandwidth. This SLA can also form the basis
for providing VPN service on a DiffServ network where the share of the ingress
nodes needs to be dynamically controlled based on customer policy.

When a customer's traffic distribution across multiple ingress nodes is dynamic, using a static allocation of bandwidth for each ingress point is waste-
ful. At any instance, either the static allocation would be too high, requiring
over-allocation of resources as well as allowing excessive bandwidth to enter the
network, or the allocation would be too low, shutting off traffic from an ingress
while other ingress nodes are not fully utilizing their share. Static allocation also
changes the semantics of the SLA to that of independent one-to-one SLAs for
every ingress node. It is more efficient to be able to specify a contract that allows the *dynamic* assignment of shares across ingress points, based on customer's
current needs. The dynamic share assignment requires gathering relevant traffic
information, based on which the share assignment will be done. This requires *coordination* among the ingress nodes. Customers have varying *customized* requirements and criteria for deciding how different ingress nodes share the bandwidth
of the SLA under different traffic conditions.

In this paper we present a simple mechanism, CCDM (Customized Coordinated Dynamic Metering), that supports the customizable dynamic distribution
of bandwidth across the different ingress nodes in the network for such an SLA.
CCDM allows the customer to specify a bandwidth distribution algorithm in
the form of code modules that periodically calculate new shares for each of the
ingress nodes based on traffic statistics and other input. We also discuss design
tradeoffs for CCDM and an analysis of its parameters. Section 2 looks at other
related work. Section 3 presents how the customers can specify policies. Section 4
describes the design of CCDM. In Section 5 we describe several example policy
sets. Details of the simulation and simulation results are presented in Section 6.
Details of our implementation and measurements are presented in Section 7 and
we conclude in Section 8.

2   Related Work

Previous work [5, 12, 14] has not addressed customized, dynamic resource sharing across ingress nodes. The egress controlled SLA is discussed in   [12, 5]. Ohlman [12] considers the idea of the receiver being able to control the shares of individual flows for low-bandwidth links. It suggests both static configuration as well as dynamic signaling solutions at the egress bottleneck link. It only considers policing at the egress and does not consider how the bandwidth can be shared among ingress links. The Hose model [5] addresses a complementary problem of resource reservation and performance guarantees for the same type of SLA. However they do not address the issue of how the customer can customize and control the shares of different ingress nodes within these aggregate constraints placed on the traffic from different VPN ingress points.

Cooperative Dropping [14] provides a number of bandwidth services, including proportional sharing using multiple drop priorities. The packets of a flow are striped across multiple priority levels. At each intermediate router, preferential dropping is done depending on the priority level of the packets and the current level of congestion. Depending on how the packets of a flow are striped across the multiple priorities, proportional sharing and a variety of other services can potentially be implemented. The granularity and accuracy of the bandwidth sharing depends on the number of priority levels used. This approach provides implicit cooperation across different traffic classes but does not solve the problem of sharing the resources of the same traffic class across different ingress nodes.

3   Specifying the Ingress Bandwidth Distribution

A multi-ingress single-egress traffic contract between a service provider and a customer will be based on an overall "policy envelope" such as *"The total traffic sent to Y does not exceed 30Mb"* and *"no ingress can send more than 10Mb to Y"*. Such rules are necessary within the SLA contract so that the service provider can engineer its network to meet the SLA guarantees. These can also be used to accept or reject customer traffic contracts. We refer to this maximum aggregate bandwidth (30 Mb) as the "SLA Limit". This type of SLA raises two questions: 1) how should the service provider provision the network to meet the requirements of the SLA, and 2) how do we control the bandwidth distribution across the ingress routers within the envelope specified by the SLA. In this paper we focus on the second problem.

Within the policy envelope, the share assignment can be customized in two ways: 1) generic solution for all the customers and 2) specific solution for every customer. In the first option, the service provider provides generic code for handling the dynamic assignment of shares based on a set of rules that capture the customer's policy. These rules can be represented in the general form

If condition then constraints

where both condition and constraints are formulas using customer provided constants and variables that are measured in the system. This form can handle both static as well as dynamic rules. An example of a static constraint is that ingress point A should always be able to send at least L_{AX} to egress router X, which can be specified as (S_{AX} represents the share, L_{AX} the limit)

$$\text{If true then } S_{AX} \geq L_{AX}$$

An example of a complex, dynamic rule is that when the aggregate traffic offered to the network for egress X is greater than the SLA limit (L_X), then each ingress should be assigned at least its weighted share (W_{iX}) of L_X while the combined share of traffic from ingress points A and C should not exceed L_{ACX} (Fig. 1). This rule can be represented as (B_{iX} - incoming traffic at ingress i)

$$\text{If } \Sigma B_{iX} \geq L_X \text{ then}$$
$$(\text{For } \forall i, S_{iX} \geq W_{iX} * L_X) \text{ and } (S_{AX} + S_{CX} \leq L_{ACX})$$

This rule-based approach of representing a bandwidth distribution policy gives the customer greater flexibility to customize the traffic distribution than using static allocation. However, it also has some drawbacks. First, determining the shares using a generic share assignment algorithm requires solving a system of possibly non-linear constraint equations given a maximization function, which in most cases is an NP-hard problem [6]. If the constraint equations are limited to be linear then it can be solved reasonably efficiently in polynomial time using the Projective Method [9] or by using the simplex method which outperforms polynomial time algorithms in most cases [3]. Another limitation of this approach is that the conditions and constraints can only be based on statistics that are "supported" by the service provider, like the bandwidth share of an ingress (S_{iX}) or the number of ingress nodes (n). Finally, specifying the desired bandwidth distribution as a set of rules may be awkward and difficult for complex policies.

In the second approach, the customer can specify the bandwidth distribution in the form of a code module. The service provider then uses this customized module to calculate bandwidth shares for the customer. While this approach creates some challenges for the service provider, it offers greater flexibility and control to the customer, for e.g. the customer can use information on the number or type of flows when calculating the share distribution, i.e. statistics that are not recognized by the service provider. This approach also allows optimized code for handling the share assignment rather than inefficiencies of a generalized distribution algorithm. It also allows the customer code to gather statistical data that enables it to improve the share assignment code. In this paper we will focus on this particular approach for dynamic ingress bandwidth distribution.

4   Design of CCDM

The bandwidth distribution is managed by a set of *ingress meter controllers* and a *meter coordinator*. Every ingress node has one meter controller that is responsible for periodically collecting traffic statistics and for distributing this information to the respective egress meter coordinators (Fig. 2). The meter coordinator receives the statistics from the meter controllers, computes the new

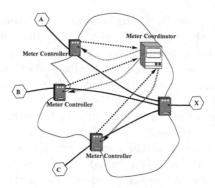

Fig. 2. CCDM using a centralized meter coordinator for a multiple-ingress single-egress SLA.

shares and then distributes them to the meter controllers. The meter controllers enforce these shares by reconfiguring the metering limits at the ingress routers. The meter coordinator acts as a centralized decision making entity for an SLA, running the share assignment algorithm.

There are many ways in which this type of mechanism can be deployed. One possibility is that customers are allowed to run fairly general code modules on the service provider's infrastructure (ingress routers and servers). This raises many security issues and such openness is not really necessary. Instead, in our design the customer provides two code modules to the service provider. The first one runs on the ingress routers and is responsible for collecting all relevant traffic statistics for the customer. These statistics can be measured using counters in the router's forwarding path, or they can be obtained using other means, e.g. higher-level signaling protocol. The second piece of code runs on the meter coordinator and computes the customized bandwidth distribution based on these statistics. The communication of statistics can be handled by the service provider, allowing it, for example, to aggregate the information for multiple SLAs.

Our proposed solution assumes a programmable network infrastructure, raising security questions with regard to the execution of customer code on service provider nodes. To deal with this issue, the customer code has limited reading and writing privileges on the host system. This follows the architecture that is used to secure control plane extensions in the Darwin extensible router [7]. We limit the statistical data that the ingress meter controller can obtain from the router allowing it to only read data belonging to "its" customer preserving privacy and limiting the ability to cheat. Similarly, the customer code can only configure its own meters and can not tamper with other customers' meters, trying to steal bandwidth. Any attempt to configure the meters is verified against the "policy envelope" to ensure compliance to the contract. The service provider can also limit the number and type of messages the meter controller is allowed to send and receive. The statistical messages are sent to the appropriate meter coordinator by service provider's code so that the customer code can not send malicious statistics to other customer's meter controllers and meter coordinators.

The meter coordinator is easier to secure since its task is to compute a bandwidth distribution based on a set of traffic statistics. It is more important to verify that the share assignment does not violate the SLA contract. Since the amount of bandwidth allowed for each ingress router is calculated by customer code, the service provider enforces the contract by making sure that the share assignment does not violate the "policy envelope". This can be done in a number of ways. One possibility is that the service provider checks and approves the meter coordinator code before it is installed. An alternative is to check the shares at runtime. It is trivial for the service provider to verify that the combined traffic shares provided by the meter coordinator module do not violate the SLA, i.e. they satisfy both per-ingress router and aggregate constraints.

We have described a centralized meter coordinator for multi-ingress single-egress SLAs. A distributed meter coordinator is more useful for a multi-ingress multi-egress SLAs where the egress nodes may also have to cooperate for assigning shares. The ingress nodes can multicast the statistics to all the other ingress nodes and each node can run the algorithm to compute their shares.

CCDM is consistent with the DiffServ philosophy of pushing complexity to the edge of the network by not requiring any information about the internal links of the service provider's network. Per update cycle, the total amount of data that has to be exchanged is proportional to the number of SLAs and the number of ingress nodes. As suggested above, aggregating the information from multiple SLAs can optimize the communication between the ingress routers and the meter coordinator. Since we don't expect a huge number of ingress routers, the centralized design is likely to be sufficient in practice.

5   Standard and Customized Policy Examples

We show three very different example policies to illustrate our solution. Table 1 shows some common parameters used in these examples.

5.1   Example 1: Fair Share Policy

Our first policy is a variant of the "Max-min fair share" policy [10] where each ingress node is given a bandwidth share proportional to its recent traffic load (demand), subject to, minimum and maximum share constraints. While this is clearly not a customized policy, we use it here to illustrate some of the features that are likely to be present in several bandwidth distribution policies. It is also likely to be an important policy in practice.

The goal is to calculate the bandwidth share S_{iX} for all ingress routers i. Informally, the bandwidth share must be proportional to the recent traffic volume (B_i) at node i, which can include past trends by using a dampening parameter (α) as follows:

$$\forall i, B_i = \alpha * B_{iX} + (1 - \alpha)B_i$$

Table 1. Notation of Parameters and Statistics Used for an SLA.

Param.	Description
SG	Statistical Gain Parameter. Allows buffer traffic ($SG \geq 1$)
K	Min. BW share. Controls the minimum share of an ingress ($0 \leq K \leq 1$)
γ	Rate gain parameter. Controls the max rate at which an ingress node can increase its rate in successive time intervals ($\gamma \geq 1$)
α	Dampening parameter. Allows longer-term traffic statistics ($0 \leq \alpha \leq 1$).
S_{iX}	Share of ingress i in bandwidth to egress X
L_X	SLA Limit i.e. total egress bandwidth
B_{iX}	Current incoming bandwidth at ingress i for egress X
U	Unused bandwidth of the SLA
n	Total number of ingress nodes
n_x	Total number of flows of type x
W_{iX}	Fractional weight of bandwidth share for ingress i
t	time interval for periodic updates

However, the bandwidth share is subject to a number of constraints. First, each ingress node is guaranteed a certain minimum bandwidth share (K). This is necessary so that idle or low bandwidth ingress nodes can increase their traffic volume in a timely way:

$$\forall i, S_{iX} \geq K * L_X$$

Second, there is a limit on how quickly a node can increase its share in successive time intervals. This is controlled using the rate gain parameter (γ):

$$\forall i, S_{iX} \leq \gamma * L_{iX}$$

The frequency of updates is based upon the required sensitivity to the traffic fluctuations. We denote this time interval parameter by t. This interval should be a multiple of the periodic interval that the meter coordinators use for exchanging the statistics so that the messages from different ingress nodes can be aggregated.

A final point is that if the sum of the bandwidth shares across the ingress routers is equal to the SLA limit L_X, the customer will never be able to fully utilize the SLA bandwidth because of shifts in traffic across ingress points leads to instantaneous entering traffic to be lower than the egress limit. In order to allow the customer to achieve a long-term average bandwidth closer to the SLA bandwidth, it may be appropriate to allow the customer to have an instantaneous bandwidth that is slightly higher than the SLA limit. We call such traffic "buffer traffic". The use of buffer traffic for performance reasons allows extra traffic to enter the network and it gets dropped at the egress meter if it exceeds the SLA limit. This can be realized by providing a statistical gain parameter (SG) that is negotiated between the customer and the service provider that limits the buffer traffic:

$$\Sigma S_{iX} \leq SG * L_X$$

The psuedocode used for the fair share policy is shown below. The traffic statistics needed are the average bandwidths B_i of the traffic arriving for this SLA at each ingress node during the period t.

1. Calculate everyone's weighted bandwidth $B_i = \alpha * B_{iX} + (1 - \alpha)B_i$
2. Calculate the assignment limit $L = L_X * SG$.
3. Find the unused bandwidth $U = min(L - \Sigma B_i, 0)$
4. Find the number of active nodes $m = \Sigma($ if $B_i > thr$ then 1 else 0$)$
5. For all the idle nodes, assign them the minimum share $S_i = K * L_X$
6. For all the active nodes sending less than the fair share $(\frac{L_X}{n})$, assign them $S_i = min(\frac{L_X}{n}, \gamma * B_i) + \frac{U}{m}$
7. For all the active nodes sending more than the fair share, assign them their fair share and a share of the unused bandwidth, $S_i = \frac{L_X}{n} + \frac{U}{m}$.
8. For all the active nodes sending more than the fair share, any remaining unallocated bandwidth (R) is distributed proportionally to their respective traffic, i.e. $S_i = \frac{L_X}{n} + \frac{U}{m} + (\frac{B_i}{\Sigma B_i}) * R$

5.2   Example 2: Strict Priority Policy

This policy is a variant of the Fair Share Policy described in Section 5.1. The difference is that instead of having fair (equal) shares for ingress nodes in times of contention, a certain ingress node has a higher priority in terms of getting bandwidth. Such a policy is useful when an ingress node carries more important traffic, e.g. traffic from strategic customers. The remaining bandwidth is distributed based on the pre-defined weighted shares of the other ingress nodes. The psuedocode for this is given below; j is the ingress node that has high priority.

1. Calculate everyone's weighted bandwidth $B_i = \alpha * B_{iX} + (1 - \alpha)B_i$
2. Calculate the assignment limit $L = L_X * SG$.
3. Assign the share to the high pty node j, $S_j = min(L_{jX}, \gamma * B_j + K * L_X)$
4. For all the remaining nodes, assign each one its weighted fair share of the remaining bandwidth (R), $S_i = min(\gamma * B_j + K * L_X, W_i * R)$
5. Any remaining unallocated bandwidth (R) is divided equally among those nodes for which $(\gamma * B_j + K * L_X > W_i * R)$

5.3   Example 3: VPN with Customized Bandwidth Management

Consider the case of a news agency that has a central server that receives live video news clips from all over the world. There are three priority types for news items. 1) Regular news items (R), 2) Special news items (S) e.g. a presidential address, and 3) Emergency news item (E) e.g. an earthquake or a war. The Emergency news items have the highest priority, followed by Special news and then regular. The news agency has a multi ingress, single egress agreement with the service provider for the aggregate EF traffic that it can send to the central server(egress) from different ingress points. The customer has a reasonable estimate, based on historical data, of what types of traffic it can expect, and

Table 2. VPN Policy Scenarios for Different Operating Ranges.

Scenario	Operating Range	Assignment Constraint
1	$n_E \leq 2$	$S_E \geq 0.1 * B_X$
2	$n_E > 2$ and $n_E \leq 5$	$S_E \leq 0.08 * n_E * B_X$
3	$n_E > 5$	$S_E \leq 0.5 * B_X$
4	$n_S \leq 3$	$S_S \geq 0.1 * B_X$
5	$n_S > 3$ and $n_S \leq 8$	$S_S \leq 0.06 * n_S * B_X$
6	$n_S > 8$	$S_S \leq 0.5 * B_X$

based on these estimates, it establishes a policy that centers on possible scenarios defined by using operating ranges. Table 2 shows an example set of operating ranges that the news agency can come up with. In the table, n_i represents the number of flows of type i for e.g. n_E represents the number of Emergency flows. Similarly, S_E represents the fraction of the traffic share for emergency traffic. Each row represents a scenario with an operating range and constraints on the share assignment. The scenarios can be revised over time as needed. Note that all flows belong to the EF service class and so the core of the network does not distinguish between them. It is only at the ingress routers that the three sub-classes are metered separately.

The traffic statistics used for this policy are specific to the customer, i.e. the number of flows where a "flow" is defined by the customer. It is the responsibility of the customer-provided ingress meter controller to collect this data, for example by asking the dataplane to monitor for certain types of special packets, or by using a signaling protocol that traffic sources can use to announce the start and end of incoming flows. Below is the psuedocode for this policy.

1. Calculate everyone's weighted bandwidth $B_i = \alpha * B_{iX} + (1 - \alpha)B_i$
2. Calculate the assignment limit $L = L_X * SG$.
3. For each type of traffic $t \in \{E, S, R\}$, find agg. traffic $b[t] += B_i[t]$
4. For each type of traffic (E, S, R) set $min[t] = 0$ and $max[t] = \gamma * b[t]$
5. Use the constraints of each applying operating range to close in on the $min[t]$ and $max[t]$ values
6. If for any t, $max[t] > min[t]$ then set $min[t] = max[t]$
7. Assign $min[t]$ share to all in the order E, S, R without exceeding L_X
8. From any remaining bandwidth, assign upto $max[t]$ share to all in the order E, S, R
9. Assign any remaining bandwidth to R.

6   Simulation

We implemented CCDM in the NS-2 network simulator [13]. We added several features of DiffServ to NS including traffic conditioning, SLA admissions control, support for different PHBs and SLA types [1] . We developed a Meter coordinator and meter controllers, which function as described in Section 4. In this section we describe our simulation results.

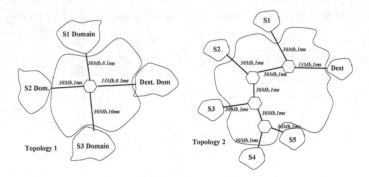

Fig. 3. Topologies Used for Simulation.

6.1   Simulation Results

We compare the performance of CCDM using the fair share policy defined in Section 5.1 with the simple approach of allowing all ingress nodes to send at the SLA rate and enforcing the SLA bandwidth limit at the egress router. While this simple approach allows the customer to send too much traffic into that network, it discourages this behavior by dropping the extra packets. Both solutions support flexible bandwidth distributions across the ingress routers.

In the first scenario we use topology 1 of Fig. 3. In each source domain (S1, S2 and S3) there are 30 TCP (ftp) traffic sources, each one attached to a separate node. The delays inside the domains are 0.01ms while the delays and bandwidths in the core are as shown in the figure. All the sources send traffic to the destination domain. We use an SLA limit of 10Mbs. Traffic from S1 starts at time 10s, from S2 at 0s and from S3 at 30s. The total simulation duration was 80s. No background traffic is used. The meters drop all the packets that exceed the metering limit.

Figure 4(left) shows how the bandwidth used at each of the three ingress routers varied over time in the absence of CCDM i.e. all ingress nodes meter at 10Mbs. We observe that the aggregate bandwidth is considerably higher than the SLA limit and we also see that during times of contention (e.g. 10-30s), the bandwidth distribution is not at all fair across the ingress nodes. The ingress nodes with the lower end-to-end delay takes more egress node bandwidth compared to the high end-to-end delay domains (S_1 vs. S_2).

We repeated the same scenario with coordinated metering. The parameter values used for the algorithm were, $\gamma = 2, t = 400$ms, $\alpha = 0.5, SG = 1.1$ and $K = 0.001$; note that the frequency of adaptation is very aggressive. This time (Fig. 4(right)) the fair share policy among the ingress nodes is adhered to more closely and during periods 10 to 70s, the difference between the throughput shares of individual source domains is small, achieving the goal of the policy. The difference that exists is because of an unequal capture of the extra bandwidth assigned by the statistical gain parameter. For an SG value of 1.0 the bandwidth difference is minimized, though the throughput is lowered. This shows that co-

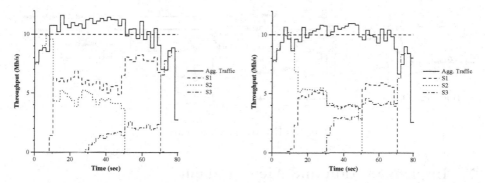

Fig. 4. SLA Traffic Entering Network Ingress Points without CCDM *(left)* and Using CCDM *(right)*. Traffic exceeding 10Mb is dropped at the egress router.

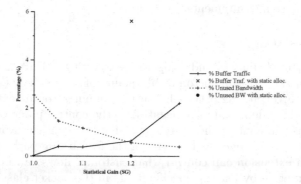

Fig. 5. Percentage Buffer Traffic and Percentage Unused Bandwidth for the SLA for different values of Statistical Gain (SG). Performance with static allocation is also shown.

ordinated metering works much better than egress metering only. We repeated similar simulations on topology 2 (Fig. 3) that provides for different levels of traffic aggregation from different ingress nodes. We saw a similar performance pattern. However there was even less unfair capture of bandwidth because of the relatively similar delays.

A comparison of the mean percentage buffer traffic and mean percentage unused bandwidth for different values of SG is shown in Fig. 5 for topology 1. The data point marked as × and as • in the graphs represents the performance without coordinated metering. Not surprisingly, we see that a higher value of SG leads to more buffer traffic and less unused bandwidth. An SG value of around 1.1 seems to be a good compromise. Both the unused bandwidth and the buffer traffic are low. Note that even with an SG value of 1 (no buffer traffic), the unused bandwidth is quite low (2.5%). It shows that share calculation tracks bandwidth very closely.

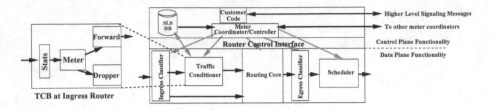

Fig. 6. Router Architecture.

7 Implementation and Measurements

We implemented CCDM and tested it over all the three policies described in section 5. In this Section we describe the details of our implementation and then present performance measurements.

7.1 Implementation

We have implemented DiffServ and Cooperative metering in Darwin [4], a system providing network support for customizable quality of service. The Darwin kernel is based on the FreeBSD 2.2.6 kernel. Darwin supports router extensions, i.e. customers can add code modules (delegates) to the control plane of the router. These extensions can control the customer traffic by using the Router Control Interface [8] (RCI) to change the behavior of the forwarding path of the router. For example, an extension can control what customer flows can benefit from a bandwidth reservation by loading specific filters in the packet classifier. Darwin restricts the actions of delegates to the traffic of the customer they represent [7].

The Darwin kernel provides a natural platform for implementing CCDM. We first added DiffServ traffic conditioning support to the input ports. Next we extended the RCI so that control plane extensions can set up traffic conditioning blocks (TCBs) and control their operation. For the SLAs used in this paper we use the TCB shown in Fig. 6, which drops all traffic exceeding the metering limit. For changing the bandwidth share on an ingress router, we use the RCI call, ds_param(tcb_id, meter_id, SET, DS_METER_LIMIT, new_value).

We provided support for dynamic metering in the form of a meter controller router extension. It runs on all the ingress routers, gathering local statistics, distributing them to the appropriate egress coordinators, and enforcing the assigned shares by dynamically configuring the ingress data path. The egress coordinators simply compute the shares of the ingress nodes based on the policy.

7.2 Experiments

We used the Darwin testbed (see Fig. 7) for evaluating our system implementation using the three policies described in Section 5. The testbed routers are connected by 100Mb Ethernet LAN. We deployed CCDM on the four routers Cozumel, Keywest, Aruba and Tobago where Cozumel acts as the egress and the others as ingress.

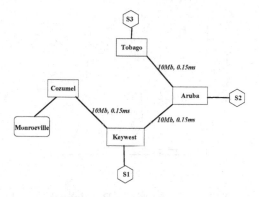

Fig. 7. Testbed Topology.

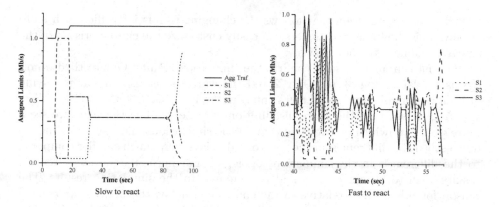

Fig. 8. Bandwidth distribution: stable with $t = 1s$ and $\alpha = 0.2$ *(left)*; more dynamic $t = 0.2s$ and $\alpha = 0.8$ *(right)*.

Fair Share Policy. We set up an SLA as defined in Section 5.1 to allow 1 Mbs from S1, S2 and S3 to Monroeville. We connected 12 TCP sources to each source domain and sent the traffic to the testbed for different time intervals. The delays are estimated by ping times. The parameters were set as follows: $SG = 1.1, \gamma = 1.3, t = 1sec, \alpha = 0.2$ and $K = 0.1$. S1 starts traffic at time 9s, S3 at 18s and S2 at 30s. Figure 8(left) shows the assigned shares for the different domains over time.

We observe that despite the different end-to-end delays, the sources share the bandwidth equally according to our fair share policy. Also, a new entrant easily and quickly captures its share. Similarly, when a source stops, its bandwidth share is quickly captured. The share assignments are very stable because of the high granularity of the update interval and the low value for α. The end-to-end times were in the 2-4ms range. If we want the share assigning strategy to capture more of the fluctuations, then we have to decrease the update time interval (t) and also increase α. In Fig. 8(right) we see a zoomed-in section of the assigned

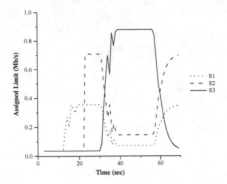

Fig. 9. Prioritized Sharing of Ingress Bandwidth: $t = 1s$ and $\alpha = 0.3$.

limits for a configuration that follows the changing trends in traffic much more closely. This shows how customers can easily customize the characteristics of the share assignment to their needs.

The measurement results confirm the simulation results. Our testbed topology is similar to Fig. 3, topology 2 and the workload is also similar in having quite a reasonable number of flows entering the network. For the policy of Section 5.1, we see that the share distribution is quite similar. However, we see a more even bandwidth distribution than in the simulations. The reason is that in our testbed the differences in the end-to-end delays were much smaller compared to the differences in the simulation topologies. Furthermore, the measurement results seem to be much more stable compared to the simulation results. The reason for this is that, relative to the end-to-end delay, the update interval (t) is larger (and more realistic!) providing stability to the metering limits. The measurement results increase our confidence in the validity of the simulation results.

Strict Priority Share Policy. Next we set up an SLA using the strict priority policy described in Section 5.2. We use the same topology and configuration. The workload was generated using 11 CBR UDP sources (`ttcp`)in every domain all sending at 10KBs (aggregate 0.88Mbs per domain). The parameters were set as follows: $SG = 1.1, \gamma = 1.3, t = 1s, \alpha = 0.3$ and $K = 0.1$. The high priority ingress gets a maximum of 80% of the share. The remaining two ingress nodes have a weighted share assignment ($\frac{1}{3}$ and $\frac{2}{3}$ in this case). In Fig. 9 we see that S3 (high priority) is able to get its high priority share when it is sending traffic (time=30-55s) and when it stops, its share is slowly given back to the other ingress nodes as controlled by α. The ratio of the shares of S1 and S2 is 1:2 throughout the run of the experiment (10-65s) showing that the share assignment is working correctly.

VPN with Customized Bandwidth Sharing. We set up the customized bandwidth sharing policy as described in Section 5.3. We use a subset of the

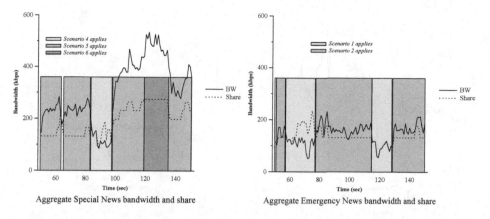

Fig. 10. Incoming bandwidth and shares assigned for Emergency and Special News with initial operating ranges.

testbed topology (Fig. 7) with aruba, keywest and cozumel. We used the Verbose_StarWarsIV_64.dat mpeg file trace data obtained from [11] for generating VBR UDP traffic. From each source domain several sources were started at different starting and ending times. Each source starts from a random location in the mpeg trace file and produces traffic accordingly. The parameters were set as follows: $SG = 1.1, \gamma = 3, t = 1s, \alpha = 1.0, K = 0$ and the SLA limit = $500Kbs$.

In Fig. 10 we see a comparison of the incoming traffic and the assigned shares for the Special and Emergency News repectively. We see that the assignment for Special News from 50 to 80s is too harsh because of scenario 5 (see Sect. 5.3, Table 2). Similarly between time 120 and 135s, Special News needs to be assigned more share which is being restricted by scenario 6. Emergency news is also unable to find enough bandwidth from 85 to 115s because of scenario 2.

We modified the constraints of these scenarios according to these observations and ran the experiment again. Now the Emergency share (Fig. 11) always meets the bandwidth requirements. The special news is also able to handle the demand much better. This example illustrates how the customer can quite easily customize and fine tune the distribution policy, based on its requirements. The meter coordinator customer code can gather such statistics and can forward them to the customer so that he can improve the share assignment code.

8   Conclusion

We have shown how our simple strategy, CCDM, can be used to handle a multi-ingress single-egress SLA in a customizable, flexible and conservative manner. The customer has the flexibility to provide its own specialized share assigning code leading to customized performance. We believe that this class of SLAs will form the basis of widely used traffic contracts by a DiffServ service provider. CCDM is necessary for the network providers for providing better service to the

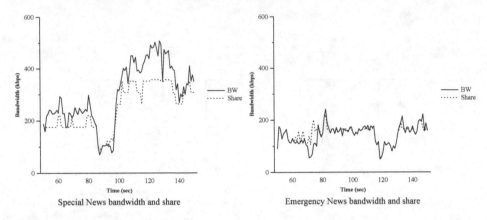

Special News bandwidth and share   Emergency News bandwidth and share

Fig. 11. Incoming bandwidth and shares assigned for Emergency and Special News with modified operating ranges.

end users as well as for efficient use of network resources from the customer's point of view. It enables the customer to control the performance of the SLA along different dimensions as we have shown by our experiments.

Further work is required to see how this basic coordination framework for a many-to-one SLA can be carried forward to a many-to-many SLA where the egress routers also share the resources using similar policies. The complementary issue of reserving network resources for such complex SLAs is another possible research direction. Finally, Another challenging issue is to handle these SLAs across different service provider domains.

References

[1] BERNET, Y., BLAKE, S., GROSSMAN, D., AND SMITH, A. An Informal Management Model for Diffserv Routers. Internet Draft, February 2001.

[2] BLACK, D., BLAKE, S., CARLSON, M., DAVIES, E., WANG, Z., AND WEISS, W. An Architecture for Differentiated Services. Network Working Group, RFC 2475, December 1998.

[3] BLELLOCH, G. Algorithms in the real world: Lecture notes (fall 1997).

[4] CHANDRA, P., FISHER, A., KOSAK, C., NG, E., STEENKISTE, P., TAKAHASHI, E., AND ZHANG, H. Darwin: Resource Management for Value-Added Customizable Network Service. ICNP'98, October 1998.

[5] DUFFIELD, N., GOYAL, P., GREENBERG, A., RAMAKRISHNAN, K., AND VAN DER MERWE, J. A Flexible Model for Resource Management in Virtual Private Networks. Proceedings of SIGCOMM, August 1999.

[6] FLOUDAS, C. A., AND VISWESWARAN, V. Quadratic optimization. Kluwer Acad. Publ. pp. 217–269, 1995.

[7] GAO, J., AND STEENKISTE, P. An Access Control Architecture for Programmable Routers. Fourth IEEE Conference on Open Architectures and Network Programming (OPENARCH'01), April 2001.

[8] GAO, J., STEENKISTE, P., TAKAHASHI, E., AND FISHER, A. A Programmable Router Architecture Supporting Control Plane Extensibility. IEEE Communications Magazine, 2000.

[9] KARMARKAR, N. A New Polynomial-time Algorithm for Linear Programming,. Combinatorica 4, pp. 373–395, 1984.

[10] KESHAV, S. An Engineering Approach to Computer Networking. Addison-Wesley, pp. 215–217, 1997.

[11] MPEG-4 AND H.263 VIDEO TRACES FOR NETWORK PERFORMANCE EVALUATION. http://www-tkn.ee.tu-berlin.de/research/trace/trace.html.

[12] OHLMAN, B. Receiver control in Differentiated services. Internet Draft, work-in-progress, expired, September 1998.

[13] THE NETWORK SIMULATOR - NS-2. http://www.isi.edu/nsnam/ns/.

[14] WEISS, W. Providing Differentiated Services through Cooperative Dropping and Delay Indication. Internet Draft, work-in-progress, expired, March 1998.

Segmented Adaptation of Traffic Aggregates

Hermann de Meer and Piers O'Hanlon

Dept. of Computer Science, University College London
{H.DeMeer,P.OHanlon}@cs.ucl.ac.uk

Abstract. Congestion control in heterogeneous Quality-of-Service (QoS) architectures remains a major challenge. The solution proposed in this paper entails three constituents. Taking the current trend towards Differentiated Services (DiffServ) as a likely candidate for future Internet QoS-architectures, our approach is based on aggregated, domain-based, and class-of-service based congestion control. The overall framework for congestion control, as suggested here, reflects essential properties of underlying QoS-architectures and their instantiations in real implementations. As such an approach calls for highly flexible architectures, we suggest the use of Active Networking, and in particular Application Level Active Networking, as an enabling technology for a seamless and rapid integration of the proposed scheme into current architectures.

1 Introduction

Congestion control in IP networks has traditionally relied upon mechanisms of the transport protocol TCP, and thus, on the cooperation of applications in the face of network congestion. With the advance of real-time multimedia communication applications on the Internet, on the other hand, a non-congestion sensitive usage of UDP has become increasingly popular for several reasons. The prospect of a massive usage of non-responsive transport protocols such as raw UDP, however, has created some concern about the viability of a "laissez-faire" approach that has traditionally been adhered to in the Internet [3].

In the same vein, issues in sharing bottlenecks and techniques of how to enforce fairness among competing elastic and non-elastic applications has triggered a hefty discussion within the Internet research community recently. While some form of fairness among competing flows appears appealing to one camp of researchers others are worried about the implied overhead of policing and enforcing control policies on a flow basis. To yet another camp, the idea of enforcing fairness in sharing bottlenecks does not appear as compelling at all. Disregarding implementation and run-time overhead, a fair sharing of bandwidth may not easily translate to a concept of fairness at the application level as QoS requirements and pricing conditions may vary vastly. While the concern about congestion control is being widely shared the jury is still out on measures of how to approach the problem [3].

QoS architectures within the confines of the DiffServ model are currently being introduced into the Internet to provide preferred handling of privileged traffic load classes [8]. Subscribers to such a privileged class expect to receive a higher

L. Wolf, D. Hutchison, and R. Steinmetz (Eds.): IWQoS 2001, LNCS 2092, pp. 342–353, 2001.
© Springer-Verlag Berlin Heidelberg 2001

level of service at a premium price. Within the confines of a DiffServ architecture, and in particular within Assured Forwarding (AF) subclasses, congestion may still arise and QoS violations may occur at times, despite traffic conditioning and resource provisioning. QoS architectures rely on class-dedicated resource provisioning of some sort and thus try to reduce the *probability of congestion occurrence* on the one hand and the *impact of congestion events* on privileged traffic on the other hand. Although the latter aspect may seem less obvious, we believe that it does reveal a most important aspect of the semantics of QoS architectures. Our approach is essentially based upon that observation.

Congestion can be seen as a possible cause (error) or "fault" in delivering a networking service with a required QoS. Since the possibility of such a QoS fault is likely to be an occasional matter-of-fact characteristic of any DiffServ architecture, we suggest that it is systematically taken into account as part of an overall model of QoS semantics of a given architecture. Such an approach is common practice in designing fault-tolerant distributed systems [2]. We therefore advocate to characterize QoS architectures not only in terms of assured transport privileges, but also in terms of adequate recovery measures in the presence of QoS faults as a result of congestion.

Assessing current trends in the design of QoS DiffServ architectures reveals inherent possibilities of QoS faults. Taking the derived fault models into account, we conclude the measures as presented in this paper. The suggested solution is based on the three elements of *aggregated congestion control, domain-based congestion control*, and *class-of-service based congestion control*.

Our approach is based on the introduction of new control and management functions into existing architectures. Introducing new services into an existing networking environment, however, is a challenge of its own, as re-programming of routers has traditionally been solely controlled by hardware vendors. While Active and Programmable Networking technologies have been pursued as a possible once-and-for-all solution to the often lamented inflexibility of networking infrastructure by the research community, there is still a long way to go for Active Networking infrastructure to be widely available [1]. Our pragmatic solution, therefore, relies on our Application Level Active Networking (ALAN) infrastructure, called FunnelWeb, that requires only a minimum support from the networking layer [4].

In Sect. 2, we first introduce the notion of QoS fault-tolerance measures in DiffServ architectures and then look at segmented adaptation as a special case of it. An illustrating test scenario completes that section. FunnelWeb, our Application Level Active Networking infrastructure, is introduced in Sect. 3, followed by presentation of an implementation of the segmented adaptation scheme based on that infrastructure. Section 4, which shows some of our simulation studies and first numerical results, is followed by the conclusion in Sect. 5.

2   Decomposition of Elasticity and Feedback

2.1   QoS Failure Semantics in Networks and Segments

Current discussions on DiffServ QoS architecting largely fall into two categories. One focus is on traffic conditioning and admission control at network edges and the other is on resource provisioning at the network nodes by defining so-called per hop behaviors (PHB). Combining proper traffic conditioning at the edges with nodes configured according to PHB specifications is designed to result in well-defined *edge-to-edge behaviors* for an *aggregate* traffic load. Such a behavior aggregate, whose scope is limited to an administrative domain as a *segment* of the network, is referred to as a per domain behavior (PDB) [6].

The ultimate goal in the current DiffServ bottom-up approach is to compose end-to-end QoS services from PDBs. Conditioning and provisioning are performed based on service level specifications (SLS) as part of more general service level agreements (SLA).

A wide spread belief is that adequately provisioned resources at the network nodes and carefully set conditioning elements at the edges can assure, or even guarantee, a targeted level of QoS. While that may be true for many conditions, there have not yet been any proofs of such procedures to reliably exist under fairly general conditions. Provisioning and conditioning are rather static and slow operations that need to account for repeated occurrences of aggregation/disaggregations at nodes of the a network segment, or to possibly deal with other hard to control stochastic events that may have an adverse impact on the delivered QoS of the network segment. While the likelihood of occurrence of congestion inducing events that may trigger a segmented QoS fault may be low, we nevertheless argue that precautions should be taken against them. This is particularly the case given the hierarchical structure of the DiffServ architecture in which end-to-end services may well be composed of a chain or mesh of PDBs. While a segmented QoS fault may be rare under ideal conditions, the whole chain could break if any segment exhibits a QoS failure, which is likely to occur an order of magnitude more often than a QoS failure of a single segment.

As a result of this analysis, we argue for the explicit extension of the QoS DiffServ model to incorporate (segmented) QoS fault tolerance. Our view is that this is in compliance with Huston's recent proposal for the next generation of DiffServ architectures [9]. Only we are suggesting a more systematic approach based on the well-understood technology of distributed fault-tolerant systems [2] and its application to the concept of (segmented) QoS faults. Expressing Huston's proposal in terms of these concepts and terminologies it could be rephrased as augmenting PDBs with QoS error state detection mechanisms in order to signal implied QoS fault events to the edge routers. The edge routers in turn may choose to trigger error recovery techniques in order to re-establish a well-defined QoS level according to a given SLA. The recovery action to perform would clearly depend on the QoS class itself as well as on other factors such as pricing. For graceful adaptation we suggest the introduction of *service curves* for definition of QoS levels within the SLA. The benefit of such an approach is that QoS guarantees can be expressed as simple functions of the service curves, as opposed to rigid definitions in traditional SLAs. This technique would allow for smooth movement between different service levels dependent upon network state and stipulated class of service through the selection of appropriate service curves.

In general, any disturbances could lead to an error such as congestion. Congestion may lead to a violation of the targeted QoS, referred to here as a QoS fault in that segment. Local error recovery methods such as adaptive QoS routing could mask the fault so that no QoS failure of the PDB becomes externally visible. But error recovery could also, and mostly will, rely on external cooperation as provided by contracted elasticity of up-stream domains to *adapt* the service-qualified load sensibly. Formally, the purpose of introducing QoS fault tolerance is to reduce to likelihood of "externally" visible end-to-end QoS failures. [1]

QoS error recovery actions would be performed on whole traffic *aggregates* in accordance with the DiffServ model and are suggested to be specified as part of (bilateral) SLAs among providers. No involvement of any flow-specific knowledge or measure is needed or wanted for the method to be effective. We see this as an argument in favor of scalability and flexibility. In analogy to TCP-based congestion control, one could refer to cooperating service providers as being *elastic*. We believe elasticity and cooperation among service providers in terms of realizing QoS fault-tolerance as vital for the success of the DiffServ model, in analogy to the viability of the best-effort model based on elasticity or TCP-friendliness of cooperating applications.

Once the DiffServ model has been adopted, we believe that the approach of segmented adaptation and other domain based error recovery methods as suggested here do not imply any further violations of the end-to-end principles [7]. On the contrary, incorporating QoS fault tolerance into the DiffServ model increases flexibility and mutual independence between applications and the network transport mechanisms as it re-establishes backward compatibility with TCP-like end-to-end congestion control [3]. It provides the means for self-protection of network segments and may even allow the introduction of *incentives* to applications for backing off, perhaps based on local access policies. We see segmented QoS error recovery as an essential means to enable the network to control its own resource utilization within the DiffServ framework. Although elasticity of applications does help significantly, the network should not *rely* on applications to cooperate in order to deliver its advertised services.

Our approach introduces a dynamic factor into SLAs and the execution of management tasks and traffic conditioning operations. Currently we limit ourselves to implementing simple QoS fault indication mechanisms and restrict the experimental setting to dynamically reconfiguring traffic conditioning operations to yield an adaptation effect on traffic aggregates. The concept, however, is by no means limited to these specific measures and is potentially open to any error state (or congestion) discovery, QoS fault notification, and QoS error recovery procedure. Depending on local policies and effectiveness of QoS error recovery, congestion signals may, however, propagate all the way back to access domains. Local access policies and strategies may become effective and non-elastic applications can, for example, be penalized. But these decisions would be made

---

[1] For more information on the the use of terms "error", "fault" and "failure" see [2].

locally by the access bandwidth brokers and would not impose any burden on the core or edge routers further down-stream.

The way adaptation of traffic aggregates is performed, as well as other mechanisms forming part of a QoS fault tolerance framework, also depends on the service class that is supported by a PDB in a given segment. This entails error detection schemes such as random early detection (RED) [3], QoS fault classification mechanisms such as identifying affected and responsible traffic aggregates, and also the adaptation strategy itself. A best-effort service, for instance, offers little clue about the nature, scope and duration of a congestion state. Entropy is reduced, however, in the case of DiffServ service classes. Here the occurrence of congestion may indicate longer term provisioning problems rather than coincidental short bursts that have an adverse effect. Adaptation may occur rarely but load reduction may have to last longer under these circumstances. We anticipate time scales to be in the order of session durations for Assured Service traffic aggregates. Expedited Service classes may require even longer down scaling of the traffic aggregate, perhaps permanently so, as a congestion state could be seen as a severe error that indicates a prohibitively poor provisioning strategy. While it seems to be obvious that service-class types have a significant impact on the best adaptation strategy for traffic aggregates, we are far from having final answers to these questions. We consider this for further research but regard a *differentiated adaptation* behavior that reflects service class differentiation as an essential constituent for segmented adaptation of traffic aggregates in the DiffServ framework.

2.2   An Economical Motivation

A definition of suitable SLAs including QoS fault tolerance mechanisms would be subject to *economically motivated* bilateral agreements and *technologically enabled* solutions. For example, imagine an up-stream provider having a contract with a down-stream provider to forward 5Mbps of a certain Assured Forwarding (AF) QoS-type of traffic aggregate through its domain. Minimizing the percentile of QoS failures by overprovisioning the down-stream networking segment may get excessively costly beyond a certain percentile threshold. In contrast, it might be much less expensive, and thus to the mutual benefit, to agree on clearly stated QoS fault models, fault notifications and recovery procedures. The resulting QoS fault tolerance may involve cooperation between neighbouring providers such as reducing the aggregated rate to 2Mpbs, say, that is sent down-stream upon receiving a congestion signal. Nothing is said of how the up-stream provider achieves the requested QoS recovery measure, be it by re-setting traffic conditioning parameters at its corresponding egress link or by applying QoS routing and traffic management techniques, or any combination thereof and further measures. The down-stream provider would rely on the potential - and rare - case of cooperative elasticity in the way it manages and runs its own network segment. Of course, the necessary QoS fault-tolerance mechanisms have to be in place as well, which is a cost factor too, so a trade-off analysis may reveal the optimum point of operation based on all given local conditions and QoS requirements.

Specifying QoS failure semantics and provisioning of QoS fault tolerance based on bilateral agreements may not only enhance the end-to-end QoS as to be experienced by applications but is also envisioned to lead to a mode of operation that allows minimization of constraining cost functions.

2.3   A Test Scenario

A test scenario is given in Fig. 1. Four autonomous systems (AS) are interconnected by properly defined DiffServ links with the capacity as indicated in the legend of Fig. 1. All indicated intra-AS links are also considered to be DiffServ capable. Edge routers are labeled by names with initial letter E, followed by a number indicating the particular AS the router belongs to and a symbol indicating the particular router in that domain. Core routers are indicated by an initial letter C. Hosts are simply designated by the source running on it.

Assured Forwarding (AF) links have been set up between the nodes with the edge routers configured for time sliding window with three color marker (TSW3CM) policier. The initial state is that four hosts are transmitting traffic. One FTP source is running over a TCP connection from end host $src1$ in AS1 to its sink in AS3. One 4Mbps UDP constant bit rate source (with randomized departure times to avoid phase effects) is running from $src2$ in AS4 to its sink in AS3. Similarly, $src3$ in AS4 is transmitting a 2Mbps CBR stream to its sink in AS3. And, finally, a third CBR UDP source $src4$ is transmitting 4Mbps UDP traffic to its sink in AS3.

We have artificially created the congested AF link c3r1–c3r2 in AS3 by defining the traffic matrix and choosing the link capacities accordingly. For the sake of demonstration we have exaggerated the congestion event as we are not as interested in an exact modeling of congestion occurrence at this point, but in the effectiveness of our congestion recovery methods in relation to a given AS.

Congestion Notification (CN) signals originating from link c3r1–c3r2 in AS3 are propagated back to the edge routers and adaptation of traffic aggregates is performed to re-condition the traffic load that enters AS3 at the particular ingress link. Two solutions are possible in this case, either the advertised capacity of egress link c2r3–e3r1 or of the ingress link e3r1–c3r1 can be reduced accordingly. As results presented later suggest, the latter action is considered as less desirable and should only be used as a intermediate step for self-protection of AS3. The goal, however, is to initiate a reduced traffic flow from AS2 to AS3 with respect to a congested traffic class. This is considered a method to recover from a segment-AS3 QoS-fault and is achieved by providing elasticity of AS2.

Once adaptation of the traffic aggregate on the identified ingress link of AS3 has been performed properly, AS3 should be well protected and be back to normal operation, with no congestion in the core nodes. The edge router of AS2 may decide to forward the CN signal to AS4 in order to trigger a reduction of advertised AF capacity of link e4r1–e2r2. If that step is performed properly, both AS2 and AS3 should have recovered from a possible segmented QoS failure and be back to normal AF forwarding mode, albeit at a potentially reduced rate. It is up to AS4 to decide on quenching source src2 or on taking other recovery actions internally. Of course, the simple scenario indicated in Fig. 1 is likely to

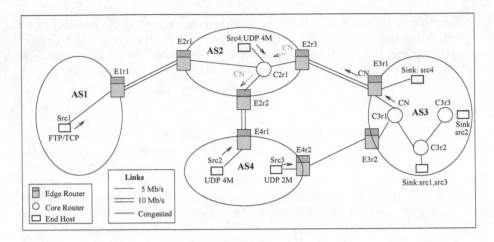

Fig. 1. Test Scenario.

be a partial representation only, in reality many more links and nodes would be available, providing more options for recovery.

Given that aggregated links may be shared by many traffic types in DiffServ we would like to mention the fact that the presented scenario implicitly entails QoS routing, as flows *src2* and *src3* follow different paths, although originating and terminating in the same ASs. While not particularly significant in the results presented in this paper, our overall approach does take QoS routing into account. However, we will not further elaborate on QoS routing in this paper.

3   An Implementation Based on ALAN

3.1   The Concept of ALAN

The concept of Application Level Active Networking (ALAN) has been developed to introduce flexibility and openness into networking architectures, at the application level, without compromising performance and security of core networking functions [4].

ALAN differs from network layer Active Networking (AN), where packets carry active code which may be executed at the network level in devices such as routers. The AN approach attempts to solve the flexibility problem while creating a number of new challenges, in particular, with respect to security, network stability and performance. In contrast, the ALAN approach leaves the basic network infrastructure unchanged and provides a framework for deployment of application level active services. General purpose computational nodes are placed inside the network at strategic locations to host execution environments that can be dynamically installed to perform application level functions that, however, interfere in a very limited way with network transport mechanisms. This concept is similar to current methods in optimization of content delivery in the Internet that have been widely adopted by the industry. The main difference is that we are deploying general purpose "boxes" that are designed and intended to be easily re-programmable.

3.2 FunnelWeb

We are using the *FunnelWeb* [4] implementation for ALAN - see Fig. 2. This system consists of deploying active elements in the network which provide an application level execution environment. In FunnelWeb the active applications, or active services, take the form of *proxylets*. Proxylets are written in Java and are loaded by reference. Proxylets may be loaded and executed on active node machines running an *Execution Environment for Proxylets* (EEP). There are interfaces for monitoring and control of the proxylet on the EEP.

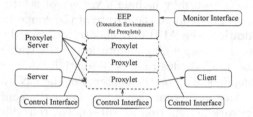

Fig. 2. FunnelWeb Architecture.

3.3 Active Segmented Adaptation

In our approach, adaptation of inter domain traffic is to be performed at the edge routers of autonomous domains according to pre-negotiated, bi-lateral, service curve oriented SLAs. Normally, traffic crossing network boundaries is handled within negotiated standard SLS operation. An SLS may specify what to do in the case of out-of-profile traffic [10], resulting in some level of traffic conditioning at the edges such as dropping or re-classification. In contrast, dynamic adaptation of traffic aggregates may well be performed on *in-profile* traffic in the *exceptional case* of congestion events. Adaptation may be implemented in any form that reduces the relevant traffic load class, be it by dropping, re-classification or re-shaping. The resulting action is dependent upon the local and class-based QoS failure semantics, the correspondingly suggested recovery methods, and further conditions such as pricing that may be included in a bi-lateral SLA.

Active services in the form of proxylets are dynamically deployed to augment and customize traffic conditioning at the edges. We refer to this approach as an active bandwidth broker (ABB). An ABB also forwards and filters congestion signals to neighbouring domains according to actively extended SLAs. A dynamic change of traffic conditioning can be accomplished easily as the required traffic conditioning modules are already in place due to the underlying DiffServ architecture. Only an interface is needed for a dynamic parameterization of the conditioning modules. The change of the parameter settings is triggered by the proxylets. Local policies can easily be incorporated and changes to policies can be realized *on demand* due to the inherent flexibilty of FunnelWeb.

Segmented adaptation is designed to utilize explicit CNs from the core of an AS to its edges. These congestion signals may be generated when certain resource utilization levels at a router exceed given thresholds. Such notifications contain information from packet headers regarding the sources of congestion. At the current point we plan to multicast notifications from the congested router to the proxylets attached to the edge routers. It may also be possible to unicast such notifications by usage of an active intermediary that would be able to specifically route the CN signal.

Once the proxylets at the edge router, as a constituent of an ABB, receives a CN they may decide to comply with it or to discard it, depending on the valid policy. Type of action taken may include a variety of strategies to control aggregates' behavior dependent upon the nature of the congestion and the service class affected. In addition, the ABB may choose to forward the CN to peering ABBs and to trigger a corresponding adaptation of the traffic aggregate at corresponding actively extended edge routers (potentially more responsible for the congesting traffic) dependent upon the contents of the CN and installed policy. Action taken would have a lifetime dependent upon the continuing congestion state of the core network as reported by subsequent refreshing CNs. If neither CN signals are received from the defective down-stream domain, nor any arriving traffic exceeds the lowered egress rate, after a "reasonable" period, recovery from an AS QoS fault is assumed to be completed and the link is opened again to full capacity. Referring to Fig. 1, when recovery has been completed for AS3 and AS2 only link e4r1–e2r2 remains scaled down as long as recovery has not been completed within AS4, all other AF links would be back to normal. Note that the current procedure of AF adaptation is based on intuition, but more systematic, preferably formal modeling, studies are needed.

We are implementing the system using two proxylets as illustrated in Fig. 3. The *SLA proxylet* provides facilities for implementation of SLS service curve behaviour which interfaces with the conditioning elements. The SLA proxylet also specifies the conditions in which peering ABBs should get involved and propagate the congestion notification signal propagate further up-stream. The *ASA proxylet*, where ASA refers to active segmented adaptation, controls operation of QoS error recovery in terms of aggregated adaptation as discussed earlier.

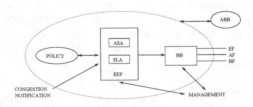

Fig. 3. Active Bandwidth Broker.

Using FunnelWeb provides high flexibility to dynamically change policies or to update adaptation and other recovery procedures upon demand. In our case, FunnelWeb allows us to rapidly and seamlessly integrate SLA extensions into existing networking infrastructures to create our customized testbed.

4   Some Simulation Results

Referring to the test scenario in Fig. 1, we performed some early simulations to study the effectiveness of adaptation of traffic aggregates. The simulations have been carried out using The Network Simulator [5]. While our focus is on inter-domain adaptation of traffic aggregates, we also present additional results on intra-domain aggregated adaptation on links local to AS3. This provides more systematic insights into the impact of adaptation steps. Indeed, the concept would also allow adaptation on local links as an immediate recovery method. The first graph in Fig. 4 depicts the flow specific throughput in the initial state of the network with the indicated congested link in AS3 of Fig. 1. Src1, src2, and src3 share the congested link. As a result, the TCP flow is almost completely starved while src2 and src3 share the remaining bandwidth but both suffer from considerable losses. Upon reception of CN signals, the total bandwidth on ingress link e3r1–c3r1 is immediately reduced to 2Mbps, shifting the congestion towards edge router e3r3. Such a measure could be useful as an immediate measure to shift a bottleneck away from a hot spot to increase the overall healthiness of a domain. The resulting situation can be seen in the middle graph of Fig. 4. The TCP session is still completely starved but at least the QoS of src3 is back to normal and the throughput is back to the targeted rate, link c3r1–c3r2 has recovered from its congested state and AF QoS is guaranteed again for that part of segment AS3. In a practical setting, this step could be seen as a realization of self-protection of the congestion affected AS.

Next, the CN signal is propagated further up-stream to edge router e2r3 in domain AS2. At which point the advertised bandwidth of link e2r3–e3r1 is reduced until no more CN signals arrive. Domain AS3 has now recovered completely and is again fully complying with a PDB specification of the AF QoS class. Note, the approach does not boil down to a hop-by-hop control mechanisms - only edge routers are involved in the adaptation process.

Finally, link capacity e4r1–e2r2 is reduced to 2Mbps and all other link restrictions are released shortly afterwards, as there is no congestion observed on any link of AS2 or AS3. The resulting effective throughput is given in the right graph of Fig. 4. AS2 and AS3 are not only loss free and fully back to AF forwarding mode, but TCP src1 has gained it's share of link c3r1–3r2.

From the simulation studies it is clear that there are difficulties in specifying reasonable adaptation strategies. So far we have only relied on our intuition, but more systematic investigations are obviously needed here. It should be noted, however, that the problem of defining strategies in back-propagation of CN signals and performing up or downscaling of QoS DiffServ virtual links appears similar in nature to matching of traffic conditioning and resource provisioning in general. Hence, this problem must be solved for the DiffServ model to

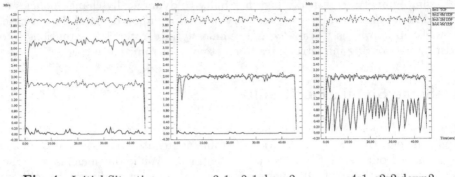

Fig. 4. Initial Situation.   e3r1–c3r1:down2.   e4r1–e2r2:down2.

become completely viable [9]. Once there are reasonable solutions for the provisioning/conditioning problem, these results should also be applicable to the related adaptation problem of traffic aggregates. But as we doubt there will ever be a perfect solution to this problem, segmented QoS fault tolerance and adaptation of traffic aggregates may be as essential to the the viability of the DiffServ model, as the solution of the provisioning/conditioning problem itself.

5   Conclusions

In this paper we have introduced the concept of segmented adaptation of traffic aggregates. The concept reflects essential characteristics of evolving DiffServ QoS architectures such as those being built on the principle of composing end-to-end QoS services from autonomous segments that are characterized by their PDBs. The entities of interest are traffic aggregates of given quality and quantity, which are negotiated between neighbouring providers Adaptation of traffic aggregates is motivated by the observation that PDBs are likely to be defective with respect to the delivered service quality at times.

Segmented QoS fault tolerance mechanisms, of which adaptation of traffic aggregates is designed to be a particular case, are to be provided to enhance end-to-end QoS and to confine or to reduce the probability of occurrences of end-to-end QoS failures. We envision segmented QoS fault tolerance as being the necessary "glue" to form end-to-end QoS services from well-defined PDBs as building blocks. Our approach has been built on an extended concept of distributed bandwidth brokers that control the adaptation of traffic aggregates according to policies provided by SLAs. An actively extended SLA specifies how, according to a service curve, service-class dependent parameters and values of an SLS are altered upon reception of a congestion signal.

Building adaptation on edge-to-edge segmentation and aggregation of traffic is believed to result in a highly scalable approach for congestion control, or, more generally, QoS fault tolerance. Decoupling applications and network transport mechanisms further within the DiffServ model is likely to increase the overall flexibility in terms of the end-to-end argument. The introduction of network self-protection as an immediate consequence of the approach eliminates the reliance on the cooperation of applications in the face of congestion, whilst being fully backward compatible with TCP-friendly schemes for congestion control.

As a proof of concept, we are working on an implementation of a testbed, based on our ALAN platform that allows the introduction of new services into an existing networking infrastructure. While early simulation results on the effectiveness of segmented adaptation of traffic aggregates have been discussed in this paper, there is large scope for exploration of the area. Some examples of open issues are finding reasonably good procedures and parameter settings for error (congestion) detection, identification and classification of segmented QoS faults of traffic aggregates or defining service-class specific adaptation procedures properly. Other pending problems are investigating the effectiveness of combining adaptation of traffic aggregates with other segmented QoS fault masking techniques such as QoS routing, or defining strategies for propagating congestion signals further up-stream, and the security issues associated with such procedures. Each mentioned aspect opens up a new potential for research within the framework of segmented QoS fault tolerance, in general, and segmented adaptation of traffic aggregates, in particular.

References

1. Campbell, A.T., H. De Meer, M.E. Kounavis, K. Miki, J.B. Vicente, and D. Villela : "A Survey of Programmable Networks"; ACM SIGCOMM Computer Communications Review, (Apr. 1999).
2. Cristian, F. : "Understanding Fault-Tolerant Distributed Systems"; Communications of the ACM, Vol. 34 (2), (Feb. 1991) 56–78.
3. Floyd, S. and K. Fall : "Promoting the Use of End-to-End Congestion Control in the Internet"; IEEE/ACM Transactions on Networking, Vol. 7(4) (1999) 458–472.
4. Ghosh, A., M. Fry and J. Crowcroft : "An Architecture for Application Layer Routing"; Yasuda, H. (Ed), Active Networks, LNCS 1942, Springer: (2000) 71–86.
5. The Network Simulator ns2, http://www.isi.edu/nsnam/ns/.
6. Nichols, K. and B. Carpenter : "Definition of Differentiated Services Per Domain Behaviors and Rules for their Specification"; Internet Draft, http://www.ietf.org/internet-drafts/draft-ietf-diffserv-pdb-def-03.txt, (Jan. 2001).
7. Saltzer, J.H., D.P. Reed, and D.D. Clark : "End-to-End Arguments in System Design"; ACM Transactions of Computer Systems, Vol. 2 (4), (Nov. 1984) 277–288.
8. Blake, S., D. Black, M. Carlson. E. Davies, Z. Wang, and W. Weiss : "An Architecture for Differentiated Services"; http://www.ietf.org/rfc/rfc2475.txt, (Dec. 1998).
9. Huston, G. : "Next Steps for the IP QoS Architecture"; http://www.ietf.org/rfc/rfc2990.txt, (Nov. 2000).
10. Goderis D., et al. : "Service Level Specification Semantics, Parameters and Negotiation Requirements", http://search.ietf.org/internet-drafts/draft-tequila-diffserv-sls-00.txt , (Jul. 2000).

Differentiated Services with Lottery Queueing

Joseph Eggleston and Sugih Jamin*

Electrical Engineering and Computer Science Department
University of Michigan, Ann Arbor, MI 48109-2122
{jeggle,jamin}@eecs.umich.edu

Abstract. This work investigates the benefits and drawbacks of using a lottery to schedule and drop packets within a router. We find that a lottery can provide many distinct levels of service without requiring that per-flow state be maintained in the routers. We also investigate the re-ordering within a flow that can occur if a lottery is used to pick packets to forward without regard to their flow order.

1   Introduction

The current Internet infrastructure provides the same level of service to all packets, namely Best Effort service. This type of service has proven to be adequate for many applications. However, given the heterogenous nature of traffic on the Internet and that network bandwidth on the Internet is a limited resource, a means for providing different levels of service to different traffic would be beneficial. Higher levels of service could be used to meet the QoS needs of certain applications (e.g., streaming audio and video), or to simply satisfy the demands of users who are willing to pay for improved service.

Before different levels of service can be provided, the network needs a way to determine the level of service packets should receive. There are two prominent methods for making this distinction. Either the router maintains enough state to identify the level of service a packet should receive based on the flow it belongs to, or packets carry the state needed to identify their service requirements. The Integrated Services (IntServ) framework [1] is a system based on the first method, while the Differentiated Services (DiffServ) model [2], [3] proposes the second.

Because it requires routers to maintain per-flow state, IntServ is generally considered less scalable than DiffServ. DiffServ preserves the stateless nature of core routers thus guarding their scalability. It does this by distinguishing between core routers and boundary routers (also known as edge routers). Boundary routers (ingress, egress, or possibly both) perform traffic shaping and policing of flows leaving the core routers to forward packets based on state carried in packet headers. This state identifies the aggregate traffic class to which that packet belongs. Core routers use this state to determine how the packet should be treated, that is, the state determines the Per-Hop Behavior

* This project is funded in part by DARPA/ITO grant F30602-97-1-0228 from the Information Survivability program. Sugih Jamin is further supported by the NSF CAREER Award ANI-9734145, the Presidential Early Career Award for Scientists and Engineers (PECASE) 1999, the Alfred P. Sloan Research Fellowship 2001, and by equipment grants from Sun Microsystems Inc. and Compaq Corp.

(PHB) the packet receives. PHB groups typically define a few traffic classes and the associated forwarding behavior those classes receive. Examples of PHBs are Assured Forwarding [4] and Expedited Forwarding [5]. Assured Forwarding prescribes four traffic classes with up to three drop precedences per class. Differentiation is achieved by assigning each class a portion of the available bandwidth and buffer space, and through use of the drop precedence. Expedited Forwarding provides one high priority class with guarantees of a certain amount of bandwidth available for packets in that class.

This work investigates the aptitude of Lottery Queueing for providing Differentiated Services. Lottery Queueing is based on the CPU scheduling mechanism presented in [6].[1] In Lottery Queueing, each packet carries a bid value set at the sender. Routers use the bid value to distinguish between packets waiting in their forwarding queues. The bid value can be used both to determine the next packet to forward and, when necessary, to choose a packet to drop. In this way, Lottery Queueing can provide differentiation in either latency, or drop rate, or both. Ideally, there will be a large range of bid values available which will allow for many levels of service. Because bid values are carried in the packet headers, it preserves the stateless nature of routers implementing it.

In this paper we show that not only can Lottery Queueing provide differentiation in both latency and drop rate between packets carrying different bid values, it does so in a manner proportional to the bid values. That is, Lottery Scheduling can provide many levels of service dependent on the number of unique bid values.

Using a taxonomy similar to that in [2], the service provided by Lottery Scheduling is best described as Probabilistically Relatively Quantified. That is, in the aggregate, packets with higher bids will receive better service than packets with lower bids. However, because it is probabilistic, individual packets with a higher bid may experience worse service than a packet with a lower bid.

Probabilistically Relatively Quantified service may be adequate for some applications. For example, assuming that Best Effort traffic is assigned a bid value of one, someone simply desiring faster than Best Effort service could use a bid value greater than one. However, some applications may have more specific service requirements. If the sender knows that a certain packet carries "important" data, it could use a higher bid value for that packet. Dynamically changing the bid value could also allow a flow to maintain a certain level of service despite changing network conditions. Because it is the receiver that knows when service requirements are not being met, a feedback mechanism from the receiver to the sender could be used. This feedback mechanism could take the form of explicit requests for a bid change, or could be inferred at the sender using Acks. The key attribute of Lottery Queueing that allows user feedback to the network for meeting QoS needs is its ability to allow fine grain changes in service dependent only on the number of bid values available.

Because allocating bandwidth is a zero-sum game, there must be some penalty that prevents users from trying to grab the largest share by always using the maximum bid value. The most obvious such penalty is charging users more for using higher bid values. Differentiated Services assumes that a customer will negotiate a contract with a network provider that specifies the guarantees provided by the network provider and

---

[1] The authors of [6] uses the term Lottery Scheduling. We also use that term, but for a specific aspect of Lottery Queueing.

the restrictions on the traffic the customer can send into the provider's network. This agreement is referred to as a Service Level Agreement (SLA). SLAs will be recursively formed between neighbor networks from the sender to the receiver so that some type of end-to-end guarantee exists. One way that Lottery Queueing could fit into this framework is for domains to agree to forward a certain amount of traffic at a set bid value. For example, there could be an agreement to forward 10 Mbps of bid 10 traffic. A more interesting use of Lottery Queueing, the one we focus on, allows senders to set and dynamically change the bid value of their flows. For this purpose, an SLA could include pricing agreements for packets carrying different bid values. If restrictions are not placed on the amount of traffic carrying each bid value, the result will be a market for available bandwidth where a larger share of the bandwidth goes to those who pay more for using higher bid values.

In Sect. 2 we describe the mechanisms behind Lottery Queueing. Section 3 relates how we implemented the test system. Section 4 presents experimental results. It includes experiments on drop rate, queueing latency, and packet re-ordering. Finally, Sect. 5 concludes the paper.

1.1 Related Work

Lottery Queueing does not provide fair share scheduling in the sense of Fair Queueing [7]. If all flows use the same bid value, the flow that sends the most will get the largest share of bandwidth. However, if flows are charged for the packets they send, Lottery Queueing could be considered to achieve a type of fairness where paying more gives you better service.

Core-Stateless Fair Queueing (CSFQ) [8] and Core-Jitter Virtual Clock (CJVC) [9] both endeavor to provide Fair Queueing while maintaining the stateless nature of the network core. CSFQ uses edge routers to mark packets with their flow's arrival rate and uses that information in the core to provide Fair Queueing at core routers. It does not provide for allocating different portions of the bandwidth to different flows. CJVC goes beyond Fair Queueing and provides guaranteed service for flows. However, it assumes a reservation based admission control policy. While Lottery Scheduling does not provide guaranteed service, its more relaxed service model can provide many levels of differentiation with less strict admission control such as Measurement-Based Admission Control [10] or no admission control.

RED [11] provides a means for congestion avoidance. RED routers track the average queue length. When the average queue length is less than a set minimum value, no packets are dropped. When the average queue length is greater than a set maximum value, all packets are dropped. Average queue lengths between the minimum and maximum values cause packets to be dropped with a certain probability. RIO [12] extends the RED system to provide two service levels by distinguishing between *in* and *out* packets. *Out* packets are dropped earlier and with a higher probability than *in* packets. It would be possible to extend Lottery Scheduling to provide a congestion control mechanism combined with service differentiation as in RIO. We already use bid values to differentiate between packets when dropping packets due to queue overflow. This could be extended to include Early-Drop with lower bid packets having a higher probability of being dropped early. We leave exploration of this possibility as future work.

2 Packet Scheduling

We describe the queueing systems we have investigated. We focus on Lottery Queueing for its potential to provide many levels of service with proportional sharing. For comparison purposes, we also examine Deterministic Queueing which uses the bid value to forward based on static priority. A queueing system consists of the scheduling discipline, which determines the next packet to be served, and a dropping policy, which determines the packet to drop when the queue is full.

2.1 Scheduling Disciplines

We study two scheduling disciplines that can differentiate based on a bid value: Lottery Scheduling and Deterministic (Static Priority) Scheduling. For comparison purposes, we also include traditional FIFO (first-in-first-out) scheduling in our study.

Lottery Scheduling is a probabilistic scheduling method. The next packet to be served is chosen by holding a lottery with the probability of an individual packet being served being proportional to its bid value:

$$\Pr[\text{packet } k \text{ is served}] = \frac{B_k}{\sum_j B_j},\qquad(1)$$

where B_i is the bid value of packet i; j represents each packet in queue. With lottery scheduling, packets with higher bids have a greater chance of being served next. Therefore, during times of congestion higher bid packets should typically experience lower queueing delay than packets with lower bids. However, since low bid packets nevertheless will have a chance of being served, they won't be starved. Even if high bid packets keep arriving low bid packets will still receive a share of the bandwidth proportional to their bid value.

One side effect of Lottery Scheduling is that packets in a flow have a higher probability of arriving out of order at the receiver. When a flow has more than one packet in a router's queue, each packet has equal probability of being forwarded next (assuming the flow has not changed the bid value carried by its packets). We explore this aspect of Lottery Scheduling in more detail in Sect. 4.4.

An alternative form of Lottery Scheduling that does not suffer from this problem would maintain the order of a flow's packets by always sending in FIFO order within a flow. This does not require a router to maintain state for every flow passing through it, it only needs to keep track of flows that currently have packets in the queue. If there are two packets from the same flow in the queue, the router would ensure that the first packt to enter the queue is forwarded first.

Deterministic Scheduling always forwards the packet with the highest bid. If the queue contains multiple packets with the same highest bid value they are served in FIFO order. As long as a flow does not change its bid value, Deterministic Scheduling will not re-order queued packets belonging to a flow.[2] The key difference between Lottery Scheduling and Deterministic Scheduling is that Lottery Scheduling provides

---

[2] However, packets can still arrive out of order at the receiver due to network topological or routing changes.

proportional scheduling. That is, with Lottery Scheduling, all flows will receive service at a level proportional to their bid value; Deterministic Scheduling allows starvation of lower bid packets if higher bid packets keep arriving.

2.2 Dropping Policies

We consider three dropping policies: Lottery Drop, Deterministic Drop, and the standard Drop Tail policies.

The mechanism behind Lottery Drop is analogous to that of Lottery Scheduling. When a new packet arrives at a full queue it is placed in the queue and a lottery is held to determine which packet to drop. For Lottery Drop the probability of an individual packet being dropped is proportional to the inverse of its bid value.

$$\Pr[\text{packet } k \text{ is dropped}] = \frac{1/B_k}{\sum_j 1/B_j} \tag{2}$$

Hence higher bid packets are less likely to be dropped than lower bid packets.

Deterministic Drop selects the lowest bid packet in the queue to drop. If there is more than one packet with the same lowest bid, the latest to arrive is dropped.

2.3 Interactions of Scheduling and Dropping Policies

As stated above, a queueing mechanism is defined by its scheduling discipline and dropping policy. When combined into a single queueing mechanism, the scheduling discipline and dropping policy are no longer independent. For example, focusing on the time spent by a single packet in the queue, the number of packets served prior to this packet influences the chances of this packet being dropped. Likewise, the number of other packets dropped influences the amount of time the packet must wait before it sees service. Hence, it is important to compare complete queueing mechanisms rather than just the forwarding or dropping policies. The following are the combinations of scheduling disciplines and dropping policies we use in this study:

1. FIFO scheduling with Drop Tail (FSDT). This is the traditional router queueing mechanism.
2. Lottery Scheduling with Drop Tail (LSDT). Lottery Scheduling provides lower latencies for higher bid packets. Drop Tail guarantees that once a packet enters the queue it will eventually be served.
3. Lottery Scheduling with Lottery Drop (LSLD). This combination tends to favor higher bid packets in both forwarding and dropping. A packet that has entered the queue is not guaranteed to be forwarded.
4. FIFO Scheduling with Lottery Drop (FSLD). This combination favors higher bid packets only when there is enough congestion to cause the queue to overflow.
5. Deterministic Scheduling with Deterministic Drop (DSDD). This combination always forwards the highest bid packet and drops the lowest bid packet.

3 Implementation

In this section we describe our implementation of the studied packet queueing systems. [3]
This implementation has not been optimized; there are certainly more efficient ways of
implementing Lottery Queueing. The implementation was intended only to allow us
to study the behavior of Lottery Queueing. We have implemented our new queueing
mechanisms in the FreeBSD 3.2 operating system kernel. The main modifications are
changes to the general Ethernet driver and the Ethernet interface device driver [13].
While two queues exist for each interface, an input queue and an output queue, we
modify only the output queue since the CPU on our machines processes packets fast
enough to avoid queueing in the input queue. There are only two modifications to the
kernel queue data structure: the addition of a variable to hold the sum of all bids in
the queue and a variable to hold the sum of all inverse bids. These variables facilitate
holding lotteries for scheduling and dropping packets as described below. Whenever a
packet is added to or removed from the queue these variables are updated.

Packet Enqueueing. Enqueueing is done by the general Ethernet driver. This is also
where packets are dropped if the queue is full. The standard Drop Tail implementation
simply adds the packet to the end of the queue. If the queue is full, the packet is dropped
and the memory used by the packet is freed.

When using Lottery Drop, an incoming packet is always added to the end of the
queue. If the queue is full a lottery is then held by pseudo-randomly picking a number
between zero and the sum of all inverse bids in the queue using the inverse bid sum
stored in the queue structure. The queue is traversed, keeping a running sum of the
inverse bid values as packets are passed, until the running sum exceeds the randomly
chosen value, in which case the corresponding packet is dropped.

For Deterministic Drop, incoming packets are inserted in bid order (high to low).
When the queue becomes full, the packet at the end of the queue is dropped.

Packet Dequeueing. Dequeueing is done by the Ethernet interface device driver. The
standard FIFO scheduling queue simply picks the packet from the front of the queue
for sending.

The Lottery Scheduling mechanism works in a manner similar to Lottery Drop. To
pick a packet for sending, a number between zero and the sum of all bids in the queue is
chosen pseudo-randomly. The queue is traversed, keeping a running sum of bid values
as packets are passed, until the running sum exceeds the randomly chosen value, in
which case the corresponding packet is forwarded.

Deterministic Scheduling forwards the packet at the front of the sorted queue.

Bid Placement. For Lottery Queueing, we require a means of carrying the bid value
inside the packet. For the purposes of the experiments in this paper we place the bid
value in the IPv4 type-of-service (TOS) field. At this time, we have made no attempt to
fit the bid field into the the Differentiated Services fields of IPv4 and IPv6 defined in

---

[3] Our reference implementation is available at
http://irl.eecs.umich.edu/jamin/papers/marx/lottery.tgz.

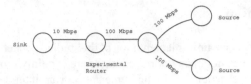

Fig. 1. Testbed Setup. The congestion point is between the Experimental Router and the Sink.

[14]. We note that Lottery Queueing will take advantage of as many bits as are made available for bids by providing more levels of service.

4   Experimental Results

We used the testbed shown in Fig. 1 in our experiments. During our tests the Source nodes send data to the Sink node. The route includes a congestion point after the Experimental Router which is running one of the experimental queueing systems. We collect data from the arriving packets at the Sink node. All packets are sent using the unreliable, connectionless User Datagram Protocol (UDP). We now show the efficacy of lottery queueing in differentiating packets with varying bids. The differentiation takes the form of both packet loss and queue waiting time. To get an idea of the basic behavior of our queueing systems, we first explore differentiation in drop rate and queue latency using CBR traffic (Sections 4.1 and 4.2). Section 4.3 describes results using more realistic traffic models. We explore the possibility of packet re-ordering due to Lottery Scheduling in Sect. 4.4.

4.1   Drop Rate

We first explore the ability of Lottery Drop and Deterministic Drop to differentiate between packets of varying bids. For this experiment, we use three source processes, each generating 374-byte packets (headers included) at a constant rate of 2000 packets per second. Through experimentation we determined that the maximum throughput of the network through the congestion point is about 3220 packets per second; each process sends at 62% of link capacity. Each process marks its packets with a different bid value: 1, 5, and 10. Packets are also marked with sequence numbers, allowing the Sink node to track which packets arrive and which are dropped at the Experimental Router. We run each experiment for 120 seconds using each of the queueing schemes described above. We allow the system to warm up and stabilize for 10 seconds before any data is collected. Figure 2 shows the percent of packets belonging to each flow that makes it to the Sink node under each of the queueing mechanism. We do not present the results of FSDT and LSDT since the drop-tail policy they use does not differentiate packets by their bid values, so each flow gets an equal share of the available bandwidth.

The DSDD results in Fig. 2a show that the high bid flow receives as much bandwidth as it needs. The bid 5 flow takes the remaining available bandwidth, leaving the low bid flow completely starved.

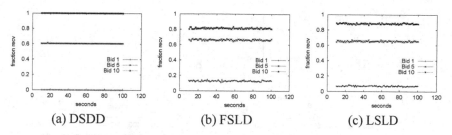

Fig. 2. Percent of packets received by each flow under three queueing mechanisms.

FSLD (Fig. 2b) and LSLD (Fig. 2c) give each flow a share of the available band-width proportional to their bid values. The differences between the FSLD and the LSLD results show that the scheduling algorithm used affects packet loss rate. LSLD shows more differentiation between the varying bid values. This is because Lottery Scheduling tends to forward the high bid packets before the low bid packets, causing the low bid packets to go through more lottery drops and increasing their chance of being dropped.

4.2 Queueing Delay

We next examine the amount of time packets with different bid values must wait in the queue before they are forwarded under the various queueing mechanisms. The Exper-imental Router keeps a count of the number of packets forwarded from a queue since the start of the experiment. By stamping each packet with the value of this counter as it enters and leaves the queue, we obtain a measure of the queueing delay for each packet. The queueing delay is thus expressed in terms of the number of other packets forwarded between the time a packet enters and leaves the queue. The Sink node col-lects the queueing delay information from all packets. It is important to note that we obtain no queueing delay data from dropped packets. To account for dropped packets in presenting the queueing delay data, we use the total number of packets sent, as opposed to the total number of packets received, to compute the cumulative distribution function (CDF) of queueing delays.

The data presented here is collected from the same experiments used to present the drop rate data in the previous section. The queue length at the Experimental Router was set to 160 packets. Packets forwarded through the FSDT queue have to wait a full queue of 159 packets 93% of the time, and they never wait less than 154 packets. This indicates that most packets that are not dropped filled the queue to capacity. Occasion-ally, the three sending processes become synchronized such that several packets arrive nearly simultaneously and are dropped, giving the queue a chance to drain slightly. This information is mostly interesting for comparison with the other queueing methods.

DSDD always forwards the highest bid packet in the queue first. In this experiment high bid packets see immediate service 98.7% of the time. Bid 5 packets wait an aver-age of 420 packets. Since the combination of the high and medium flows completely saturates the congested link, bid 1 packets are all dropped, leaving the queue full of bid 5 packets. Not only do the bid 5 packets have to wait for the full queue to drain before

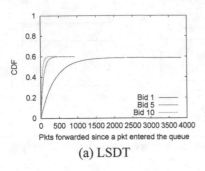
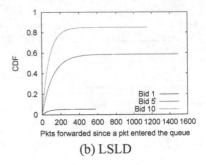

(a) LSDT   (b) LSLD

Fig. 3. Queueing delay under two queueing mechanisms. The CDFs do not reach 1 reflecting the amount of packets dropped.

they see service, service is interrupted every time a high bid packet arrives at the queue, making the wait longer.

We present the CDFs of queueing delay under LSDT (Fig. 3a) and LSLD (Fig. 3b). The CDFs do not go up to 1, reflecting the amount of packets dropped. While the drop policy of LSDT does not differentiate packets by their bid values in making drop decisions, the use of Lottery Scheduling does take the bid values into account in forwarding packets. This is reflected in the difference in queueing delay CDFs of the various bid values. The LSLD CDFs demonstrate an interesting and non-intuitive interaction between the scheduling and dropping policy. The tail for the Bid 1 CDF is much shorter than the tails for the higher bid CDFs suggesting that the low bid packets see faster service. Because Lottery Drop is being used, the low bid packets are dropped preferentially over the high bid packets. If a low bid packet is not forwarded quickly from the queue, it is likely it will be dropped. Therefore, the sink only counts the low bid packets that happen to be selected quickly for forwarding. The maximum queueing delay in the distributions is much higher than the queue size of 160 packets because packets are served at random. Packets with the highest bid value must contend with others of the same bid value for service, hence they also can see queueing delay longer than the queue size.

We do not present the queueing delay CDFs for FSLD because the FIFO scheduling discipline does not differentiate service by bid value, hence all packets see the same queueing delay distribution.

4.3   Long-Range Dependent Traffic

All the flows in the experiment described above transmitted at constant bit rate. Researchers in network traffic characterization have observed long-range dependency in aggregate network traffic [15]. To study the effectiveness of our queueing mechanisms on long-range dependent traffic, we conduct a similar experiment on sources generating On/Off traffic with Pareto distributed On and Off times. Aggregate traffic from such sources has been shown to exhibit long-range dependency [16].

For this experiment we run several sources at each bid value of 1, 5, and 10. Each source uses Pareto distributed On/Off times. The mean and shape parameter for the

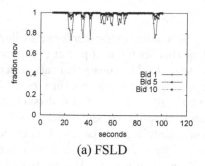
(a) FSLD

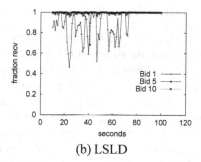
(b) LSLD

Fig. 4. Fraction of packets received by each flow under two queueing mechanisms for Pareto On/Off sources.

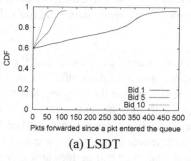
(a) LSDT

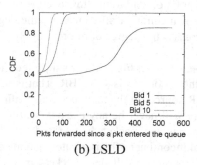
(b) LSLD

Fig. 5. Queueing delay under two queueing mechanisms for Pareto On/Off sources. The CDFs do not reach 1 reflecting the amount of packets dropped.

both the On and Off distributions are 400 ms and 1.1, respectively. When On, sources transmitt 374 byte packets at a rate of 100 packets per second. The data collected for each bid value is from the aggregate of all sources sending at that bid value.

Figures 4a and 4b show that in the face of long-range dependent (LRD) traffic, while higher bid flows continue to receive preferential treatment under FSLD, and an exaggerated preferential treatment under LSLD, lower bid flows do not suffer as much as in the previous case. The high variance of long-range dependent traffic allows lower bidding traffic to continue to be served at network switches, albeit with a longer delay. Hence, when network traffic is very bursty, lower bidding traffic experiences longer queueing delay but not higher loss rate. This effect can be seen by comparing Fig. 3a against Fig. 5a. Note that the CDFs of all bid values are higher in the latter graph where the traffic is generated by Pareto On/Off sources; however, the percentage of lower bid packets staying longer in the queue is much higher. This effect is even more pronounced under the LSLD queueing mechanism, as can be seen by comparing Fig. 2c against Fig. 4b for drop rates of CBR and Pareto On/Off sources, respectively, and the corresponding queueing delays shown in Figures 3b and 5b, respectively.

4.4   Out-of-Order Packets

In this section we analyze the proclivity of Lottery Scheduling to re-order packets within the network. As shown in (1), Lottery Scheduling chooses the next packet to forward without any knowledge of the flow that packet belongs to. This means that each packet a flow has in the queue has an equal opportunity of being the next packet forwarded (assuming each of those packets carry the same bid value). To assist in understanding how serious this effect will be, we have analyzed a simplified model of our queueing schemes that include Lottery Scheduling as their scheduling mechanism. This model gives some insight into what will affect the possibility of re-ordering and the extent to which re-ordering will take place if Lottery Scheduling is used. The most significant result found using this model is that the main parameters governing the degree of re-ordering a flow experiences at a router are the percentage of incoming traffic to the router belonging to the flow and the bid value used by the flow. Increasing the percentage of incoming traffic increases the degree of re-ordering, while increasing bid value decreases the degree of re-ordering.

Model Description. In our simplified model all sources are CBR sources and the sum of those sources is still CBR. That is, there is no burstiness in the traffic.[4] With these CBR sources, if the total incoming traffic to a queue is less than the outgoing link speed there will never be any packets waiting in the queue to be sent, and so there is never any chance of packets being reordered. Therefore, the only interesting case is when the total incoming traffic is greater than the outgoing link speed. In this case, the queue will always remain full since packets are arriving faster than they can be forwarded. Since the queue length is constant (always full) the total number of incoming packets must equal the total number of outgoing packets, where a packet is outgoing when it is either forwarded or dropped. Since the queue is in equilibrium, it follows that the number of packets that a single flow has in the queue will remain constant.

The following system of equations helps describe the behavior of this model using the LSLD queueing scheme:

$$\frac{n_i \cdot b_i}{\sum_{\text{all flows}} n_j \cdot b_j} \cdot f_{forw} + \frac{\frac{n_j}{b_j}}{\sum_{\text{all flows}} \frac{n_j}{b_j}} \cdot f_{drop} = f_i \qquad (3)$$

$$\sum_{\text{all flows}} n_i = qlen \qquad (4)$$

n_x: the number of packets flow x has in the queue.
b_x: the bid value that flow x marks its packets with.
f_x: the fraction of incoming packets belonging to flow x.
f_{forw}: the fraction of total incoming packets forwarded.
f_{drop}: the fraction of total incoming packets dropped. This is $1 - f_{forw}$.
$qlen$: the length of the queue in packets.

---

[4] While the model assumes CBR traffic, we believe the results give insight into more realistic traffic as well. For example, packets from On/Off traffic can only be re-ordered while the source is On, and while the source is On it is sending at a constant bit rate.

Note that (3) actually represents a whole system of equations, one for each unique flow using the queue. Equation (4) provides a constraint that guarantees the solution to the system is one where the queue is full. Equation (3) represents how flow i is treated in the queue. The right side of the equation gives the percentage of the incoming traffic belonging to flow i. The left side of the equation represents the fact that each packet of flow i must either be forwarded or dropped. The left half of the left side is the portion of forwarded packets belonging to flow i. The right half of the left side is the portion of dropped packets that belong to flow i.

The equations representing LSDT are similar. In this case, because Drop Tail is used, the fraction of dropped packets belonging to a flow is equal to that flow's fraction of the total incoming traffic. Therefore, a flow's fraction of the total incoming traffic is equal to the flow's fraction of packets entering the queue after dropping. This, in turn, is equal to the flow's portion of all forwarded packets. From this we get the following equation replacing (3) in the system above:

$$\frac{n_i \cdot b_i}{\sum_{\text{all flows}} n_j \cdot b_j} = f_i \tag{5}$$

Using these systems of equations we can predict the number of packets each flow will have in the queue when the respective queueing scheme is used. This is useful because the probability that the next packet belonging to a flow forwarded from a router is out-of-order is determined by the number of packets that flow has in the queue as seen in the following equation.

$$\text{Pr(Out of Order)} = 1 - \frac{1}{n_x} \tag{6}$$

Where Pr(Out of Order) is the probability that flow x's next packet sent from the queue is out-of-order. Intuitively, the more packets a flow has in the queue, the greater the chance that the next forwarded packet belonging to that flow will be out-of-order.

Predicted Behavior. Figure 6a is helpful in understanding the behavior predicted by the LSLD system of equations. The x-axis of the graph marks the number of packets a flow has in the queue, n_i; the y-axis is the percentage of total incoming packets belonging to that flow, f_i. Each line represents a different bid value, b_i, for the flow. In all cases, all the incoming traffic not belonging to flow i, $1 - f_i$ percent of the total incoming traffic, has a bid value of 1. f_{forw} and f_{drop} are 0.8 and 0.2 respectively, and $qlen$ is 100. When b_i is 1 the function is a straight line. Intuitively, this makes sense because the background traffic also carries a bid of 1 so LSLD degenerates to randomly picking packets from the queue to forward and drop which implies that the number of packets a flow has in the queue should be directly proportional to the percentage of incoming traffic belonging to that flow. When b_i is 1000 the function is approximately a step-function with the shift in n_i taking place when f_i equals f_{forw}. The step represents the transition from flow i needing less than the outgoing link bandwidth to needing more. Because b_i is so much greater than the bid value of the background traffic, 1000 versus 1, when f_i is less than the outgoing link bandwidth, flow i's packets are forwarded immediately from the queue keeping n_i near 0. When f_i becomes greater than f_{forw}

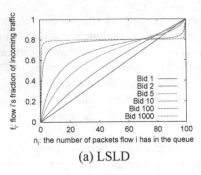

(a) LSLD

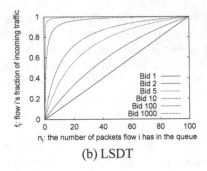

(b) LSDT

Fig. 6. The behavior predicted by the LSLD (equation (3) and LSDT (equation 5) models. Each line represents the function relating the fraction of incoming packets belonging to a flow using a certain bid value to the number of packets that flow should expect in the queue at any one time.

flow i's packets start to compete with each other for the outgoing link bandwidth. This leaves the queue full of flow i's packets since the background traffic packets have a much greater chance of being dropped.

Figure 6b uses the same parameters as Fig. 6a to show the behavior of a LSDT system. The figure shows that when using LSDT a high bid flow does not take over the queue until f_i approaches 100% of the incoming bandwidth. This is because all flows are dropped in proportion to their send rate regardless of their bid value, so whatever percentage of the incoming bandwidth that flow takes, it will receive the same percentage of the outgoing bandwidth.

Experimental Verification. To convince ourselves of the accuracy of our equations we ran experiments in a setup approximating our model. Fig. 7 shows the results of those experiments. The line shows the calculated value of n_i on the x-axis for different values of f_i on the y-axis. The parameters used are the same as the bid 10 analysis for LSLD as described above. The individual points show experimental results. f_{forw}, f_{drop}, $qlen$, and b_i were kept constant throughout the experiments matching the values used in the calculations. Based on the observed maximum outgoing link bandwidth we could simply calculate the total amount of incoming traffic needed to get the desired f_{forw} and f_{drop}. Each experiment used a different value of f_i for the bid 10 traffic with bid 1 traffic making up the remaining amount of total incoming traffic. For the purposes of this experiment, we had the router keep track of how many packets each flow had in the queue. Each outgoing packet was marked with the number of packets each flow had in the router's queue at that time. The points in the chart show the average number of packets in the queue while the error bars show the standard deviation. As Fig. 7 shows, the calculated values predicted very well what was observed.

Preventing Out-of-Order Packets. In this section we look at the maximum fraction of incoming bandwidth, f_i, a flow can take while keeping the number of packets it has in the queue less than two. By keeping the number of packets in the queue less than two, a flow is guaranteed not to suffer from re-ordering due to Lottery Scheduling. Of

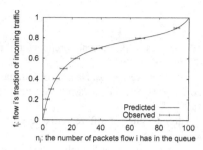

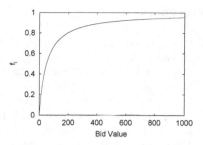

Fig. 7. Calculated vs. experimental packets in queue. The solid line shows the values predicted by (3). The individual points show experimental results.

Fig. 8. The maximum fraction of incoming traffic to a router a flow can have with various bid values which keeps the number of packets that flow has in the queue less than 2. This defines the amount of traffic a flow can send through a router without experiencing re-ordering.

course, this f_i is going to vary depending on other flows using the queue and the bid values of all the flows. In this section we look this fraction for different b_i's assuming the remaining traffic carries a bid value of one.

Figure 8 shows the maximum f_i a flow can use while keeping its n_i less than two when the queueing scheme is LSDT or LSLD with f_{forw} equal to 1. (For LSLD with f_{forw} other than 1, the graph will asymptotically approach that value of f_{forw}.) This result is presented for evaluation purposes only. We are not advocating keeping per-flow state in core routers for the purpose of allowing flows to track their f_i's.

5   Conclusion

In this paper, we have examined the possibility of providing different levels of service using a lottery to schedule and drop packets. We have shown that both Lottery Scheduling and Lottery Drop have the ability to provide many distinct levels of service. Lottery Drop provides each flow a share of the bandwidth proportional to the bid value that flow is using. Likewise, flows with higher bids see proportionally faster service when Lottery Scheduling is used. Trials with bursty traffic are especially encouraging. They show that Lottery Scheduling can still provide service differentiation when the queue length fluctuates.

The high volume of traffic usually associated with core routers suggests that Lottery Queueing may be best suited for use in the core. Since it preserves the stateless nature of the core, it scales well with the number of flows. In fact, a large number of flows will benefit Lottery Scheduling. The likelihood of packet re-ordering due to Lottery Scheduling decreases as the share of traffic belonging to any one flow decreases. If any one flow is an insignificant portion of the core traffic, the likelihood of re-ordering will be low.

References

1. S. Shenker, R. Braden, and D. Clark, "Integrated services in the internet architecture: an overview," RFC 1633, Internet Engineering Task Force, June 1994.
2. Y. Bernet, J. Binder, S. Blake, M. Carlson, B. Carpenter, S. Keshav, E. Davies, B. Ohlman, D. Verma, Z. Wang, and W. Weiss, "*A Framework for Differentiated Services*," Internet Draft, draft-ietf-diffserv-framework-02.txt, Feb. 1999.
3. S. Blake, D. Black, M. Carlson, and E. Davies, "An architecture for differentiated services," RFC 2475, Internet Engineering Task Force, December 1998.
4. J. Heinanen, F. Baker, W. Weiss, and J. Wroclawski, "Assured forwarding phb group," RFC 2597, Internet Engineering Task Force, June 1999.
5. V. Jacobson, K. Nichols, and K. Poduri, "An expedited forwarding phb," RFC 2598, Internet Engineering Task Force, June 1999.
6. C.A. Waldspurger and W.E. Weihl, "Lottery scheduling: Flexible proportional-share resource management," *Proc. of the USENIX Symposium on Operating Systems Design and Implementation (OSDI) '94*, pp. 1–12, 1994.
7. A. Demers, S. Keshav, and S.J. Shenker, "Analysis and simulation of a fair queueing algorithm," *Proc. of ACM SIGCOMM '89*, pp. 1–12, Sept. 1989.
8. I. Stoica, S. Shenker, and H. Zhang, "Core-stateless fair queueing: Achieving approximately fair bandwidth allocations in high speed networks," *Proc. of ACM SIGCOMM '98*, pp. 118–130, Sept. 1998.
9. I. Stoica and H. Zhang, "Providing guaranteed services without per flow management," *Proc. of ACM SIGCOMM '99*, pp. 81–94, Aug. 1999.
10. S. Jamin, P.B. Danzig, S.J. Shenker, and L. Zhang, "A measurement-based admission control algorithm for integrated services packet networks (extended version)," *ACM/IEEE Transactions on Networking*, vol. 5, no. 1, pp. 56–70, Feb. 1997,   Available from http://netweb.usc.edu/jamin/admctl/ton96.ps.Z.
11. S. Floyd and V. Jacobson, "Random early detection gateways for congestion avoidance," *ACM/IEEE Transactions on Networking*, vol. 1, no. 4, pp. 397–413, Aug. 1993.
12. D. Clark. and W. Fang,   "Explicit allocation of best effort packet delivery service," *ACM/IEEE Transactions on Networking*, vol. 6, no. 4, pp. 362–373, Aug. 1998.
13. G. Wright and W. Stevens, *TCP/IP Illustrated, Volume 2: The Implementation*, Addison-Wesley, 1995.
14. K. Nichols, S. Blake, F. Baker, and D. Black, "Definition of the differentiated services field (ds field) in the ipv4 and ipv6 headers," RFC 2474, Internet Engineering Task Force, December 1998.
15. W.E. Leland, M.S. Taqqu, W. Willinger, and D.V. Wilson, "On the self-similar nature of ethernet traffic (extended version)," *ACM/IEEE Transactions on Networking*, vol. 2, no. 1, pp. 1–15, Feb. 1994.
16. W. Willinger, M.S. Taqqu, R. Sherman, and D.V. Wilson, "Self-similarity through high-variability: Statistical analysis of ethernet lan traffic at the source level," *Proc. of ACM SIGCOMM '95*, pp. 100–113, Aug. 1995.

On Creating Proportional Loss-Rate Differentiation:

Predictability and Performance

Ulf Bodin, Andreas Jonsson, and Olov Schelén

Luleå University of Technology, SE-971 87 Luleå, Sweden
{uffe,aj,olov}@cdt.luth.se

Abstract. Recent extensions to the Internet architecture allow assignment of different levels of drop precedence to IP packets. This paper examines differentiation predictability and implementation complexity in creation of proportional loss-rate (PLR) differentiation between drop precedence levels. PLR differentiation means that fixed loss-rate ratios between different traffic aggregates are provided independent of traffic loads. To provide such differentiation, running estimates of loss-rates can be used as feedback to keep loss-rate ratios fixed at varying traffic loads. In this paper, we define a loss-rate estimator based on average drop distances (ADDs). The ADD estimator is compared with an estimator that uses a loss history table (LHT) to calculate loss-rates. We show, through simulations, that the ADD estimator gives more predictable PLR differentiation than the LHT estimator. In addition, we show that a PLR dropper using the ADD estimator can be implemented efficiently.

1 Introduction

Today's Internet supports a wide spectrum of applications with different demands on forwarding quality. The Internet community has recognized that one service only (i.e., best-effort) may not be enough to meet these demands. The Internet Engineering Task Force (IETF) is therefore designing architectural extensions to support service differentiation on the Internet. The Differentiated Services (DiffServ) architecture [2][9] includes router mechanisms for differentiated forwarding.

With DiffServ, levels of drop precedence can be assigned to IP packets. Differentiation between drop precedence levels is part of the Assured Forwarding (AF) per-hop behavior (PHB) group [8]. AF can be used to offer differentiation among rate adaptive applications that respond to packet loss (e.g., applications using TCP). The traffic of each user is tagged as being *in* or *out* of their service profiles. Packets tagged as *in*-profile are assigned lower drop precedence than those tagged as *out*-of-profile. In addition, a packet within a user's profile may be tagged with one out of several levels of drop precedence. For now, there are three levels of drop precedence defined for AF.

For AF, it is required that the levels of drop precedence are ordered so that for levels $x < y < z$, $P_{drop}(x) \leq P_{drop}(y) \leq P_{drop}(z)$ and $P_{drop}(x) < P_{drop}(z)$[1]. To further refine the differentiation, it can be defined in quantitative terms. For example, the loss-rate ratios (i.e., $P_{drop}(x)/P_{drop}(y)$) can in case of congestion be set to a target value ≤ 1.

---

[1] $P_{drop}(x)$ is the drop probability for traffic at drop precedence level x.

L. Wolf, D. Hutchison, and R. Steinmetz (Eds.): IWQoS 2001, LNCS 2092, pp. 372–386, 2001.

This paper examines proportional loss-rate (PLR) differentiation [5] in terms of predictability and implementation complexity. We grade the predictability of a PLR differentiation by studying short-term variations in loss-rate ratios between drop precedence levels at changing traffic loads and load distributions. In addition, we study if long-term loss-rate ratios achieve target loss-rate ratios at changing traffic loads. We consider a PLR differentiation *robust* if short-term loss-rate ratios have negligible variations and *capable* if long-term loss-rate ratios approximate target loss-rate ratios reasonable well at changing traffic load conditions.

A PLR differentiator can be divided into two modules. First, a *drop controller* decides when a packet needs to be dropped. A drop controller can, for example, perform early congestion signaling with Random Early Detection (RED) [7], or just drop packets when no buffer space is available for queuing, etc. When the drop controller has decided that a packet needs to be dropped, a *PLR dropper* selects a drop precedence level from which a packet will be dropped (to maintain the PLR differentiation), selects a packet at the victim level and removes it from the queue.

Changing traffic loads and load distributions between drop precedence levels can cause loss-rate ratios to deviate from the target loss-rate ratios. To create PLR differentiation under such conditions, a PLR dropper can use running estimates of loss-rates as feedback to adjust towards the target loss-rate ratios. For example, when the drop controller triggers a drop, a packet at the drop precedence level with the minimum normalized loss-rate (NLR) can be selected and dropped from the queue. The NLR is in [5] defined as: $\bar{l}_i/\sigma_i$ where $\bar{l}_i$ is the loss-rate and σ_i is the differentiation constant in class i. In this paper, we adopt this method of selecting the victim level when dropping packets. Using the NLR selector, we study how properties of the loss-rate estimator influence PLR differentiation predictability.

The proportional loss-rate model is proposed and motivated in [4] and [5]. In [5], a loss-rate estimator is proposed, which estimates the loss-rate by counting the number of losses at each class during a time window of M packet arrivals. One implementation of this uses a cyclic queue named a Loss History Table (LHT). The problem with this is that an appropriate value of M has to be chosen. Firstly, it has to be at least one dropped packet at all drop precedence levels in the last M packets arrived. Otherwise, the measured loss-rate will be zero for levels at which no packet is dropped in the last M arrivals. This leads to inaccurate loss-rate estimation which in turn leads to loss-rate relations which differs from the configured ones. Unfortunately, a large M makes the PLR dropper less adaptive to changing traffic loads. Hence, there is a trade-off in chosing an appropriate value of M. Large M gives capable but not robust PLR differentiation, while small M gives robust but not capable PLR differentiation.

In this paper, we define an estimator that uses average drop distances (ADDs) as estimates of loss-rates. For each drop precedence level, an ADD covers a history which length is defined in number of drops. This makes the estimator adapt the history length at each level to changing load distributions. The history covered by the ADD estimator can be set short without risking estimated loss-rates to be zero for some traffic loads. With the ADD estimator, the history length simply determines how fast changed traffic load conditions are detected. Hence, the ADD estimator does not have the same trade-off in choosing history length as the LHT estimator.

We evaluate through simulations the PLR differentiation predictability of the ADD and the LHT estimator for two levels of drop precedence. These simulations show that the trade-off in choosing M disables the LHT estimator from providing both robust and capable PLR differentiation with one single M. For the ADD estimator, weights can be found that gives both robust and capable PLR differentiation.

When designing forwarding mechanisms for Internet routers, their performance is important. Computation and/or memory intensive mechanisms make routers more expensive, which can make deployment in routers handling high bit-rates unfeasible. To examine the performance of the ADD estimator together with the NLR selector, we have implemented these mechanisms in the kernel of FreeBSD. With this implementation, only 131 clock cycles in average are needed to update three ADDs on an Intel Pentium II 350Mhz. Selecting from which of drop precedence level to drop needs 59 clock cycles in average for three levels.

The rest of the paper is structured as follows. In Sect. 2, we define the ADD estimator, discuss its properties and compare these with properties of the LHT estimator. Next, in Sect. 3, the need for differentiation predictability at both long and short time-scales is discussed. In Sect. 4, simulations comparing the LHT estimator and the ADD estimator are presented. In Sect. 5, effective implementations of the ADD estimator and the NLR selector are described. Moreover, performance measurements for these mechanisms are presented in this section. Finally, in Sect. 6, we conclude our work.

2   Estimating Loss-Rates

In this section, we define a loss-rate estimator that uses average drop distances (ADDs). The basic properties of this estimator is discussed and compared with the properties of the loss history table (LHT) estimator [5].

Table 1. Symbols Used in this Paper.

l	Aggregate loss-rate.	$\overline{d}_i$	Estimated average drop distance at L_i
L_i	Drop precedence level i.	$\overline{d}_{i_{old}}$	Old estimated average drop distance at L_i
λ_i	Arrival rate at L_i.	σ_i	Differentiation constant for class i.
d_i	Drop distance counter at L_i.	$B(t)$	Set of backlogged L_i at time t.
$d_{i_{old}}$	Old drop distance counter at L_i.	g_i	EWMA filter constant for L_i.
$\overline{l}_i$	Estimated loss-rate at L_i.	w_i	EWMA filter weigth for L_i.

2.1   The Average Drop Distance (ADD) Estimator

An ADD estimator calculates an average drop distance for each drop precedence level. The drop distance is the number of successfully transferred packets between two lost packets. We denote the estimated ADD at L_i as $\overline{d}_i$ and the estimated loss-rate at L_i as $\overline{l}_i = 1/\overline{d}_i$. We denote the estimated loss-rate ratio between L_i and L_j as $\overline{l}_i/\overline{l}_j$ and the target loss-rate ratio between L_i and L_j as σ_i/σ_j. The definition of normalized loss-rate (NLR) and the method for selecting precedence level when dropping a packet are both adopted from [5].

ADD estimation can be performed by computing the average drop distance over a certain number of packet drops. It is however hard to pick an appropriate number of drops to consider, especially when arrival rates and drop rates vary frequently. For quick response to changing conditions a small number of drops should be considered and for stability a large number of drops. For this reason, we instead use exponential weighted moving averages (EWMAs) to give higher weigth to recent drop distances. Although EWMA suits the purpose and can be implemented efficiently, we do not claim that it is the optimal averaging function. With EWMAs (1), the ADD estimator covers a configurable history length, $g_i, 0 \leq g_i \leq 1$, coupled to the number of rescent drops for each drop precedence level, L_i. We limit g_i to integer powers of two for efficient implementation through shift operations: $g_i = 2^{-w_i}$, where the weight w_i is a positive integer.

$$\bar{d}_{i,n} = \bar{d}_{i,n-1} \cdot (1 - g_i) + d_i \cdot g_i \tag{1}$$

Larger w_i results in more stable (i.e., more capable) estimations of $\bar{d}_i$ and longer detection times for changed trafficload conditions (i.e., less robust estimations). However, our experiments indicate that estimations of $\bar{d}_i$ is stable enough even with small values of w_i. In Sect. 4, we show that robust and capable PLR differentiation can be obtained with very small values of w_i.

The detection time is determined by both w_i and the number of drops per time unit. For robustness, changed traffic load conditions should be detected equally fast at all drop precedence levels. If $\bar{d}_i$ is only updated upon packet drops at L_i, different arrival rates and loss-rates between drop precedence levels causes different update fequency and detection times between these levels. For example, assume that the actual loss-rate is higher at L_k than at L_j. Then, $\bar{d}_k$ is updated more often than $\bar{d}_j$ and changed traffic load conditions is thus detected faster at L_k than at L_j. Moreover, if the load suddenly decreases and stays at the lower level, $\bar{d}_j$ may not be updated at all. This is because $\bar{l}_j/\sigma_j$ becomes larger than $\bar{l}_k/\sigma_k$, which causes drops to be strictly given to traffic at L_k. We refer to this problem as *update locking*.

To avoid the risk for update locking and to make detection times similar between drop precedence levels, we also recalculate $\bar{d}_j$ at drop precedence levels, j, which was not targeted for a drop. We do this by restoring $\bar{d}_j$ to the value it had before at the time of the previous drop at L_j and then recalculate $\bar{d}_j$ with all new arrivals at L_j added to the last drop distance at L_j. Not only do we get a more up to date estimate of $\bar{d}_j$, but also we solve the update locking problem since $\bar{d}_j$ goes towards infinity with $d_j + d_{j_{old}}$ (see Fig. 1).

Equal weights, w_i, for all drop precedence levels makes the detection time at L_i shorter (i.e., $\frac{\sigma_i \cdot \lambda_i}{\sigma_j \cdot \lambda_j}$ times shorter) than at L_j. To compensate for this, separate weights for each drop precedence level can be applied such that (2) is met as closely as possible[2]. Equation (2) is based on the closed form expression for EWMA (1).

$$(1 - 2^{-w_i}) = (1 - 2^{-w_j})^{\frac{\sigma_j \cdot \lambda_j}{\sigma_i \cdot \lambda_i}} \tag{2}$$

---

[2] Equation (2) cannot be met exactly since w_i is a positive integer.

The algorithm for the ADD estimator and the NLR selector is shown in Fig. 1. Note that the inverse NLR, $\overline{d}_i \cdot \sigma_i$, instead of the NLR, $\overline{l}_i/\sigma_i$, is used to select from which L_i to drop. Hence, at congestion we drop a packet at the drop precedence level with the *maximal* inverse NLR instead of the *minimal* NLR as done in [5].

With the algorithm shown in Fig. 1, $\overline{d}_i$ does not change if no packets arrive at L_i and will therefore be invalid after a idle period at this drop precedence level. This becomes a problem if $\overline{d}_i \cdot \sigma_i$ is larger than the inverse NLR for other levels. The first packets arriving at L_i immediately after the idle period will then be dropped until $\overline{d}_i \cdot \sigma_i$ becomes smaller than the inverse NLR for some other level. Similarly, the first packets arriving at L_i immediately after an idle period will not be dropped if $\overline{d}_i \cdot \sigma_i$ is smaller than the inverse NLR for some other level. Hence, loss-rate ratios can temporarily be larger or less than the target loss-rate ratios. We refer to this problem as *invalid ADDs*.

Dropping the first packets arriving after an idle period can be devastating since TCP sources perform an exponential back-off when loosing SYN packets. Due to the exponential back-off, it can take considerable time for $\overline{d}_i$ to decrease since no packets arrive at L_i. We solve the invalid ADDs problem by updating $\overline{d}_i$ to a value calculated from known ADDs at other drop precedence levels. The update is made if no packet has arrived after *maxidle* updates of ADDs for other levels (Fig. 2).

Packet arrival at L_k:
$\quad d_k{+}{+}$
Packet drop at L_k:
$\quad \overline{d}_{k_{old}} \leftarrow \overline{d}_k$
$\quad \overline{d}_k \leftarrow \overline{d}_k \cdot (1 - 2^{-w_k}) + d_k \cdot 2^{-w_k}$
$\quad d_{k_{old}} \leftarrow d_k$
$\quad \forall i \in B(t) \setminus \{k\}:$
$\qquad \overline{d}_i \leftarrow \overline{d}_{i_{old}}$
$\qquad d_{i_{old}} \leftarrow d_{i_{old}} + d_i$
$\qquad \overline{d}_i \leftarrow \overline{d}_i \cdot (1 - 2^{-w_k}) + d_{i_{old}} \cdot 2^{-w_k}$
$\quad \forall i \in B(t):$
$\qquad d_i \leftarrow 0$
Selecting L_k for next drop:
$\quad k \leftarrow \arg\max_{i \in B(t)} \overline{d}_i \cdot \sigma_i$

Packet arrival at L_k:
$\quad idle_k \leftarrow 0$
Packet loss at L_k:
$\quad \forall i \in B(t) \setminus \{k\}:$
$\qquad$ if $d_i = 0$ and $idle_i{+}{+} > maxidle$:
$\qquad\quad a \leftarrow \arg\min_{j \in B(t)} \overline{d}_j \cdot \sigma_j$
$\qquad\quad \overline{d}_i \leftarrow \overline{d}_a \cdot \frac{\sigma_a}{\sigma_i}$

Fig. 1. The Algorithm for the ADD Estimator and the NLR Selector.

Fig. 2. Method for Detecting and Updating Idle Drop Precedence Levels.

The method for detecting and updating idle drop precedence levels shown in Fig. 2 can cause deviations of loss-rate ratios from target loss-rate ratios. For example, say that L_i is frequently idle for periods long enough to trigger an update and that each update decreases $\overline{d}_i \cdot \sigma_i$. Moreover, say that the active periods are short and that $\overline{d}_i \cdot \sigma_i$ therefore does not reach the actual inverse NLR before L_i gets idle again (i.e., although very few packets are dropped at L_i, $\overline{d}_i$ does not increase enough to reflect the loss-rate at L_i). This may cause the loss-rate ratio between L_i and the next lower level to be less than the target loss-rate ratio between these levels. We refer to this problem as *frequent updates*. To avoid the frequent updates problem, *maxidle* should be set large. We recommend setting *maxidle* to trigger updates after idle periods of several seconds. This mechanism is disabled in the simulations presented in Sect. 4.

Distributions in arrival rates between drop precedence levels (i.e., λ_j/λ_i) at congested links is usually unknown and may change rapidly for bursty traffic patterns. However, different arrival rates is a severe problem only if the loss-rate changes rapidly with changing traffic loads. This is often the case for pure FIFO queues, but not for Random Early Congestion (RED) [7] managed queues. RED smoothes the loss-rate using a low pass filter (e.g., EWMA). We take advantage of the smooth changes in loss-rates provided by RED and do not compensate for different arrival rates (i.e., we set $\lambda_j/\lambda_i = 1$). When loss-rates are controlled with RED, differences in detection time caused by different arriving rates have limited effect on the PLR differentiation offered. This is shown in Sect. 4 where simulations are presented.

A consequence of using RED to smooth loss-rates is that the ADD estimator depends on proper operation of RED. Recent studies of RED have shown that the average queue length and thus the loss-rate can oscillate under certain conditions (the discontinuity in the standard RED drop function[3] and/or some combinations of link bandwidth, average packet size and load levels can cause such oscillations [11]). Based on these studies, a new active queue management (AQM) mechanism is developed that gives more robust loss-rates than RED [3]. The ADD estimator should gain from the smoother loss-rates provided by this new AQM mechanism. However, in this paper we evaluate the ADD estimator with RED without the gentle modification.

2.2   The Loss History Table (LHT) Estimator

The loss history table (LHT) estimator is defined in [5]. The estimated loss-rate $\bar{l}_i$ is the number of drops at L_i in the last M arrivals divided by M. The cyclic queue used to count drops is named loss history table (LHT). M has to be large enough to always cover at least one drop at all drop precedence levels. Otherwise, acceptable estimation accuracy is not obtained since $\bar{l}_i$ occasionally becomes zero. Equation (3) gives a lower bound on M [5]. N is the number of drop precedence levels supported and $m = \arg\min_{0 \leq i \leq N-1} \lambda_i \cdot \sigma_i$.

$$M_{min} = \frac{\sum_{i=0}^{N-1} \frac{\lambda_i \cdot \sigma_i}{\lambda_m \cdot \sigma_m}}{l} \tag{3}$$

M should be larger than the lower bound given by (3) in order to to provide capable PLR differentiation. This is shown through simulations in [5]. For instance, bursty traffic can give considerable variations in the arrival rate and the loss-rate over short time-scales, which will degrade the differentiation if M too small. Unfortunately, large M makes the detection of changed traffic load conditions slow. Hence, there is a trade-off in selecting M. Larger M gives more capable, but less robust PLR differentiation and smaller M less capable, but more robust PLR differentiation.

---

[3] The standard drop function in RED jump from the maximal drop probability (e.g., 0.1) to 1 instantly. This discontinuity is however removed with the gentle modification of RED [6].

2.3 Comparison

The ADD estimator provides both robust and capable PLR differentiation with one configuration. In contrast to the LHT estimator, it provides accurate loss-rate estimation by always encountering several drops at every drop precedence level. Hence, without risking inaccurate loss-rate estimation with incapable PLR differentiation as result, the ADD estimator can be configured to encounter few drops to detect changed traffic load conditions rapidly. The LHT estimator cannot be configured to provide both accurate loss-rate estimation and rapid detection of changed traffic load conditions. Fast detection of changed traffic load conditions is needed to provide robust PLR differentiation.

If a small number of drops is covered by the loss history, the loss-rate estimation becomes unstable at short time-scales. Such unstable loss-rate estimation can cause variations in actual loss-rate ratios at short time-scales and deviations of actual loss-rate ratios from target loss-rate ratios at long time-scales. The EWMA makes it hard to give the parameters w_i a clear physical interpretation, as opposed to the LHT estimator, where M corresponds to the number of packet arrivals. However, this is not necessary to configure the ADD estimator appropriately. We show in Sect. 4 that by using $w_0 = 1$ and $w_1 = 4^4$, the ADD estimator provides robust and capable PLR differentiation between L_0 and L_1 for $\sigma_0 = 1$ and $\sigma_1 = 10$.

3 Measuring Loss-Rates

In this section, we discuss over which time-scales loss-rate ratios are likely to be measured by network operators and to be perceived by users. Network operators may monitor loss-rates by polling routers periodically using SNMP or command line interfaces. The overhead associated with periodic polling makes it appropriate to monitor loss-rates over time-scales in order of minutes rather than seconds. However, users are likely to perceive loss-rate ratios over time-scales spanning from few seconds to several minutes.

PLR differentiation allows individual users to choose a service that provides an appealing balance between forwarding quality and cost. With PLR differentiation, a user can dynamically switch between levels of drop precedence to find a level with a loss-rate low enough for the application used. A user can begin tagging all the packets with a high drop precedence level. If the loss-rate at this level is considered unacceptably high after a period, the user can switch to a lower drop precedence level. Eventually, the user should find a level that provides a loss-rate adequate for the user's needs. Hence, the user does not have to pay for additional and unneeded forwarding quality.

To make the result of switching from one level of drop precedence to another level predictable, the PLR differentiation needs to be robust and capable. Loss-rate ratios measured over several minutes need to closely approximate target loss-rate ratios. Otherwise, users cannot predict the result of switching drop precedence level. Moreover, loss-rate ratios measured over a few seconds need to have negligible variations. This is to make the result of switching drop precedence level immediately notable to users.

---

[4] With $w_0 = 1$ and $w_1 = 4$ the equality in (2) is approximately satisfied when arrival rates are equal for L_0 and L_1.

4   Simulations

In this section, we present simulations evaluating the predictability of the PLR differentiation created with the LHT estimator and the ADD estimator respectively. The simulations are made with the network simulator (ns) [10]. Sect. 4.1 describes the simulation setup. We study PLR differentiation at a time-scale of two minutes in Sect. 4.2 and at a time-scale of five seconds in Sect. 4.3.

4.1   Simulation Setup

For the simulations, a topology with ten hosts (s0, …, s9), ten receivers (r0, …, r9) and two routers (A and B) is used. The routers are connected via a 50 Mbps link with 20 ms delay (Fig. 3). A PLR dropper supporting two levels of drop precedence is used to differentiate traffic at this link. The target loss-rate ratio σ_1/σ_0 is set to 10 times (i.e., the loss-rate at drop precedence level 1 is targeted to be 10 times higher than the loss-rate at level 0). The drop controller is RED and the drop strategy is Drop-Tail[5]. The configuration of RED is: min threshold 70 packets, max threshold 210 packets and max drop probability 10 percent.

The bit-rates of links connecting hosts and receivers to routers are reconfigured with uniformly distributed random values between 22 and 32 Mbps once every two simulated seconds. The delays of these links are reconfigured equal often with uniformly distributed random values between 0.1 and 0.9 ms. Similar values are used in [1] to emulate switched Ethernet. A positive consequence of making these reconfigurations is that synchronization affects among TCP connections get reduced[6].

Each host (s0, …, s9) has three TCP Reno connections with each receiver (r0, …, r9) (i.e., 300 connections are established over link A-B). The receivers use delayed ACKs. MTU is 1460 bytes. The TCP connections are established randomly within the first 10 simulated seconds. These random variables are uniformly distributed. Using these connections, the receivers download data from Pareto distributed ON-OFF sources at the hosts. The scale parameter for the Pareto distribution is 1.5, the average length of *ON* periods is set to 50 ms and the average length of *OFF* periods is set to 950 ms. The rate of *ON* periods for each source is set to 490 kbps. This generates a highly variable traffic load causing loss-rates in between 1.72 and 5.15 percent at link A-B when measured over two minutes (the simulations presented in Sect. 4.2). When measured over five seconds (the simulations presented in Sect. 4.3), loss-rates are in between 1.15 and 5.78 percent at this link.

For all simulations, a warm-up period of 60 simulated seconds is used to let the congested queue at router A and the loss-rate estimator examined stabilize. After these 60 seconds, counters for the number of packet arrivals and drops at each of the two levels of drop precedence are initialized to measure loss-rate ratios. In Sect. 4.2 and 4.3, loss-rate ratios are plotted with a log 10 scale at the y-axis. The log scale is chosen to view deviations of loss-rate ratios equally independent on whether they are larger or less than the target loss-rate ratio.

---

[5] Packets are removed from the end of the queue for the drop precedence level.
[6] The random drops made by RED also reduce the risk of having TCP flows synchronize.

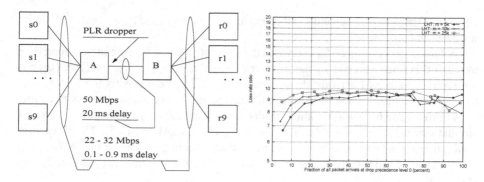

Fig. 3. Simulated Topology.

Fig. 4. Actual Loss-Rate Ratios Measured over Two Minutes (The LHT Estimator).

4.2   Long-Term PLR Differentiation

For each of the loss-rate estimators, we examine their long-term PLR differentiation predictability with 19 simulations. The distribution in number of TCP connections at the two levels of drop precedence is changed between simulations. At drop precedence level 0 (L_0), the number of TCP connections is varied between 15 and 285 in steps of 15. At L_1, the number of TCP connections is varied between 285 and 15 in steps of 15. The x-axis is graded with the fraction of all packet arrivals at L_0. Each simulation is 120 seconds long.

   Figure 4 shows simulation results when using the LHT estimator. For these simulations, (3) gives $M_{min} \approx 5000$ packets when 15 TCP connections are at L_0[7]. With this distribution, about 5 percent of all packet arrivals are at L_0. As discusseded in Sect. 2.2, M should however be set larger than M_{min}. We present simulations with $M = 5000$, $M = 10000$ and $M = 25000$ packets. At a bit-rate of 50 Mbps, 10000 packets of size 1460 bytes are forwarded in 2.336 seconds. Hence, with $M = 10000$ packets, the LHT estimator can be expected to adapt to changing traffic load conditions faster than in five seconds. With $M = 25000$ packets, this adaptation should be slower than in five seconds (25000 packets of size 1460 bytes are forwarded in 5.84 seconds).

   In Fig. 4, it can be seen that at low arrival fractions at L_0, loss-rate ratios is less than 9 for all M simulated. As expected, higher M gives higher loss-rate ratios at such low fractions. With too few packets at L_0, the LHT occasionally falsely estimates $\bar{l}_i$ to zero. The dropper will then select L_0 for a packet drop. At high arrival fractions at L_0, loss-rate ratios varies and go below 9 for all M simulated. When no packets arrive at L_1, packets can only be dropped at L_0 since the queue at L_1 is empty. Consequently, when a packet should be dropped at L_1 to increase the loss-rate ratio, it will have to be dropped at L_0 instead. With close to 100 percent of all packet arrivals at L_0, packet drops are forced to L_0 because of an empty queue at L_1 in 4.4 percent of all drops for

---

[7] $M_{min} = 3767$ packets for $M = 5000$, $M_{min} = 5997$ packets for $M = 10000$ and $M_{min} = 5059$ packets for $M = 25000$.

$M = 5000$ packets, 8.3 percent of all drops for $M = 10000$, and 22 percent of all drops for $M = 25000$.

A forced packet drop at L_0 decreases the loss-rate ratio. It takes relatively long time to repair this since a drop at L_0 has a larger impact than a drop at L_1. If a forced drop is not repaired before M arrivals, the loss-rate ratio will be permanently too low. As can be observed in Fig. 4, larger fractions of forced drops at L_0 decreases the loss-rate ratio.

Figures 5 and 6 show simulation results when using the ADD estimator. We present simulations with (w_0, w_1) = (1,2), (1,3), (1,4), (2,3), (2,4) and (2,5). The three first configurations are shown in Fig. 5 and the last three configurations in Fig. 6. (w_0, w_1) = (1,4) and (2,5) approximates the equality in (2) when arrival rates are equal at L_0 and at L_1.

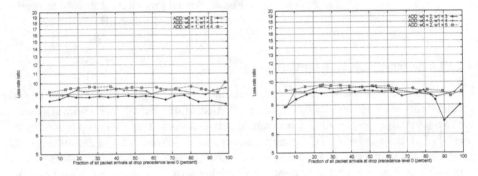

Fig. 5. Loss-rate ratios measured over two minutes (the ADD estimator, configuration set 1).

Fig. 6. Loss-rate ratios measured over two minutes (the ADD estimator, configuration set 2).

Loss-rate ratios are degraded for low arrival fractions at L_0 with the ADD estimator (Figs. 5 and 6). This is because the ADD estimator detects an increasing loss-rate more rapidly for L_1 when there are more arrivals at L_1 than at L_0. This property of the ADD estimator is discussed in Sect. 2.1. Without RED smoothing actual loss-rates, the problem of different detection times cause severe degradations in loss-rate ratios.

For configurations not satisfying (2) (i.e., (w_0, w_1) = (1,2), (1,3), (2,3) and (2,4)), loss-rates are lower than for configurations that do (Figs. 5 and 6). Lower w_1 than given by (2) implies that changes in loss-rates are detected faster at L_1 than at L_0 except for high arrival fractions at L_0. For example, with (w_0, w_1) = (1,2), λ_0/λ_1 needs to be 4.15 to satisfy (2). This means that loss-rates are detected faster at L_1 than at L_0 for all arrival fractions at L_0 up to 80 percent. This percentage is similar with (w_0, w_1) = (2,3) as with (w_0, w_1) = (1,2). With (w_0, w_1) = (1,3) and (2,4), it is about 65 percent.

For high arrival fractions at L_0, the fraction of forced drops from L_0 gets high for all configurations of the ADD estimator (i.e., (w_0, w_1) = (1,2) gives up to 6.4 percent forced drops, (1,3): 13 percent, (1,4): 21 percent, (2,3): 12 percent, (2,4): 24 percent and (2,5): 32 percent). This suggests that loss-rate ratios should be degraded for high arrival fractions at L_0. However, since the invalid ADDs problem cause increases in loss-rate ratios (Sect. 2.1), the degradation expected from the high fractions of forced drops from L_0 get balanced out so that loss-rate ratios approximates the target loss-rate ratio of 10 (Figs. 5 and 6). This explains the increase in loss-rate ratios with larger w_1 in these simulations.

In Figs. 5 and 6, it can be seen that the ADD estimator provides capable PLR differentiation for low arrival fractions at L_0 with the configurations that satisfy (2) (i.e., for $(w_0, w_1) = (1,4)$ and $(2,5)$). The LHT estimator needs large M to provide capable PLR differentiation for arrival fractions less than 15 percent (i.e., $M = 25000$ packets).

The $(1,4)$ configuration of the ADD estimator is preferable before the $(2,5)$ configuration since it gives fewer forced drops. Moreover, the invalid ADDs problem is less severe with small weights. Fewer forced drops and a less severe invalid ADDs problem should make the PLR differentiation more robust.

4.3 Short-Term PLR Differentiation

For each of the two loss-rate estimators evaluated, we examine their short-term PLR differentiation predictability. The simulations run for 360 seconds after the warm-up period. Loss-rate ratios are measured 5 seconds interval. At 180 and 300 seconds, the distribution in number of TCP connections between L_0 and L_1 is changed. In the first 60 seconds after the warm-up, there are 15 TCP connections at L_0 and 285 TCP connections at L_1. In the next 120 seconds of the simulations, there are 150 TCP connections at each drop precedence level. Finally, in the last 120 seconds of the simulations, there are 285 TCP connections at L_0 and 15 TCP connections at L_1.

For this scenario, we have used the same parameters for the LHT estimator and the ADD estimator as for the simulations presented in Sect. 4.2 (i.e., $M = 5000$, 10000, 25000 and $(w_0, w_1) = (1,2)$, $(1,3)$, $(1,4)$, $(2,3)$, $(2,4)$ and $(2,5)$). Fig. 7 shows simulation results for the LHT estimator with $M = 5000$ and 10000 packets and Fig. 8 for the LHT estimator with $M = 10000$ and 25000 packets. Thereafter, Fig. 9 through Fig. 12 show simulation results for the different configurations of the ADD estimator.

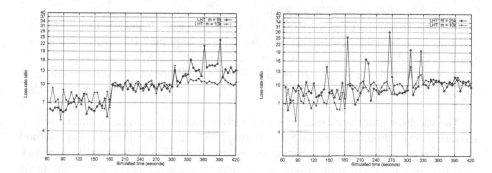

Fig. 7. Loss-Rate Ratios Measured over Five Seconds ($M = 5000$ and 10000).

Fig. 8. Loss-Rate Ratios Measured over Five Seconds ($M = 10000$ and 25000).

When the arrival fraction at L_0 is low or high, loss-rate ratios are closer to the target loss-rate ratio with larger M (Figs. 4, 7 and 8). Using small M, chances are that some levels have not ben targeted for a drop in the last M arrivals, causing the estimated loss-rate to be zero for levels with low arrival rate. With $M = 5000$ packets, this happens frequently for both low and high arrival fractions at L_0, but only for low arrival fractions at L_0 with $M = 10000$ packets. With $M = 25000$ packets, estimated loss-rates become seldom zero for levels with low arrival rate and loss-rate ratios therefore better approximate the target loss-rate ratio. Figures 7 and 8 also shows that large M causes large variation in loss-rate ratios. This is because larger M gives slower detection of changed traffic load conditions.

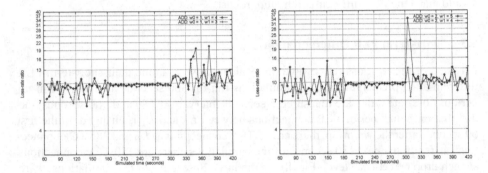

Fig. 9. Loss-Rate Ratios Measured over Five Seconds ($(w_0,w_1) = (1,4)$ and $(1,3)$).

Fig. 10. Loss-Rate Ratios Measured over Five Seconds ($(w_0,w_1) = (2,5)$ and $(2,4)$).

For low arrival fractions at L_0, the variation in loss-rate ratios is larger for higher weights (before 180 seconds in Figs. 9 and 10). Nevertheless, this variation is similar for the ADD estimator with $(w_0,w_1) = (1,4)$ and for the LHT estimator with $M = 5000$ packets (Fig. 7). The variation in loss-rate ratios is smaller with $M = 5000$ packets than with $M = 10000$ or 25000 packets.

For high arrival fractions at L_0 (after 300 seconds in Figs. 9 and 10), the variation in loss-rate ratios is higher with the ADD estimator than with the LHT estimator if a configuration satisfying (2) is used. The high variation with the ADD estimator is caused by the invalid ADDs problem described in Sect. 2.1. Although all arriving packets at L_1 are dropped, the loss-rate is increased slowly due to very few packet arrivals at this level.

The variation in loss-rate ratios for high arrival fractions at L_0 can be decreased by activating the method for detecting and updating idle levels shown in Fig 2. This method cannot however eliminate the variation in loss-rate ratios for high arrival fractions at L_0. Moreover, the method for detecting and updating idle levels can cause long-term loss-rate ratios to deviate from the target loss-rate ratio. This is because of the frequent updates problem described in Sect. 2.1.

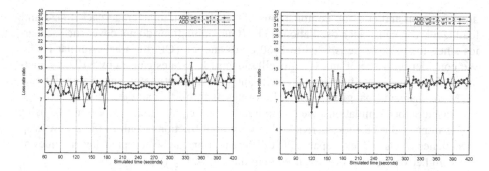

Fig. 11. Loss-Rate Ratios Measured over Five Seconds $((w_0,w_1) = (1,2)$ and $(1,3))$.

Fig. 12. Loss-Rate Ratios Measured over Five Seconds $((w_0,w_1) = (2,3)$ and $(2,4))$.

4.4   Summary of Simulation Results

In Sect. 4.2, we examine differentiation predictability at a time-scale of two minutes for the ADD estimator and the LHT estimator respectively. We show that the LHT estimator needs large M to provide capable PLR differentiation for arrival fractions less than 15 percent (i.e., $M = 25000$ packets). The ADD estimator provides capable PLR differentiation for low arrival fractions at L_0 with configurations that satisfy (2) (i.e., for $(w_0, w_1) = (1,4)$ and $(2,5)$).

Next, in Sect. 4.3, we examine differentiation predictability at a time-scale of five seconds for the two loss-rate estimators evaluated. When configured for equal arrival rates at both drop precedence levels (i.e, for $(w_0, w_1) = (1,4)$ and $(2,5)$), the variation in loss-rate ratios is similar or lower with the ADD estimator than with the LHT estimator (i.e., for $M - 25000$ packets). With $M = 10000$ packets, this variation is higher with the ADD estimator than with the LHT estimator. Such a configuration of the LHT estimator does not however give a capable PLR differentiation for low arrival fractions at L_0. Hence, the LHT estimator cannot provide both capable and robust PLR differentiation with one configuration. Since the ADD estimator can be both capable and robust with one configuration, we consider it more predictable than the LHT estimator.

4.5   Configuration Recommendations

The robustness of the PLR differentiation can be improved with the ADD estimator by setting $(w_0,w_1) = (1,3)$ or $(w_0,w_1) = (2,4)$. Such configurations give robust PLR differentiation at high fractions of all traffic at L_0, but less capable PLR differentiation at most traffic distributions.

Simulations with different link speeds, RTTs and number of TCP flows indicate that the above given configurations are not particularly sensitive to these parameters [8]. The ADD estimator may however be sensitive to scenarios in which the EWMA averaging function of RED gives an oscillating loss-rate (the discontinuity in the standard RED

---

[8] Due to limited space, we do not show simulations with different link speeds, RTTs and number of TCP flows.

drop function and/or some combinations of link bandwidth, average packet size and load levels can cause such oscillations [11]). Our simulations do not cover such scenarios since the packet size is fixed to 1460 bytes and all TCP flows have similar RTTs. We consider the issue of analyzing oscillations of loss-rates caused by RED, evaluating different averaging functions for RED and the ADD estimator, and examining new AQM mechanisms that gives smother loss-rates as for further studies.

Based on our simulations, we recommend to set $w_0 = 1$ and w_i for other L_i using (2). If improved robustness is required at very high fractions of all traffic at low drop precedence levels and less capable PLR differentiation can be accepted, lower values of w_i or higher values of w_0 can be used than those given by (2).

5 Implementation Complexity

In this section, we describe an efficient implementation of the ADD estimator and the NLR selector. We also include an evaluation of the computational cost of an implementation on a test platform. This evaluation show that the overhead introduced by an implementation of ADD is small compared to other tasks a router need to perform. The computational cost of these mechanisms $\in O(n)$ (linear complexity), where n is the number of drop precedence levels. Since n is expected to be small (e.g., $n = 3$ for DiffServ AF), we do not consider this to be significant.

The ADD estimator is designed to allow implementation without floating-point arithmetics, divisions, or multiplications.

To further improve the performance of the differentiation dropper, the drop distance counter, d_i, is increased with σ_i instead of 1 upon packet drops at drop precedence level i. This gives the inverse NLR, $\overline{d}_i \cdot \sigma_i$, without multiplications (4).

$$\overline{d}_{i,n} \cdot \sigma_i = \overline{d}_{i,n-1} \cdot \sigma_i \cdot (1 - g_i) + d_i \cdot \sigma_i \cdot g_i \qquad (4)$$

The differentiation constants, σ_i, can be treated as fixed point decimal numbers so that relations with decimal precision can be configured by scaling σ_i with a factor of 10^p, where p is the desired number of decimal positions.

6 Conclusions

In this paper we define a loss-rate estimator based on average drop distances (ADDs). The ADD estimator is designed to offer *robust* and *capable* proportional loss-rate (PLR) differentiation at varying traffic loads. We consider a PLR differentiation *robust* if short-term loss-rate ratios have negligible variations and *capable* if long-term loss-rate ratios approximate target loss-rate ratios reasonable well at changing traffic load conditions.

We evaluate, through simulations, the PLR differentiation predictability of the ADD estimator and an estimator implemented with a loss history table (LHT) for two levels of drop precedence. These simulations show that the LHT estimator cannot provide both robust and capable PLR differentiation with one single M. For large M, the target loss-rate ratio is well approximated by the loss-rate ratio at long time-scales. However, for such M, the short-term loss-rate ratio can vary appreciably when traffic load varies.

For small M, low variation in the short-term loss-rate ratio is obtained at varying traffic loads, but it does not reach the target loss-rate ratio at long lime-scales. For the ADD estimator, weights can be found that gives both robust and capable PLR differentiation (i.e. the short-term loss-rate ratio has low variation and the long-term loss-rate ratio approximates the target loss-rate ratio). The ADD estimator requires however that the actual loss-rate is smooth (e.g. by using RED). Without proper smoothing of the actual loss-rate, the ADD estimator may not give predictable PLR differentiation.

To evaluate the performance of the ADD estimator, we have implemented a PLR dropper using the ADD estimator in the kernel of FreeBSD. With three levels of drop precedence supported, this dropper needs only 131 clock cycles in average to update ADDs and 59 clock cycles in average to select from which precedence level to drop on an Intel Pentium II 350Mhz.

References

1. Polly Huang Anja Feldmann, Anna C. Gilbert and Walter Willinger. Dynamics of ip traffic: A study of the role of variability and the impact of control. *ACM Computer Communications Review*, October 1999.
2. S. Blake, D. Black, M. Carlson, E. Davies, Z. Wang, and W. Weiss. An architecture for differentiated service. Request for Comments (Informational) 2475, Internet Engineering Task Force, December 1998.
3. Don Towsley C.V. Hollot, Vishal Misra and Wei-Bo Gong. On designing improved controllers for aqm routers supporting tcp flows. *Proceedings of Infocom 2001*, April 2001.
4. Constantinos Dovrolis and Parameswaran Ramanathan. A case for relative differentiated services and the proportional differentiation model. *IEEE Network*, 13(5):26–34, September/October 1999.
5. Constantinos Dovrolis and Parameswaran Ramanathan. Proportional differentiated services, part II: Loss rate differentiation and packet dropping. In *Proceedings of IWQoS 2000*, Pittsburgh, June 2000.
6. Sally Floyd. Recommendations on using the gentle variant of red. Notes, March 2000.
7. Sally Floyd and Van Jacobson. Random early detection gateways for congestion avoidance. *IEEE/ACM Transactions on Networking*, 1(4):397–413, August 1993.
8. J. Heinanen, F. Baker, W. Weiss, and J. Wroclawski. Assured forwarding PHB group. Request for Comments 2597, Internet Engineering Task Force, June 1999.
9. K. Nichols, S. Blake, F. Baker, and D. Black. Definition of the differentiated services field (DS field) in the IPv4 and IPv6 headers. Request for Comments (Proposed Standard) 2474, Internet Engineering Task Force, December 1998.
10. UCB/LBNL/VINT. Network simulator—ns (version 2.1b5). http://www-mash.cs.berkeley.edu/ns/, 1999.
11. Wei-Bo Gong Vishal Misra and Don Towsley. Fluid-based analysis of a network of aqm routers supporting tcp flows with an application to red. *ACM Computer Communications Review*, October 2000.

A Novel Scheduler for a Low Delay Service within Best-Effort

Paul Hurley[1]*, Mourad Kara[2], Jean-Yves Le Boudec[1], and Patrick Thiran[3]

[1] Institute for Computer Communication and Applications (ICA)
EPFL, Switzerland
[2] School of Computing, University of Leeds, United Kingdom
[3] Sprint ATL, Burlingame, CA, USA

Abstract. We present a novel scheduling algorithm, Duplicate Scheduling with Deadlines (DSD). This algorithm implements the ABE service [5] which allows interactive, adaptive applications, that mark their packets green, to receive a low bounded delay at the expense of maybe less throughput. ABE retains the best-effort context by protecting flows that value higher throughput more than low bounded delay, whose packets are marked blue. DSD optimises green traffic performance while satisfying the constraint that blue traffic must not be adversely affected. Using a virtual queue, deadlines are assigned to packets upon arrival, and green and blue packets are queued separately. At service time, the deadlines of the packets at the head of the blue and green queues are used to determine which one to serve next. It supports any mixture of TCP, TCP Friendly and non TCP Friendly traffic. We motivate, describe and provide an analysis of DSD, and show simulation results.

1 Introduction

We describe and analyse Duplicate Scheduling with Deadlines (DSD), a novel scheduling algorithm which implements ABE [5], an enhancement to the IP best-effort service to provide low queueing delay at the expense of maybe less throughput. In order to understand the reasoning behind and advantages of DSD, we provide a brief overview of ABE. A full description and discussion of ABE can be found in [5]. ABE does not need to police how much traffic opts to use low delay, and retains the operational simplicity of a single class best-effort network. ABE packets are marked either green or blue, with green packets receiving a low, bounded delay at every hop. For the service to remain best-effort with no overall advantage to either traffic type, sources which choose not to avail of the lower delay must receive at least as good a service as they would as if all packets had been blue. The introduction of ABE must be transparent to them.

As such, ABE requires that *green does not hurt blue*. If some source decides to mark some of its packets green rather than blue, then the quality of service received by sources that mark all their packets blue must remain the same or become better. The delay and throughput of blue sources must not deteriorate

---

* Part of this work was done while working at Sprint ATL. Contact author. paul.hurley@epfl.ch, Ph. +41-21-693-6626

L. Wolf, D. Hutchison, and R. Steinmetz (Eds.): IWQoS 2001, LNCS 2092, pp. 389–403, 2001.

and be at least as good as it was if a "flat best-effort" network, namely if all packets were blue. As a consequence, green packets are more likely to be dropped during periods of congestion than blue ones. This requirement applies whether green traffic originates from TCP Friendly sources, those which receive no more throughput than a TCP flow would, or from non TCP Friendly flows. Despite the mandate that non TCP sources be TCP Friendly [3], it is still the case that many multimedia flows are not TCP Friendly. It is worth mentioning that, with or without ABE, non TCP Friendly sources may, in some cases, severely hurt other, TCP Friendly sources. The ABE requirement simply means that giving low delay to such sources does not make things worse.

The overall design goal of DSD is to provide green with the best possible service while still ensuring protection for blue, namely that green does not hurt blue. Any significant extra gain by blue packets would be at the expense of green ones, reducing the incentive to choose green. In ABE, the first part of the green does hurt blue requirement is *local transparency*, which is satisfied, if, for each blue packet in the ABE scenario:

1. the delay is not larger than it would have been in a flat best-effort scenario, where a node would treat all ABE packets as one single best effort class;
2. it is dropped only if it would have been in the flat best-effort scenario.

DSD is a solution to the optimisation problem, which minimises the number of green losses subject to the following constraints:

- Green packets receive a no larger queueing delay than d (thus satisfying the low delay requirement).
- Local transparency to blue holds.
- The scheduling is work conserving.
- No reordering: Blue (respectively green) packets are served in the order of arrival.

DSD sends a duplicate of each packet arrival to a virtual queue. A blue packet is only dropped if its duplicate was in the virtual queue. Otherwise it receives a deadline equivalent to the service time its duplicate has in the virtual queue. A green packet is accepted if it passes what is called the *green acceptance test*, and then assigned a deadline equal to its arrival time plus the maximum time it can wait in the queue d. Green and blue packets are queued separately, and the deadlines of the packets at the head of blue and green queues are used to determine which one is to be served next; blue packets served at the latest their deadline permits and green served in the meantime if they have been in the queue for less than d seconds, and are dropped otherwise.

The virtual queue is not restricted to drop-tail queueing (although in the simulation results we show it is). An Active Queue Management scheme such as RED [7] can be supported for blue traffic by applying it to the virtual queue, and using those results in assigning losses and deadlines. Some of the building blocks in DSD are similar to those in other scheduling techniques. The calculation and tagging of deadlines to each arriving packet is also performed by Earliest Deadline First (EDF)[8] schedulers and its variants. However, EDF sorts packets according to deadlines, whereas DSD remains FIFO within each of its two queues, and the deadlines are used at service to determine whether the head of the green or the head of the blue queue should be served. The use of a virtual queue has

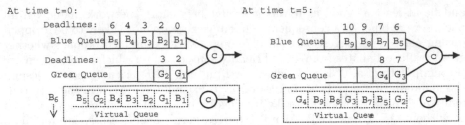

Fig. 1. Two snapshots as an example of DSD, at time $t = 0$ (left) and $t = 5$ (right). For this example, all packets are the same size and "packet" time is used. To facilitate understanding, we consider first the case where green packets do not undergo the green acceptance test and where $g = 1$. The maximal buffer size is $Buff = 7$ packets. The maximum green queue wait is $d = 3$ packets. B and G denote blue and green packets respectively. In the first snapshot, B_1 is served at time $t = 0$ in order to meet its deadline, then G_1, B_2, B_3, B_4. G_2 has to be dropped from the green queue because it has to wait for more than $d = 3$, whereas B_6 had to be dropped because the virtual queue length was $Buff$ when it arrived. At time $t = 5$, we reach the situation of the second snapshot. As no blue packet has reached its deadline yet, G_3 can be served, followed by $B_5, B_7,$ $G_4, B_8,$ and B_9.

been used many times, for example in an admission control context [9].

The second part of green does not hurt blue is throughput transparency. If some sources sending green traffic are rate-adaptive (TCP Friendly), and greedy, local transparency to blue may not be sufficient. It is quite possible that, by becoming green, a TCP Friendly source would achieve a higher data rate, due to the reduction in round-trip time (which is a direct result of the known bias in TCP in favour of flows with shorter round-trip times). To provide throughput transparency, an ABE node must ensure that an entirely green flow gets a lesser or equal throughput than if it were blue. Unlike local transparency, throughput transparency seems impossible to implement exactly, since it requires knowledge of the round-trip time for every flow, which is not feasible in practice and the rate adaptation algorithm implemented by a source may significantly deviate from strict TCP friendliness. Indeed, the dependency of rate on round-trip time is not necessarily a desirable feature of a rate adaptation algorithm.

To provide throughput transparency in the DSD scheduler, a controller, as described in Section 4, acts upon a parameter g to control the service received by green packets. g is the probability of serving the green queue first in the event that the deadlines of the packets at the head of each queue can both be met if the other queue was served beforehand, when there is a tie so to speak. The delay and loss ratio are monitored, and the controller adjusts g to make sure throughput transparency is maintained.

In the next section, a full description for DSD is given, followed in Section 3 by a presentation of some of its properties. A control loop for DSD is described in Section 4, followed by simulations of DSD in Section 5.

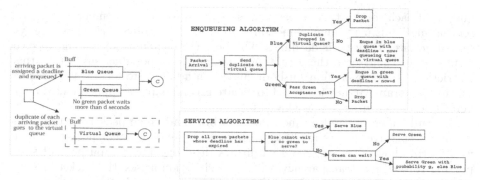

Fig. 2. Outline of the DSD Algorithm.

2 Scheduler Description

An example of how DSD works is given in Figure 1, while Figure 2 provides an overview of the algorithm. It is assumed the router has only output port queueing. Duplicates of all incoming packets are sent to a virtual queue with a buffer size *Buff*. A duplicate is admitted if the virtual buffer is not full. Packets in the virtual queue are served according to FCFS at rate c, as they would be in a flat best-effort. The times at which duplicates will be served are used to assign blue packets *deadlines* at which they would have been served in flat best-effort. The original arriving packets are fed according to their colour into a green and a blue queue. Blue packets are always served at the latest their deadline permits subject to work conservation. Green packets are served in the meantime if they have been in the queue for less than d seconds, and are dropped otherwise.

A blue packet is dropped if its duplicate was not accepted in the virtual queue. Otherwise, it is tagged with a deadline, given by the time at which its duplicate will be served in the virtual queue, and placed at the back of the blue queue. A green packet is accepted if it passes what is called the *green acceptance test* and dropped otherwise. A green packet arriving at time t fails the test if the sum of the length of the green queue at time t (including this packet), and of the length of the first part of the blue queue that contains packets tagged with a deadline less than or equal to $t + d + pg_{new}$, where pg_{new} is the transmission delay for the incoming green packet, is more than $c(d + pg_{new})$, and passes otherwise. The use of the test ensures the total buffer occupancy, namely the sum of the green and blue queue lengths, does not exceed *Buff*, which is discussed in Section 3. Consider again the example in Figure 1, except green packets are now enqueued only if they pass the green acceptance test. This amounts here to accepting a green packet at time t if the number of green packets in the queue at time t, augmented by the number of blue packets in the queue with a deadline between $[t, t + 4]$ is no more than 4. The only difference from Figure 1 is that G_2 is no longer enqueued. Indeed, when it arrived, the green queue already contained packet G_1, and the blue queue contained packets B_1, B_2 and B_3. The total queue length at time was 5 packets (including G_2), and so G_2 fails the test.

An accepted green packet is then assigned a deadline which is the sum of its arrival time plus its maximum waiting time d, and placed at the back of the green queue. At each service time, a decision is made as to which queue to serve. The serving mechanism's primary function is to ensure that blue packets are always served no later than their deadlines. The best performance green could receive would be to then serve the green queue as much as possible, subject to this restriction. However, as previously discussed, in addition to local transparency, throughput transparency is needed to ensure green adaptive applications do not benefit too much from lower delay.

It can happen at service time that both blue and green packets at the head of their respective queues are able to wait, as letting the other packet go first would still allow it to be served within its deadline. For the purpose of supporting throughput transparency, when this situation arises, the packet serving algorithm uses the current value of the *green bias* g, a value in the range $[0, 1]$, to determine the extent to which green is favoured over blue. More precisely, when both blue and green packets can wait, g is the probability that the green packet is served first. The value $g = 1$ corresponds to the case where green is always favoured. Conversely, the value $g = 0$ corresponds to the systematic favouring of blue packets. In the example in Figure 1, the packets served would have thus been, successively, B_1, B_2, G_1, B_3, B_4, B_5, B_7, G_3, G_4, B_8 and B_9.

Table 1. Pseudocode of DSD. *now* is the current time, *p*.deadline denotes the latest time a packet *p* can remain in the queue (whose value is tagged onto packet *p*), and *p*.transmissionDelay denotes its transmission delay.

Packet Enqueueing Algorithm	**Packet Serving Algorithm**
packet p arrives at the output port $dup = p$ Add dup to the virtual queue if p is blue 　if dup was dropped from virtual queue 　　drop p 　else 　　vd = queueing delay received by dup 　　　in virtual queue 　　p.deadline $= now + vd$ 　　add p to blue queue else // p is green 　if p fails "green acceptance test" 　　drop p 　else 　　p.deadline $= now + d$ 　　add p to green queue	drop stale green packets, those packets from green queue who cannot be served within their deadline headGreen = packet at head of green queue headBlue = packet at head of blue queue if headGreen = 0 // no green to serve 　if headBlue !=0 　　serve headBlue else if headBlue = 0 // no blue to serve 　serve headGreen else // both queues contain packets 　p_g = headGreen.transmissionDelay 　$dead_g$ = headGreen.deadline 　p_b = headBlue.transmissionDelay 　$dead_b$ = headBlue.deadline
"Green Acceptance Test" pg_{new} = transmission delay for p l_g = length of green queue l_b = length of packets in blue queue with 　　deadlines $<= now + d + pg_{new}$ if $l_g + l_b > c * (d + pg_{new})$ 　return "p fails test" else 　return "p passes test"	if $now > dead_b - p_g$ 　serve headBlue // because it cannot wait else if $now > dead_g - p_b$ 　serve headGreen // because it cannot wait else with probability g 　serve headGreen else 　serve headBlue

A value of g less than 1 causes the delay for green traffic to be increased. This increase in delay for green TCP Friendly traffic reduces their throughput, thereby enabling blue traffic to increase its throughput. Increasing the delay of non TCP Friendly traffic may not reduce their throughput, but blue flows are, in the worst case, as equally protected from this type of traffic as they would have been in a flat best-effort service. The value of g chosen is made according to a control loop which is described in Section 4.

All green packets who miss their deadline, by waiting for more than d seconds (these packets are said to have become *stale*), are removed from the green queue. Pseudocode for DSD is given in Table 1. Removing stale green packets, those packets from the green queue whose deadline cannot be met, involves a search of this queue up to the first alive green packet. In practice, these stale green packets can be cleaned up between service times, as was done in our dummynet implementation, and it has proven sufficiently fast. However, for really high speed networks this search may prove expensive. As such, further optimisations of this algorithm may be needed, and are the subject of on-going work.

3   Scheduler Properties

Some important properties of DSD are:

1. Buffer space constraint: the total buffer occupancy for real packets (green and blue counted together) is always less than *Buff*, the size of the virtual queue used for duplicates.
2. All accepted blue packets will be served by their deadlines. Accepted blues are thus served at the same time as, or earlier than, they would have been in flat best-effort.
3. All green packets are served before d, or are otherwise dropped. Low bounded (per hop) delay for the green packets is enforced by dropping a green packet that waits or would have to wait d seconds in the queue.
4. The green acceptance test does not unnecessarily drop green packets in the following sense. If all enqueued green packets are to be served, then it is impossible to serve, within d seconds, an incoming green packet that arrived at time t and would violate the green admission test. Also, if $g = 1$, the green admission test is optimal in the sense that it accepts exactly the green packets that will be served within d seconds. Note that if $g < 1$, some green packets may become stale and be dropped by the packet serving algorithm.

Items 2 and 3 are obvious consequences of the DSD algorithm. Item 1 is Theorem 1, proven in the appendix. Item 4 is Theorem 2, also proven in the appendix.

4   Control Loop for DSD

For the reasons described in the Introduction (Section 1), unlike local transparency, maintaining throughput transparency is by its nature approximate. g is used as a control parameter to balance the throughputs of green and blue, which are estimated by the formula,

$$\theta = \frac{s}{R\sqrt{\frac{2p}{3}} + 3t_1\sqrt{\frac{3p}{8}}p(1 + 32p^2)} \tag{1}$$

where R is the round-trip time, p the loss rate, t_1 the TCP retransmit time (roughly proportional to the round-trip time), and s the packet size [4].

A fixed value is used to represent the non-queueing delay portion of the round-trip time of a flow. This value is chosen to be small, since this favours blue traffic. For the purposes of the control, flows are assumed to be greedy, since this also increases the protection to blue flows. Estimates for the delay and loss ratio for both green and blue traffic are monitored, and throughput estimated by Equation (1). Let $\theta_b(t)$ and $\theta_g(t)$ be these estimates for the blue and green throughput respectively at time t. The value of g is chosen so that their ratio is close to a desired value γ, which is slightly larger than 1, to provide blues with a small advantage in throughput, and to offer a safety margin for protection from errors in throughput estimation. g is updated every T seconds according to the control law,

$$g(t + T) = (1 - \alpha)g(t) + \frac{\alpha}{1 + (\gamma\theta_g(t)/\theta_b(t))^K}, \qquad (2)$$

where $\alpha \in (0, 1)$ and $K > 0$ are two control parameters. T is a chosen parameter of the system which determines the rate of update of g. The initial value of g upon commencement of control can be chosen to be 1, namely, $g(0) = 1$.

Let us briefly explain the rationale behind this choice of control law, which we do not claim to be optimal. In the ideal case where $\theta_b = \gamma\theta_g$, there should not (a priori) be any bias against blue nor green, and the value of g should be $1/2$. If θ_b is larger than $\gamma\theta_g$, then g must be increased, and vice versa if θ_b is smaller than $\gamma\theta_g$. We wish to maintain symmetry in the amount by which we increase or reduce g: the amount by which g is increased if $\theta_b/\gamma\theta_g$ is multiplied by some factor A should be the same amount by which g is decreased if $\theta_b/\gamma\theta_g$ is divided by the same factor A. Denoting by $\xi = \ln(\theta_b/\gamma\theta_g)$, the targeted g should therefore be an increasing function F of ξ with central symmetry around 0, and such that $F(0) = 1/2$, $F(\xi) = 0$ for $\xi \to -\infty$ and $F(\xi) = 1$ for $\xi \to +\infty$. Such a function is the sigmoid function $F(\xi) = \frac{1}{1+\exp(-K\xi)}$ where K is the slope of the function at the origin. The larger K, the closer the sigmoid function to the step (Heaviside)

function $\overline{F}(\xi) = \begin{cases} 1 & \text{if } \xi > 0 \\ 1/2 & \text{if } \xi = 0 \\ 0 & \text{if } \xi < 0. \end{cases}$ The control law $g(t+T) = g(t)+\alpha(F(\xi)-g(t))$

where α is the adaptation gain, will therefore bring g to the targeted value. If $0 \leq \alpha \leq 1$, this control law keeps $g(t)$ between 0 and 1 at all times t. Replacing ξ by $\ln(\theta_b/\gamma\theta_g)$ in this equation, we get the control loop equation for the green bias as given in Equation (2).

5   Simulation of DSD

In this section, we show simulations, using ns-2 [1], of DSD run on the topology shown in Figure 3. There are $n_{b,1}$ blue sources and $n_{g,1}$ green sources with an outgoing link propagation delay of 20ms (sources of type 1), and $n_{b,2}$ blue and $n_{g,2}$ green sources with an outgoing 10Mbps link of propagation delay 50ms (sources of type 2). All sources pass the 5Mbps link L of propagation delay 20ms, and terminate via a 10Mbps link of propagation delay 10ms. These blue

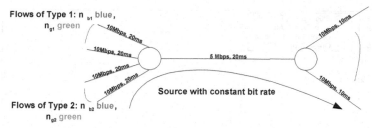

Fig. 3. Simulation Topology.

sources are TCP Reno, and the green sources are the TCP Friendly algorithm as described in [6]. There is also green traffic which sends a constant rate r (CBR) and passes through the link L.

The router buffer size was 60 packets (i.e. $Buff = 60$) and the maximum delay green can queue for, d, was 0.04s. For simplicity, the size of all packets is fixed at 1000 bytes. The control loop updates its value of g every 0.5s (i.e. $T = 0.5$), the gain parameter α was 1.1, and the conservative value of 20ms was taken to be the round-trip time used for estimating throughput. The router distinguishes green and blue by a bit in the packet header. Each simulation ran for 300 seconds of simulated time.

The first goal of this simulation study was to show that green does not hurt blue, under a variety of conditions; namely when there are flows of various round-trip times, where green flows may be either TCP Friendly or non TCP Friendly, may be greedy or not. In addition, we illustrate that green flows benefit from low delay (at the expense of less throughput), and show the effect DSD has on the loss rates of each traffic type. ABE results are compared to the flat best-effort scenario, where all packets are treated equally at the router.

We first examine some scenarios when there are only TCP and TCP Friendly flows. For the case where there are 5 blue TCP and 5 green TCP Friendly flows of each type ($n_{b,1} = n_{b,2} = n_{g,1} = n_{g,2} = 5$), Figure 4 shows the average transfer rate for each blue and green connection, of both types at each time t. Figure 5 shows the end-to-end delay distributions received for green packets under ABE and flat best-effort. Blue flows of each type receive more throughput with ABE than the did in flat best-effort, thus benefiting from the use of ABE. Green flows receive less, and in exchange, the green queueing delay is small and bounded by $d = 0.04s$. The green loss ratio was 4.97% when using ABE, and 3.3% in the flat best-effort, while the blue loss ratio decreased from 3.2% to 2.5% when moving to ABE. The extra throughput that blue flows of type 1 receive over type 2 flows follows from the lower round trip-time they experience.

ABE is designed to work independently of asymmetry in the amount of green and blue traffic. For the case where there are 5 blue TCP and 3 green TCP Friendly flows of type 1 ($n_{b,1} = 5$, $n_{g,1} = 3$) and 3 blue TCP and 5 green TCP Friendly flows of type 2 ($n_{b,2} = 3$, $n_{g,2} = 5$), Figure 6 shows that again green does not hurt blue. The situation where blue traffic is TCP, and green traffic is no longer TCP Friendly, but a constant bit-rate source is now examined. Here there are 5 blue TCP flows of each type ($n_{b,1} = n_{b,2} = 5$) and CBR green traffic which sends at 1Mbps. The number of packets received for each blue traffic type

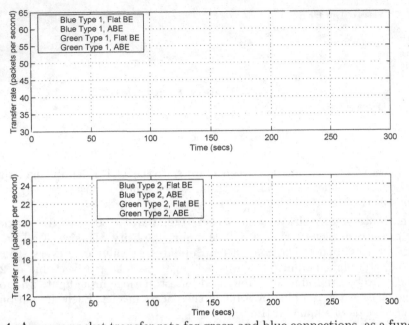

Fig. 4. Average packet transfer rate for green and blue connections, as a function of time t, when the router implemented DSD and when it implemented flat best-effort. The results are obtained by simulating the network described on Figure 3, with 5 blue flows and 5 green flows of each type, namely $n_{b,1} = n_{b,2} = n_{g,1} = n_{g,2} = 5$, and no CBR traffic.

and for the CBR source is shown as a function of time in Figure 7. What we see is that the blue traffic receives more slightly throughput with DSD than with flat best-effort, due to the local transparency property, and the non TCP Friendly CBR traffic receives less.

We now look at the scenario where there is blue TCP traffic $(n_{b,1} = n_{b,2} = 5)$, and green traffic is composed both of TCP Friendly sources $(n_{g,1} = n_{g,2} = 5)$ and CBR traffic of rate 1Mbps. The average packet transfer rate for the blue and green of type 1, and for the CBR source as a function of time is shown in Figure 8. The results for type 2 traffic is omitted for ease of reading.

6 Conclusions

We described and analysed a new scheduler, DSD, to enable ABE, a best-effort low-delay service, and thus facilitate multimedia adaptive applications to, in some cases, increase their utility. The freshness in approach involves the assignment of deadlines based on a virtual queue and maximum tolerable queueing delay, and the deadline decision based serving algorithm.

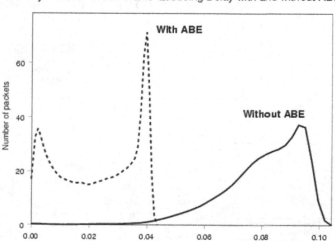

Density Plots for Green Traffic Queueing Delay with and without ABE

Fig. 5. Density plot of ueueing delay received by green packets under ABE/DSD and flat best-effort. 5 blue TCP and 5 green TCP Friendly flows of each type.

Appendix: Proof of DSD Properties

We prove items 1 and 4 from Section 3. The notation is illustrated in Figure 9. Denote by $q_g(t)$ the length of the green queue at time t, by $q_b(t)$ the total length of the blue queue (regardless of packet deadlines), and by $q_v(t)$ the length of the virtual queue, at time t (thus $q_g(t) = l_g$ in the pseudocode in Table 1).

Theorem 1 (Buffer Space Constraint). *The sum of the blue and green queue lengths does not exceed the maximum virtual queue size Buff: at any time t,* $q_b(t) + q_g(t) \leq \text{Buff}$.

Proof. Consider the system at any time $t \geq 0$.

(i) Suppose first that $q_g(t) = 0$: this means that there are no green packets in the queue at that time. Since a blue packet is accepted if and only if its corresponding duplicate is also admitted, and since it is always served no later than its duplicate, a blue packet will be present in the blue queue if and only if its duplicate is present in the virtual queue. Therefore $q_b(t) \leq q_v(t)$. Since the virtual queue length is always less than *Buff*,

$$q_b(t) + q_g(t) = q_b(t) \leq q_v(t) \leq \text{Buff}.$$

(ii) Suppose from now on that $q_g(t) > 0$. Let s be the latest time before t when an incoming green packet has arrived and been admitted, and let pg_{new} denote its processing time.

Let $q_b^{\text{head}}(t, s)$ denote the length of the portion of the blue queue with packets having deadlines in $[t, s + d + pg_{new}]$, (We have thus $q_b^{\text{head}}(s, s) = l_b$ in the

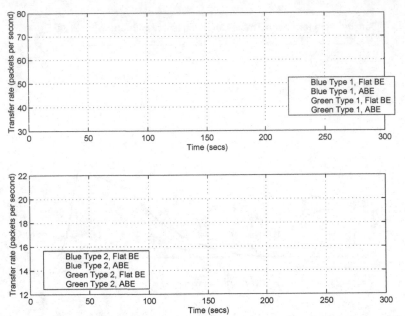

Fig. 6. Average packet transfer rate per green and blue connection, as a function of time t, when the router implemented ABE/DSD and when it implemented flat best-effort. The results are obtained by simulating the network described on Figure 3, with $n_{b,1} = 5, n_{g,1} = 3, n_{b,2} = 3, n_{g,2} = 5$.

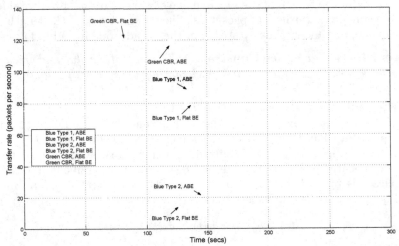

Fig. 7. Average packet transfer rate per green and blue connection, as a function of time t, when the router implemented ABE/DSD and when it implemented flat best-effort. There are 5 blue flows of each type and a CBR flow of 1Mbps which is green.

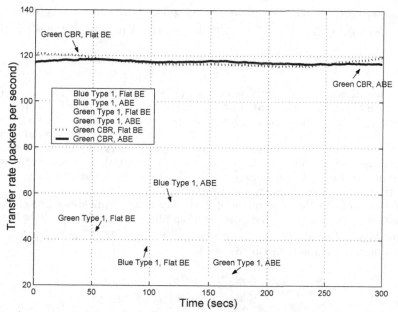

Fig. 8. Average packet transfer rate per green and blue connection of Type 1, and for the CBR source as a function of time t, when the router implemented ABE/DSD and when it implemented flat best-effort. The CBR source sends at 1Mbps and there are 5 of each other type of source.

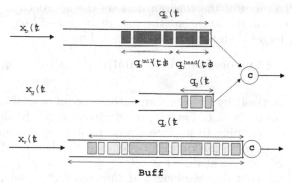

Fig. 9. Notation used in proofs in Appendix.

pseudocode in Table 1). It contains the bits that are counted in $q_b^{\text{head}}(s,s)$, and that have not been served yet. Likewise, since there are no new green arrivals in $(s,t]$, $q_g(t)$ contains the bits that are counted in $q_g(s)$, and that have not been served yet. Because $q_g(t) > 0$, the green queue is never empty in $[s,t]$, and therefore the server is never idle during this interval. Therefore

$$q_b^{\text{head}}(t,s) + q_g(t) = q^{\text{head}}(s,s) + q_g(s) - c(t-s).$$

At time s, since the latest green packet passed the green acceptance test, $q_b^{\text{head}}(s, s) + q_g(s) \leq c(d + pg_{new})$. Combining this with the previous equation,

$$q_b^{\text{head}}(t, s) + q_g(t) \leq c(d + pg_{new}) - c(t - s). \tag{3}$$

On the other hand, let $q_b^{\text{tail}}(t, s) = q_b(t) - q_b^{\text{head}}(t, s)$ denote the length of the portion of the blue queue at time t with packets having deadlines larger than $s + d + pg_{new}$. The deadline of an enqueued blue packet is larger than $s + d + pg_{new}$ if there are at least $c(d + pg_{new} + s - t)$ bits to be served by the virtual server of rate c, before the corresponding duplicate begins its service. As the total buffer space of the virtual queue is $Buff$, the maximum number of bits at time t in the virtual queue that can belong to duplicates of blue packets with deadline larger than $s + d + pg_{new}$ is thus at most $Buff - c(d + pg_{new} + s - t)$. As a blue packet is present in the blue queue if and only if its duplicate is present in the virtual queue, the length of the portion of the blue queue with packets having a deadline larger than $s + d + pg_{new}$, satisfies $q_b^{\text{tail}}(t, s) \leq Buff - c(d + pg_{new} + s - t)$. Combining this inequality with (3), we get

$$\begin{aligned}
q_b(t) + q_g(t) &= q_b^{\text{tail}}(t, s) + q_b^{\text{head}}(t, s) + q_g(t) \\
&\leq Buff - c(d + pg_{new}) + (s - t) \\
&\quad + c(d + pg_{new}) - c(t - s) = Buff.
\end{aligned}$$

Using the fact that the amount of bits admitted in the virtual queue is the same as the incoming fresh traffic as long as $q_v < Buff$, Theorem 1 can be refined and the following lemma established, which will be used in the proof of Theorem 2. The proof of the theorem follows the approach of the proof of Lemma 3, p. 20, in [10]. It states that the sum of green and blue queues at any time is never any larger than the size of the virtual queue.

Lemma 1 (Virtual Queue Bounds Actual). *At any time t, $q_b(t) + q_g(t) \leq q_v(t)$.*

Proof. Denote respectively by $a(t)$ the cumulative amount of traffic (in bits) that has arrived in the router in $[0, t]$. Let $x(t)$ (respectively $x_v(t)$) be the cumulative amount of bits corresponding to packets (resp. duplicates) that have been admitted in the router (resp. in the virtual queue) in $[0, t]$. The sum of the backlogs in the blue and green queues at time t is $q_g(t) + q_b(t) = \sup_{0 \leq s \leq t}\{x(t) - x(s) - c(t - s)\}$ whereas the virtual queue length at time t is $q_v(t) = \sup_{0 \leq s \leq t}\{x_v(t) - x_v(s) - c(t - s)\}$.

Let t be a given time, and let $0 \leq v \leq t$ be the smallest time such that the virtual queue was not full during the time interval $[v, t]$. Because of the duplicates scheme, the traffic that actually entered the virtual system during $[v, t]$ is therefore identical to the traffic that arrived during this time interval: $x_v(t) - x_v(v) = a(t) - a(v)$. If $v = 0$, the backlogged data in the actual system at time t is given by

$$\begin{aligned}
q_g(t) + q_b(t) &= \sup_{0 \leq s \leq t}\{x(t) - x(s) - c(t - s)\} \leq \sup_{0 \leq s \leq t}\{a(t) - a(s) - c(t - s)\} \\
&= \sup_{0 \leq s \leq t}\{x_v(t) - x_v(s) - c(t - s)\} = q_v(t).
\end{aligned}$$

If $v > 0$, it means that the virtual system was full at time v, and hence that $q_v(v) = Buff$. Then using Theorem 1 to obtain the first inequality below,

$$
\begin{aligned}
q_g(t) + q_b(t) &= \sup_{0 \le s \le t} \{x(t) - x(s) - c(t - s)\} \\
&= \sup_{0 \le s \le v} \{x(t) - x(s) - c(t - s)\} \vee \sup_{v \le s \le t} \{x(t) - x(s) - c(t - s)\} \\
&= \sup_{0 \le s \le v} \{x(t) - x(v) - c(t - v) + x(v) - x(s) - c(v - s)\} \\
&\qquad \vee \sup_{v \le s \le t} \{x(t) - x(s) - c(t - s)\} \\
&= \{x(t) - x(v) - c(t - v) + \sup_{0 \le s \le v} \{x(v) - x(s) - c(v - s)\}\} \\
&\qquad \vee \sup_{v \le s \le t} \{x(t) - x(s) - c(t - s)\} \\
&= \{x(t) - x(v) - c(t - v) + q_g(v) + q_b(v)\} \vee \sup_{v \le s \le t} \{x(t) - x(s) - c(t - s)\} \\
&\le \{x(t) - x(v) - c(t - v) + Buff\} \vee \sup_{v \le s \le t} \{x(t) - x(s) - c(t - s)\} \\
&\le \{a(t) - a(v) - c(t - v) + Buff\} \vee \sup_{v \le s \le t} \{a(t) - a(s) - c(t - s)\} \\
&= \{x_v(t) - x_v(v) - c(t - v) + Buff\} \vee \sup_{v \le s \le t} \{x_v(t) - x_v(s) - c(t - s)\} \\
&= \{x_v(t) - x_v(v) - c(t - v) + q_v(v)\} \vee \sup_{v \le s \le t} \{x_v(t) - x_v(s) - c(t - s)\} \\
&= \{x_v(t) - x_v(v) - c(t - v) + \sup_{0 \le s \le v} \{x_v(v) - x_v(s) - c(v - s)\}\} \\
&\qquad \vee \sup_{v \le s \le t} \{x_v(t) - x_v(s) - c(t - s)\} \\
&= \sup_{0 \le s \le v} \{x_v(t) - x_v(s) - c(t - s)\} \vee \sup_{v \le s \le t} \{x_v(t) - x_v(s) - c(t - s)\} \\
&= \sup_{0 \le s \le t} \{x_v(t) - x_v(s) - c(t - s)\} = q_v(t).
\end{aligned}
$$

Theorem 2. 1. *If all enqueued green packets are to be served, then the green acceptance test does not unnecessarily drop green packets that could otherwise have been served within d seconds.*

2. *If $g = 1$, then there are no stale green packets.*

Proof. Suppose a green packet, with transmission time pg_{new} arrives at time t.

(Item 1) Call $q_g(t-)$ the green queue size just before time t. We show that if all enqueued green packets in $q_g(t-)$ are to be served, then it is impossible to serve an incoming green packet that arrived at time t and would violate the green admission test within d seconds. To be able to complete the service of the green packet before $t + d$, one must be able to serve all packets currently in the green queue, which takes $q_g(t-)/c$ seconds, all blue packets whose deadline falls in $[t, t + d]$, which takes $q_b^d(t)/c$ seconds, and the incoming green packet itself, which takes pg_{new}. The sum of these three times must not exceed d, which shows that the green acceptance test is indeed necessary.

(Item 2) Suppose $g = 1$. Clearly there is enough time to serve all packets currently in the green queue, all blue packets present at time t and those whose deadline falls in $[t, t+d]$, and the incoming green packet itself. Because the queue is FIFO, any green packet that arrives after t will be served after the green packet that arrived at time t is served, and hence does not delay the considered green packet which arrived at time t. One has to check that any accepted blue packet that has arrived after t does not prevent the green packet that arrived at time t to be served either. Call u the smallest time larger than or equal to t such that $q_v(u) > c(d + pg_{new})$ (if no such time exists, set $u = \infty$). Then $q_v(v) \leq c(d + pg_{new})$ for all $t \leq v < u$. Because of Lemma 1, the sum of the blue and green queue lengths is less than $c(d + pg_{new})$: $q_g(u) + q_b(u) \leq q_v(u) \leq c(d + pg_{new})$, which means that all packets that arrived at any time $t \leq v < u$, including the green packet that arrived at time t, will begin their service within d seconds from their arrival time, in the FIFO order. Conversely, for all $v \geq u$, $q_v(v) \geq q_v(u) - c(v - u) > c(d + pg_{new}) - c(v - u)$ and hence that any blue packet that arrived at any time $v \geq u$ will have a deadline to begin its service such that

$$v + q_v(v)/c > v + (d + pg_{new}) - (v - u).$$
$$= u + d + pg_{new} \geq t + d + pg_{new},$$

i.e. after the green packet under consideration will have completed its service.

References

1. ns v2 simulator. `http://www.isi.edu/nsnam/ns`
2. S. Floyd, K. Fall. Promoting the Use of End-to-End Congestion Control in the Internet. IEEE/ACM Transactions on Networking, August 1999.
3. TCP Friendly web site. `http://www.psc.edu/networking/tcp_friendly.html`
4. J. Padhye, V. Firoiu, D. Towsley, J. Kurose. Modeling TCP Throughput: A Simple Model and its Empirical Validation. *Proceedings of SIGCOMM'98*.
5. P. Hurley, M. Kara, J.Y. Le Boudec, P.Thiran. ABE: Providing a Low Delay Service Within Best-Effort. IEEE Network Magazine, May 2001. Available at `http://www.abeservice.org`
6. C. Boutremans, J.Y. Le Boudec. Adaptive delay aware error control for internet telephony. Technical Report Research Report DSC 2000/31, EPFL-DSC, `http://dscwww.epfl.ch`, 2000.
7. S. Floyd, V. Jacobson. Random Early Detection Gateways for Congestion Avoidance. *IEEE/ACM Transactions on Networking*, V.1 N.4, August 1993, p.397-413.
8. H. Zhang. Service Disciplines for Guaranteed Performance Service in Packet-Switching Networks. Proceedings of the IEEE, Vol. 83 No. 10, October 1995.
9. T. Ferrari, W. Almesberger, J. Y. Le Boudec. SRP: a Scalable Resource Reservation Protocol for the Internet. Computer Communications, September 1998, Vol 21. No. 14. Special issue on 'Multimedia networking', 1200-1211.
10. J.-Y. Le Boudec, P. Thiran. Network calculus viewed as a Min-plus System Theory applied to Communication Networks. Technical report SSC/1998/016, EPFL, Lausanne, Switzerland, April 1998.

JoBS: Joint Buffer Management and Scheduling for Differentiated Services*

Jörg Liebeherr and Nicolas Christin

Computer Science Department, University of Virginia, Charlottesville, VA 22904, USA
{jorg,nicolas}@cs.virginia.edu

Abstract. A novel algorithm for buffer management and packet scheduling is presented for providing loss and delay differentiation for traffic classes at a network router. The algorithm, called JoBS (Joint Buffer Management and Scheduling), provides delay and loss differentiation independently at each node, without assuming admission control or policing. The novel capabilities of the proposed algorithm are that (1) scheduling and buffer management decisions are performed in a single step, and (2) both relative and (whenever possible) absolute QoS requirements of classes are supported. Numerical simulation examples, including results for a heuristic approximation, are presented to illustrate the effectiveness of the approach and to compare the new algorithm to existing methods for loss and delay differentiation.

1 Introduction

Quality-of-Service (QoS) guarantees in packet networks are often classified according to two criteria. The first criterion is whether guarantees are expressed for individual end-to-end traffic flows (*per-flow QoS*) or for groups of flows with the same QoS requirements (*per-class QoS*). The second criterion is whether guarantees are expressed with reference to guarantees given to other flows/flow classes (*relative QoS*), or whether guarantees are expressed as absolute bounds (*absolute QoS*).

Efforts to provision for QoS in the Internet in the early and mid-1990s, which resulted in the *Integrated Services* (IntServ) service model [3], focused on per-flow absolute QoS guarantees. However, due to scalability issues and a lagging demand for per-flow absolute QoS, the interest in Internet QoS eventually shifted to relative per-class guarantees. Since late 1997, the *Differentiated Services* (DiffServ) [2] working group has discussed several proposals for per-class relative QoS guarantees, e.g., [4,17].

With the exception of the Expedited Forwarding service [11], proposals for relative per-class QoS discussed within the DiffServ context define the service differentiation qualitatively, in the sense that some classes receive lower delays and a lower loss rate than others, but without quantifying the differentiation. Recently, research studies have tried to strengthen the guarantees of relative per-class QoS, and have proposed new buffer management and scheduling algorithms which can support stronger notions of relative QoS [6,7,15,16]. Probably the best known such effort is the *proportional service*

---

* This work is supported in part by the National Science Foundation through grants NCR-9624106 (CAREER), ANI-9730103, and ANI-0085955.

L. Wolf, D. Hutchison, and R. Steinmetz (Eds.): IWQoS 2001, LNCS 2092, pp. 404–418, 2001.
© Springer-Verlag Berlin Heidelberg 2001

differentiation model [6,7], which attempts to enforce that the ratios of delays or loss rates of successive priority classes be roughly constant. For two priority classes such a service could specify that the delays of packets from the higher-priority class be half of the delays from the lower-priority class, but without specifying an upper bound on the delays.

In this paper, we express the provisioning of per-class QoS within a formalism inspired by the network calculus [5]. We present a rate allocation and dropping algorithm for a single output link, called *Joint Buffer Management and Scheduling (JoBS)*, which is capable of supporting a wide range of relative, as well as absolute, per-class guarantees for loss and delay, without assuming admission control or traffic policing. The algorithm operates as follows. Upon each arrival, a prediction is made on the delays of the currently backlogged traffic. Then, the service rates allocation to classes are adjusted to meet delay requirements. If necessary, traffic from certain classes is selectively dropped. A unique feature of the presented algorithm is that rate allocation for link scheduling and buffer management are approached together in a single step. The JoBS algorithm provides delay and loss differentiation independently at each node. End-to-end delays and end-to-end loss rates are thus dependent on the per-node guarantees of traffic and on the number of nodes traversed.

This paper is organized as follows. In Section 2 we give an overview of the current work on relative per-class QoS guarantees. In Sections 3 and 4, we specify our algorithm for buffer management and rate allocation. In Section 5 we propose a heuristic approximation of the algorithm. In Section 6 we evaluate the effectiveness of our algorithm via simulation. In Section 7 we present brief conclusions.

2   Related Work

Due to space considerations, we limit our discussions to the relevant work on scheduling and buffer management algorithms for relative service differentiation.

Scheduling. The majority of work on per-class relative service differentiation suggests to use well-known fixed-priority, e.g., [17], or rate-based scheduling algorithms, e.g., [9]. Only a few scheduling algorithms have been specifically designed for relative delay differentiation. The Proportional Queue Control Mechanism (PQCM, [15]) and Backlog-Proportional Rate scheduler (BPR, [6]) are variations of the GPS algorithm [18]. Both schemes use the backlog of classes to determine the service rate allocation, and bear similarity to the scheduling component of JoBS, in the sense that they dynamically adjust service rate allocations to meet relative QoS requirements.

Different from the rate-based schedulers discussed above, the Waiting-Time Priority scheduler (WTP, [7]) implements a well-known scheduling algorithm with time-dependent priorities ([12], Ch. 3.7). Likewise, the Mean-Delay Proportional scheduler (MDP, [16]) uses a dynamic priority mechanism, but sets priorities based on the average experienced delay of packets. Finally, the Hybrid Proportional Delay scheduler (HPD, [6]) uses a combination of time-dependent priorities and average experienced delay to set the priority of a given packet.

The Alternative Best-Effort service (ABE, [10]) provides service differentiation for two traffic classes. The first class is provided with absolute delay guarantees, and the second class has guarantees for a lower loss rate. The delay guarantees for the first class are enforced by dropping all traffic that has exceeded the delay bound.

In contrast to the schedulers presented in this section, the scheduling algorithm presented in this paper not only considers the current state and past history of the link, but, in addition, makes predictions on future delays to improve the performance of its scheduling decisions.

Buffer Management. For a discussion of buffer management algorithms, we refer to a recent survey article [13]. Many proposals for buffer management in IP networks are motivated with the need to improve TCP performance (e.g., RED [8], REM [1]). Techniques specifically targeted for class-based service differentiation include RIO [4] and multiclass RED [19]. Of these schemes, REM is closest in spirit to the dropping algorithm presented in this paper, since REM treats the problem of marking (or dropping) arrivals as an optimization problem.

The Proportional Loss Rate (PLR) dropper [7] is specifically designed to support proportional differentiated services. PLR enforces that the ratio of the loss rates of two successive classes remains roughly constant at a given value. There are two variants of PLR. PLR(M) uses only the last M packets for estimating the loss rate of a class, whereas PLR(∞) has no such memory constraints.

With the possible exception of [10], the work on relative per-class service differentiation generally considers delay and loss differentiation as orthogonal issues, which are handled by separate algorithms.

3   An Approach to Joint Buffer Management and Scheduling

In this section, we introduce the key concepts of *Joint Buffer Management and Scheduling* (JoBS). Before we provide a detailed description, we first give an informal overview of the operations.

3.1   Overview

We assume that each output link performs per-class buffering of arriving traffic and that traffic is transmitted from the buffers using a rate-based scheduling algorithm [21] with a dynamic, time-dependent service rate allocation for classes. Traffic from the same class is transmitted in a First-Come-First-Served order. There is no admission control and no policing of traffic. The set of performance requirements are specified to the algorithm as a set of per-class QoS constraints. As an example, for three classes, the QoS constraints could be of the form:

– Class-1 Delay $\approx 2 \cdot$ Class-2 Delay,
– Class-2 Loss Rate $\approx 10^{-1} \cdot$ Class-3 Loss Rate, or
– Class-3 Delay $\leq 5\ ms$.

Here, the first two constraints are relative constraints and the last one is an absolute constraint. The set of constraints can be any mix of relative and absolute constraints. Since absolute constraints may render a system of constraints infeasible, some constraints may need to be relaxed. We assume that all QoS constraints are prioritized, so that an order is provided in which constraints are relaxed in case the system of constraints is infeasible.

The time-dependent service rate allocation operates as follows. For every arrival, a prediction is made on the delays of all backlogged traffic. Then, the service rate allocation to traffic classes is modified so that all QoS constraints will be met. If no feasible rate allocation for meeting all constraints exists, traffic is dropped, either from a new arrival or from the current backlog.

We find it convenient to view the service rate allocation in terms of an optimization problem. The constraints of the optimization problem are relative or absolute bounds on the loss and delay as given in the example above (*QoS constraints*) and constraints on the link and buffer capacity (*system constraints*). The objective function of the optimization primarily aims at minimizing the amount of traffic to be dropped, and, as a secondary objective, aims at maintaining the current service rate allocation. The first objective prevents traffic from being dropped unnecessarily, and the second objective tries to avoid frequent fluctuations of the service rate allocation. The solution of the optimization problem yields a service rate allocation of classes and determines how much traffic must be dropped.

To explore the principal properties of the optimization, we will, at first, assume that sufficient computing resources are available to solve the optimization problem for each arrival to the link. In a later section, we will approximate the optimization with a heuristic which incurs less computational overhead.

3.2   Formal Description

Next we describe the basic operations of the service rate allocation and the dropping algorithms at a link with capacity C and total buffer space B. We assume that all traffic is marked to belong to one of Q traffic classes. In general, we expect Q to be small, e.g., $Q = 4$. Classes are marked by an index. We use a convention, whereby a class with a smaller index requires a better level of QoS. We use $a_i(t)$ and $l_i(t)$ to denote the traffic arrivals and amount of dropped traffic from class i at time t. We use $r_i(t)$ to denote the service rate allocated to class i at time t. We assume that $r_i(t) > 0$ only if there is a backlog of class-i traffic in the buffer (and $r_i(t) = 0$ otherwise), and we assume that scheduling is work-conserving, that is, $\sum_i r_i(t) = C$, if there is at least one backlogged class at time t.

Remark. Throughout this paper, we take a fluid-flow interpretation of traffic, that is, the output link is regarded as serving simultaneously traffic from several classes. Since actual traffic is sent in discrete-sized packets, a fluid-flow interpretation of traffic is idealistic. However, scheduling algorithms that closely approximate fluid-flow schedulers with rate guarantees are available [18,21].

We now introduce the notions of *arrival curve*, *input curve*, and *output curve* for a traffic class i in the time interval $[0, t]$. The arrival curve A_i and the input curve R_i^{in} of class i are defined as

$$A_i(t) = \int_0^t a_i(x)dx \ , \quad R_i^{in}(t) = A_i(t) - \int_0^t l_i(x)dx \ . \tag{1}$$

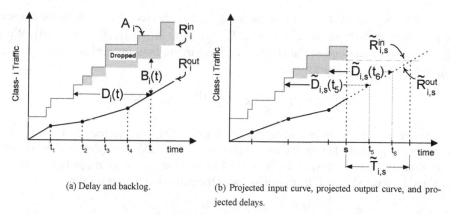

(a) Delay and backlog.   (b) Projected input curve, projected output curve, and projected delays.

Fig. 1. Delay, Backlog, and Projections. In Figure 1(b), the projection is performed at time s for the time interval $[s, s + \tilde{T}_{i,s}]$.

So, the difference between the arrival and input curve is the amount of dropped traffic. The output curve R_i^{out} of class-i is the transmitted traffic in the interval $[0, t]$, given by

$$R_i^{out}(t) = \int_0^t r_i(x)dx \ . \tag{2}$$

We refer to Figure 1(a) for an illustration. In the figure, the service rate is adjusted at times t_1, t_2, and t_4, and packet drops occur at times t_2 and t_3.

The vertical and the horizontal distance between the input and output curves from class i, respectively, are the backlog B_i and the delay D_i. This is illustrated in Figure 1(a) for time t. The delay D_i at time t is the delay of an arrival which is transmitted at time t. Backlog and delay at time t are defined as

$$B_i(t) = R_i^{in}(t) - R_i^{out}(t) \ , \quad D_i(t) = \max_{x<t}\{x \mid R_i^{out}(t) \geq R_i^{in}(t - x)\} \ . \tag{3}$$

Upon a traffic arrival, say at time s, the new service rates $r_i(s)$ and the amount of traffic to be dropped $l_i(s)$ for all classes are set such that all QoS and system constraints can be met at times greater than s. If all constraints cannot be satisfied at the same time, then some QoS constraints are relaxed in a predetermined order.

To determine the rate allocation, the scheduler makes a projection of the delays of all backlogged traffic. For the purpose of the projection, it is assumed that the current state of the link will not change after time s. Specifically, indicating projected values by a tilde (˜), for times $t > s$, we assume that (1) service rates remain as they are (i.e., $\tilde{r}_i(t) = r_i(s)$), (2) there are no further arrivals (i.e., $\tilde{a}_i(t) = 0$), and (3) there are no further packet drops (i.e., $\tilde{l}_i(t) = 0$).

With these assumptions, we now define the notions of projected input curve $\tilde{R}_{i,s}^{in}$, projected output curve $\tilde{R}_{i,s}^{out}$, and projected backlog $\tilde{B}_{i,s}$, for $t > s$ as follows:

$$\tilde{R}_{i,s}^{in}(t) = R_i^{in}(s) \ , \quad \tilde{R}_{i,s}^{out}(t) = R_i^{out}(s) + (t - s)r_i(s) \ , \quad \tilde{B}_{i,s}(t) = \tilde{R}_{i,s}^{in}(t) - \tilde{R}_{i,s}^{out}(t) \ . \tag{4}$$

We refer to the *projected horizon* for class i at time s, denoted as $\tilde{T}_{i,s}$, as the time when the projected backlog becomes zero, i.e., $\tilde{T}_{i,s} = \min_{x>0}\{x \mid \tilde{B}_{i,s}(s+x) = 0\}$. With this notation, we can make predictions for delays in the time interval $[s, s + \tilde{T}_{i,s}]$. We define the projected delay $\tilde{D}_{i,s}$ as

$$\tilde{D}_{i,s}(t) = \max_{t-s<x<t} \{x \mid \tilde{R}_{i,s}^{out}(t) \geq R_i^{in}(t-x)\}. \tag{5}$$

If there are no arrivals after time s, the delay projections are correct. In Figure 1(b), we illustrate the projected input curve, projected output curve, and projected delays for projections made at time s. In the figure, all values for $t > s$ are projections and are indicated by dashed lines. The figure includes the projected delays for times t_5 and t_6.

4   Service Rate Adaptation and Drop Algorithm

In this section we discuss an algorithm to perform the service rates allocation to classes and the decision to drop traffic in terms of an optimization problem.

Each time s at which an arrival occurs, a new optimization is performed. The optimization variable is a time-dependent vector $\mathbf{x}_s = (r_1(s)\ldots r_Q(s)\, l_1(s)\ldots l_Q(s))^T$, which contains the service rates $r_i(s)$ and the amount of traffic to be dropped $l_i(s)$. The optimization problem has the form

$$\begin{aligned}
\textbf{Minimize} \quad & F(\mathbf{x}_s) \\
\textbf{Subject to} \quad & g_j(\mathbf{x}_s) = 0,\ j = 1,\ldots,M \\
& h_j(\mathbf{x}_s) \geq 0,\ j = M+1,\ldots,N,
\end{aligned} \tag{6}$$

where $F(.)$ is an objective function, and the g_j's and h_j's are constraints. The objective function, which will be presented in Subsection 4.2, will be chosen so that the amount of dropped traffic and the changes to the current service rate allocation are minimized. The constraints of the optimization problem are QoS constraints and system constraints. The optimization at time s is done with knowledge of the system state before time s, that is the optimizer knows R_i^{in} and R_i^{out} for all times $t < s$, and A_i for all times $t \leq s$.

In the remainder of this section we discuss the constraints and the optimization function. The optimization can be used as a reference system against which practical scheduling and dropping algorithms can be compared.

4.1   System and QoS Constraints

There are two types of constraints. *System constraints* describe constraints and properties of the output link, and *QoS constraints* define the desired service differentiation.

System Constraints.   The system constraints specify physical limitations and properties at the output link. The first such constraint states that the total backlog cannot exceed the buffer size B, that is, $\sum_i B_i(t) \leq B$ for all times t. The second system constraint enforces that scheduling at the output link is work-conserving. At a work-conserving link, $\sum_i r_i(t) = C$ holds for all times t where $\sum_i B_i(t) > 0$. Other system constraints enforce that transmission rates and loss rates are non-negative. Also, the amount of traffic that can be dropped is bounded by the current backlog. So we obtain $r_i(t) \geq 0$ and $0 \leq l_i(t) \leq B_i(t)$ for all times t.

QoS Constraints. We consider two types of QoS constraints, relative constraints and absolute constraints. QoS constraints are either constraints on delays or constraints on the loss rate. The number and type of QoS constraints is not limited. Since absolute QoS constraints may result in an infeasible system of constraints, one or more constraints may need to be relaxed at certain times. We assume that the set of QoS constraints is assigned some total order, and that constraints are relaxed in the given order until the system of constraints becomes feasible. In addition, QoS constraints for classes which are not backlogged are simply ignored.

Absolute delay constraints (ADC) enforce that the projected delays of class i satisfy a worst-case bound d_i. That is,

$$\max_{s < t < s + \tilde{T}_{i,s}} \tilde{D}_{i,s}(t) \leq d_i \,, \tag{7}$$

for all $t \in [s, s + \tilde{T}_{i,s}]$. If this condition holds for all s, the delay bound d_i is never violated.

Relative delay constraints (RDC) specify the proportional delay differentiation between classes. As an example, for two classes 1 and 2, the RDC enforces a relationship

$$\frac{\text{Delay of Class 2}}{\text{Delay of Class 1}} \approx \text{constant} \,.$$

Since, in general, there are several packets backlogged from a class, each likely to have a different delay, the notion of 'delay of class i' needs to be further specified. For example, the delay of class i could be specified as the delay of the packet at the head of the class-i queue, the maximum projected delay as in Eqn. (7), or via other measures. We choose a measure, called *average projected delay* $\overline{D}_{i,s}$, which is the time average of the projected delays from a class, averaged over the horizon $\tilde{T}_{i,s}$. We obtain

$$\overline{D}_{i,s} = \frac{1}{\tilde{T}_{i,s}} \int_s^{s + \tilde{T}_{i,s}} \tilde{D}_{i,s}(x) dx \,. \tag{8}$$

To provide some flexibility in the scheduling decision, we do not enforce relative delay constraints strictly, but allow for some slack. Using the metric defined in Eqn. (8), and translating the notion of slack into a tolerance level, we can write the relative delay constraints as

$$k_i(1 - \varepsilon) \leq \frac{\overline{D}_{i+1,s}}{\overline{D}_{i,s}} \leq k_i(1 + \varepsilon) \,, \tag{9}$$

where $k_i > 1$ is a constant defining the proportional differentiation desired, and ε ($0 \leq \varepsilon \leq 1$) indicates a tolerance level. If relative constraints are not specified for some classes, the constraints are adjusted accordingly. Note that in the delay constraints in Eqs. (7) and (9), all values with exception of the components of the optimization variable $\mathbf{x}_s$ are known at time s.

Next we discuss constraints on the loss rate. Similar to delays, there are several sensible choices for defining 'loss'. Here, we select a loss measure, denoted by $p_{i,s}$, which expresses the fraction of lost traffic since the beginning of the current busy period at time t_0.[1] So, $p_{i,s}$ expresses the fraction of traffic that has been dropped in the time interval $[t_0, s]$, that is,[2]

$$p_{i,s} = \frac{\int_{t_0}^{s} l_i(x)dx}{\int_{t_0}^{s} a_i(x)dx} = 1 - \frac{R_i^{in}(s^-) + (a_i(s) - l_i(s)) - R_i^{in}(t_0)}{A_i(s) - A_i(t_0)}. \qquad (10)$$

In the last equation, all values except $l_i(s)$ are known at time s. With this definition we now specify absolute and relative constraints on the loss rates.

An *absolute loss constraint (ALC)* specifies that the loss rate of class i, as defined above, never exceeds a limit L_i, that is,

$$p_{i,s} \leq L_i . \qquad (11)$$

Relative loss constraints (RLC) specify the desired proportional loss differentiation between classes. Similar to the RDCs, we provide a certain slack within these constraints. The RLC for classes $(i + 1)$ and i has the form

$$k_i'(1 - \varepsilon') \leq \frac{p_{i+1,s}}{p_{i,s}} \leq k_i'(1 + \varepsilon') , \qquad (12)$$

where $k_i' > 1$ is the target differentiation factor, and ε' ($0 \leq \varepsilon' \leq 1$) indicates a level of tolerance.

4.2 Objective Function

Provided that the QoS and system constraints can be satisfied, the objective function will select a solution for $\mathbf{x}_s$. Even though the choice of the objective function is a policy decision, we select two specific objectives, which, we believe, have general validity: (1) *avoid dropping traffic* , and (2) *avoid changes to the current service rate allocation*. The first objective ensures that traffic is dropped only if there is no alternative way to satisfy the constraints. The second objective tries to hold on to a feasible service rate allocation as long as possible. We give the first objective priority over the second objective.

The following formulation of an objective function expresses the above objectives in terms of a cost function:

$$F(\mathbf{x}_s) = \sum_{i=1}^{Q}(r_i(s) - r_i(s^-))^2 + C^2 \sum_{i=1}^{Q} l_i(s) , \qquad (13)$$

where C is the link capacity. The first term expresses the changes to the service rate allocation and the second term expresses the losses at time s. Note that, at time s, $r_i(s)$ is part of the optimization variable, while $r_i(s^-)$ is a known value. In Eqn. (13) we use the quadratic form $(r_i(s) - r_i(s^-))^2$, since $\sum_i(r_i(s) - r_i(s^-)) = 0$ for a work-conserving link with a backlog at time s. The scaling factor C^2 in front of the second

---

[1] A busy period is a time interval with a positive backlog of traffic. For time x with $\sum_i B_i(x) > 0$, the beginning of the busy period is given by $\sup_{y < x}\{\sum_i B_i(y) = 0\}$.

[2] $s^- = s - h$, where $h > 0$ is infinitesimally small.

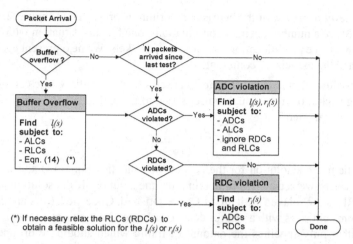

Fig. 2. Outline of the Heuristic Algorithm.

sum of Eqn. (13) ensures that traffic drops are the dominating term in the objective function.

This concludes the description of the optimization process in JoBS. The structure of constraints and objective function makes this a *non-linear optimization problem*, which can be solved with available numerical algorithms [20].

5   Heuristic Approximation

We next present a heuristic that approximates the optimization presented in the previous section, with significantly lower computational complexity. The presented heuristic should be regarded as a first step towards a router implementation.

Approximating a non-linear optimization problem such as the one presented in Section 4 can be performed by well-known techniques such as fuzzy systems, or neural networks. However, these techniques are computationally too expensive if a high accuracy in the approximation is desired. Therefore, we choose a different approach, which decomposes the optimization problem into several computationally less intensive problems. The heuristic algorithm presented here maintains a feasible rate allocation until a buffer overflow occurs or a delay violation is predicted. At that time, the heuristic picks a new feasible rate allocation and/or drops traffic. Unless there is a buffer overflow, the tests for violations of ADCs and RDCs are not performed for every packet arrival, but only periodically.

A set of constraints, which contains absolute constraints (ALCs or ADCs), may be infeasible at certain times. Then, some constraints need to be relaxed. In our heuristic algorithm, the constraints are prioritized in the following order: system constraints have priority over absolute constraints, which in turn have priority over relative constraints. If the system of constraints becomes infeasible, the heuristic relaxes the relative constraints (RLCs or RDCs). If this does not yield a feasible solution, the heuristic relaxes one or more absolute constraints.

A high-level overview of the heuristic algorithm is presented in Figure 2. The algorithm consists of a number of small computations, one for each situation which requires to adjust the service rate allocation and/or to drop packets. We next present each of these situations and the associated computation.

Buffer Overflow. If an arrival at time s causes a buffer overflow, one can either drop the arriving packet or free enough buffer space to accommodate the arriving packets. Both cases are satisfied if

$$\sum_i l_i(s) = \sum_i a_i(s) . \tag{14}$$

The heuristic picks a solution for the $l_i(s)$ which satisfies Eqn. (14) and the RLCs in Eqn. (12), where we set $\varepsilon' = 0$ to obtain a unique solution. If the solution violates an ALC, the RLCs are relaxed until all ALCs are satisfied. Once the $l_i(s)$'s are determined the algorithm continues with a test for delay constraint violations, as shown in Figure 2. The algorithm only specifies the amount of traffic which should be dropped from a particular class, however, the algorithm does not select the position in the queue from which to drop traffic. In the present paper, we assume a Drop-Tail dropping policy.

If there are no buffer overflows, the algorithm makes projections for delay violations only once for every N packet arrivals. The selection of N represents a tradeoff between the runtime complexity of the algorithm and performance of the scheduling with respect to satisfying the constraints. Simulation experiments, as described in Section 6, show that the value $N = 100$ provides good performance.

The tests use the current service rate allocation to predict future violations. For delay constraint violations, the heuristic distinguishes three cases.

Case 1: No violation. In this case, the service rates are unchanged.

Case 2: RDC violation. If some RDC (but no ADC) is violated, the heuristic algorithm determines new rate values. Here, the RDCs as defined in Eqn. (9) are transformed into equations by setting $\varepsilon = 0$. Together with the work-conserving property, one obtains a system of equations, for which the algorithm picks a solution. If the solution violates an ADC, the RDCs are relaxed until the ADCs are satisfied.

Case 3: ADC violation. Resolving an ADC violation is not entirely trivial as it requires to recalculate the $r_i(s)$'s, and, if traffic needs to be dropped to meet the ADCs, the $l_i(s)$'s. To simplify the task, our heuristic ignores all relative constraints when an ADC violation occurs, and only tries to satisfy absolute constraints.

The heuristic starts with a conservative estimate of the worst-case delay for the class-i backlog at time s. For this, the heuristic uses the fact that for all $x \in [s, s + \tilde{T}_{i,s}]$, $\tilde{D}_{i,s}(x) \leq D_i(s) + \frac{B_i(s)}{r_i(s)}$, which can be verified by referring to Figures 1(a) and 1(b). Then, using $B_i(s) = B_i(s^-) + a_i(s) - l_i(s)$, we can write a sufficient condition for satisfying the ADC of class i with delay bound d_i at time s,

$$\underbrace{\frac{1}{r_i(s)} \frac{B_i(s^-) + a_i(s) - l_i(s)}{d_i - D_i(s)}}_{\rho_i} \leq 1 . \tag{15}$$

The heuristic algorithm will select the $r_i(s)$ and $l_i(s)$ such that Eqn. (15) is satisfied for all i. Initially, rates and traffic drops are set to $r_i(s) = r_i(s^-)$ and $l_i(s) = 0$. Since at

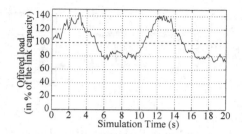

Fig. 3. Offered Load.

least one ADC is violated, there is at least one class with $\rho_i > 1$, where ρ_i is defined in Eqn.(15). Now, we apply a greedy method which tries to redistribute the rate allocations until $\rho_i \leq 1$ for all classes. This is done by reducing $r_i(s)$ for classes with $\rho_i < 1$, and increasing $r_i(s)$ for classes with $\rho_i > 1$. If it is not feasible to achieve $\rho_i \leq 1$ for all classes by adjusting the $r_i(s)$'s, the $l_i(s)$'s are increased until $\rho_i \leq 1$ for all i. To minimize the number of dropped packets, $l_i(s)$ is never increased to a point where an ALC is violated.

6   Evaluation

We present an evaluation of the algorithms developed in this paper via simulation. Our goals are (1) to determine if and how well the desired service differentiation is achieved; (2) to determine how well the heuristic algorithm from Section 5 approximates the optimization from Section 4; and (3) to compare our algorithm with existing proposals for proportional differentiated services.

We present two simulation experiments. In the first experiment, we compare the relative differentiation provided by the optimization algorithm described in Section 4, *JoBS (optimization)*, the heuristic approximation of Section 5, *JoBS (heuristic)*, and *WTP/PLR(∞)* [7], which provided uniformly the best results among previously proposed schemes for relative service differentiation. In the second experiment, we augment the set of constraints by absolute loss and delay constraints on the highest priority class, and show that JoBS can effectively provide both relative and absolute differentiation.

6.1   Experimental Setup

We consider a single output link with capacity $C = 1$ Gbps and a buffer size of 6.25 MB. We assume $Q = 4$ classes. The length of each experiment is 20 seconds of simulated time, starting with an empty system. In all experiments, the incoming traffic is composed of a superposition of Pareto sources with $\alpha = 1.2$ and average interarrival time of 300 μs. The number of sources active at a given time oscillates between 200 and 550, following a sinusoidal pattern. All sources generate packets with a fixed size of 125 bytes. The resulting offered load is plotted in Figure 3. At any time, each class contributes 25% of the aggregate load, yielding a symmetric load. In a realistic environment, one would expect to have "less" high priority traffic than low priority traffic.

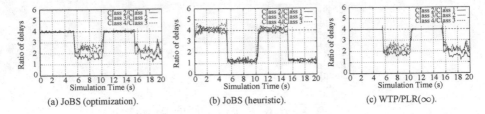

(a) JoBS (optimization).   (b) JoBS (heuristic).   (c) WTP/PLR(∞).

Fig. 4. Experiment 1: Relative Delay Differentiation. The graphs show the ratios of the delays for successive classes. The target value is $k = 4$.

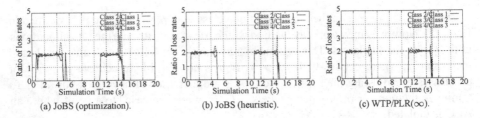

(a) JoBS (optimization).   (b) JoBS (heuristic).   (c) WTP/PLR(∞).

Fig. 5. Experiment 1: Relative Loss Differentiation. The graphs show the ratios of loss rates for successive classes. The target value is $k' = 2$.

Therefore, a symmetric load can be regarded as a realistic worst-case that can occur during bursts of high-priority traffic.

6.2 Simulation Experiment 1: Relative Differentiation Only

The first experiment focuses on relative service differentiation, and does not include absolute constraints. The objectives for the relative differentiation are so that we want to have a ratio of four between the delays of two successive classes, and a ratio of two between the loss rates of two successive classes. Thus, for JoBS, we set $k_i = 4$ and $k'_i = 2$ for all i. The tolerance levels are set to $(\varepsilon, \varepsilon') = (0.001, 0.05)$ in JoBS (optimization), and to $\varepsilon = 0.01$ in JoBS (heuristic). The results of the experiment are presented in Figures 4 and 5, where we graph the ratios of delays and loss rates, respectively, of successive classes for JoBS (optimization), JoBS (heuristic), and WTP/PLR(∞). The plotted delay and loss values are averages over moving time windows of size 0.1 s.

When the link load is above 90% of the link capacity, that is, in time intervals $[0\ s, 6\ s]$ and $[10\ s, 15\ s]$, all methods provide the desired service differentiation. The oscillations around the target values in JoBS (optimization) and JoBS (heuristic) are mostly due to the tolerance values ε and ε'. The selection of the tolerance values ε and ε' in JoBS presents a tradeoff: smaller values for ε and ε' reduce oscillations, but incur more work for the algorithms. When the system load is low, that is, in time intervals $[6\ s, 10\ s]$ and $[16\ s, 20\ s]$, only JoBS (optimization) and WTP/PLR(∞) manage to achieve some delay differentiation, albeit far from the target values. However, at an underloaded link, the absolute values of the delays are very small for all classes.

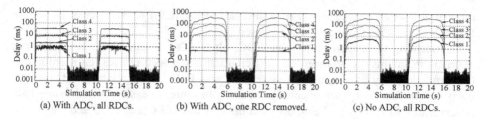

(a) With ADC, all RDCs.   (b) With ADC, one RDC removed.   (c) No ADC, all RDCs.

Fig. 6. Experiment 2: Absolute Delay Differentiation. The graphs show the delays of all packets. All results are for JoBS (heuristic).

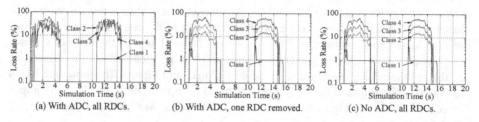

(a) With ADC, all RDCs.   (b) With ADC, one RDC removed.   (c) No ADC, all RDCs.

Fig. 7. Experiment 2: Absolute Loss Differentiation. The graphs show the loss rates of all classes. All results are for JoBS (heuristic).

Finally, one should note that the total loss rate is of interest, as a scheme may provide excellent proportional loss differentiation, but have an overall high loss rate. Additional plots provided in [14] show that the loss rates and the absolute values for the delays are very similar in all schemes.

6.3 Simulation Experiment 2: Relative and Absolute Differentiation

In this second experiment, we evaluate how well our algorithm can satisfy a mix of absolute and relative constraints on both delays and losses. Here, we only present results for JoBS (heuristic). WTP/PLR(∞) does not support absolute guarantees.

We consider the same simulation setup and the same relative delay constraints as in Experiment 1, but add an absolute delay constraint (ADC) for Class 1 such that $d_1 = 1$ ms, and we replace the relative loss constraint (RLC) between Classes 1 and 2 by an absolute loss constraint (ALC) for Class 1 such that $L_1 = 1\%$. We call this scenario "with ADC, all RDCs". With the given relative delay constraints from Experiment 1, the other classes have implicit absolute delay constraints, which are approximately [3] 4 ms for Class 2, 16 ms for Class 3, and 64 ms for Class 4. Removing the RDC between Class 1 and Class 2, we avoid the 'implicit' absolute constraints for Classes 2, 3, and 4, and call the resulting constraint set "with ADC, one RDC removed". We also include the results for JoBS (heuristic) from Experiment 1, with the ALC on Class 1 replacing the RLC between Classes 1 and 2, and refer to this constraint set as "no ADC, all RDCs". In Figure 6 we plot the absolute delays of all packets, and in Figure 7 we plot the loss

---

[3] Due to the tolerance value ε, the exact values are not integers.

rates of all classes, averaged over time intervals of length 0.1 s. We discuss the results for each of the three constraint sets proposed.

Concerning the experiment "with ADC, all RDCs", Figure 6(a) shows that the heuristic maintains the relative delay differentiation between classes, thus, enforcing the 'implicit' delay constraints for Classes 2, 3, and 4. With a large number of absolute delay constraints, the system of constraints easily becomes infeasible, which brings two observations. First, Figure 7(a) shows that the loss rates of Classes 2, 3 and 4 are similar. This result illustrates that the heuristic relaxes relative loss constraints to meet the absolute delay constraints. Second, Figure 6(a) shows that the absolute delay constraint d_1 is sometimes violated. However, such violations are rare (over 95% of Class-1 packets have a delay less than 900 μs), and Class-1 packet delays always remain reasonably close to the delay bound d_1. For the experiment "with ADC, one RDC removed", Figure 6(b) shows that, without an RDC between Classes 1 and 2, the ratio of Class-2 delays and Class-1 delays can exceed a factor of 10 at high loads. With this constraint set, the absolute delay constraint d_1 is never violated, and Figure 7(b) shows the RLCs are consistently enforced during periods of packet drops. Finally, for the experiment "no ADC, all RDCs", Figure 6(c) shows that, without the ADC, the delays for Class 1 are as high as 5 ms.[4]

7 Conclusions

We proposed an algorithm, called JoBS (Joint Buffer Management and Scheduling), for relative and absolute per-class QoS guarantees without information on traffic arrivals. At times when not all absolute QoS guarantees can be satisfied simultaneously, JoBS selectively ignores some of the QoS guarantees. The JoBS algorithm reconciles rate allocation and buffer management into a single scheme, thereby acknowledging that scheduling and dropping decisions at an output link are not orthogonal issues, but should be addressed together. JoBS implements the desired service differentiation based on delay predictions of backlogged traffic. The predictions are used to update service rate allocations to classes and the amount of traffic to be dropped. We showed in a set of simulation experiments, that JoBS can provide relative and absolute per-class QoS guarantees for delay and loss.

In future work, we will extend the approach presented in this paper to TCP congestion control. As a point of departure, we will attempt to express existing active queue management schemes, e.g., RED [8] and RIO [4], within the formal framework introduced in this paper.

---

[4] The delay values for Classes 2, 3, and 4 in Figures 6(b) and (c) appear similar, especially since we use a log-scale. We emphasize that the values are *not* identical, and that the results are consistent.

References

1. S. Athuraliya, D. Lapsley, and S. Low. An enhanced random early marking algorithm for internet flow control. In *Proceedings of IEEE INFOCOM 2000*, pages 1425–1434, Tel-Aviv, Israel, April 2000.
2. S. Blake, D. Black, M. Carlson, E. Davies, Z. Wang, and W. Weiss. An architecture for differentiated services. IETF RFC 2475, December 1998.
3. R. Braden, D. Clark, and S. Shenker. Integrated services in the internet architecture: an overview. IETF RFC 1633, July 1994.
4. D. Clark and W. Fang. Explicit allocation of best-effort packet delivery service. *IEEE/ACM Transactions on Networking*, 6(4):362–373, August 1998.
5. R. Cruz, H. Sariowan, and G. Polyzos. Scheduling for quality of service guarantees via service curves. In *Proceedings of the International Conference on Computer Communications and Networks (ICCCN)*, pages 512–520, Las Vegas, NV., September 1995.
6. C. Dovrolis. *Proportional Differentiated Services for the Internet*. PhD thesis, University of Wisconsin-Madison, December 2000.
7. C. Dovrolis and P. Ramanathan. Proportional differentiated services, part II: Loss rate differentiation and packet dropping. In *Proceedings of IWQoS 2000*, pages 52–61, Pittsburgh, PA., June 2000.
8. S. Floyd and V. Jacobson. Random early detection for congestion avoidance. *IEEE/ACM Transactions on Networking*, 1(4):397–413, July 1993.
9. S. Floyd and V. Jacobson. Link-sharing and resource management models for packet networks. *IEEE/ACM Transactions on Networking*, 3(4):365–386, August 1995.
10. P. Hurley, M. Kara, J.-Y. Le Boudec, and P. Thiran. ABE: Providing a low delay service within best-effort. Technical Report DSC/2000/34, EPFL-DI-ICA, September 2000.
11. V. Jacobson, K. Nichols, and K. Poduri. An expedited forwarding PHB. IETF RFC 2598, June 1999.
12. L. Kleinrock. *Queueing Systems. Volume II: Computer Applications*. John Wiley & Sons, New York, NY, 1976.
13. M. A. Labrador and S. Banerjee. Packet dropping policies for ATM and IP networks. *IEEE Communications Surveys*, 2(3), 3rd Quarter 1999.
14. J. Liebeherr and N. Christin. Buffer management and scheduling for enhanced differentiated services. Technical Report CS-2000-24, University of Virginia, August 2000.
15. Y. Moret and S. Fdida. A proportional queue control mechanism to provide differentiated services. In *Proceedings of the International Symposium on Computer and Information Systems (ISCIS)*, pages 17–24, Belek, Turkey, October 1998.
16. T. Nandagopal, N. Venkitaraman, R. Sivakumar, and V. Bharghavan. Delay differentiation and adaptation in core stateless networks. In *Proceedings of IEEE INFOCOM 2000*, pages 421–430, Tel-Aviv, Israel, April 2000.
17. K. Nichols, V. Jacobson, and L. Zhang. Two-bit differentiated services architecture for the Internet. IETF RFC 2638, July 1999.
18. A. K. Parekh and R. G. Gallagher. A generalized processor sharing approach to flow control in integrated services networks: the single-node case. *IEEE/ACM Transactions on Networking*, 1(3):344–357, June 1993.
19. S. Sahu, P. Nain, D. Towsley, C. Diot, and V. Fioroiu. On achievable service differentiation with token bucket marking for TCP. In *Proceedings of ACM SIGMETRICS 2000*, pages 23–33, Santa Clara, CA, June 2000.
20. K. Schittkowski. NLPQL: A FORTRAN subroutine solving constrained nonlinear programming problems. *Annals of Operations Research*, 5:485–500, 1986.
21. L. Zhang. Virtual clock: A new traffic control algorithm for packet switched networks. *IEEE/ACM Trans. Comput. Syst.*, 9(2):101–125, May 1991.

Optimal Call Admission Control under Generalized Processor Sharing Scheduling*

Antonis Panagakis and Ioannis Stavrakakis

Department of Informatics, University of Athens, 15784 Athens, Greece
{grad0260,istavrak}@di.uoa.gr

Abstract In this paper the problem of Call Admission Control is considered for leaky bucket constrained sessions with deterministic service guarantees (zero loss and finite delay bound), served by a Generalized Processor Sharing scheduler at a single node in the presence of best effort traffic. Based on an optimization process a CAC algorithm capable of determining the (unique) optimal solution is derived. The derived algorithm is applicable, under slight modification, in a system where the best effort traffic is absent and capable of guaranteeing that a solution to the CAC problem does not exist, if not found. The provided numerical results indicate that the presented algorithm can achieve, under certain conditions, a significant improvement on bandwidth utilization compared to a (deterministic) effective bandwidth based CAC scheme.

1 Introduction

The Generalized Processor Sharing (GPS) scheduling discipline has been widely considered to allocate bandwidth resources to multiplexed traffic streams. Its effectiveness and capabilities in guaranteeing a certain level of Quality of Service (QoS) to the supported streams in both a stochastic ([5,6,4]) and deterministic ([1,2,3,7]) sense have been investigated. Traffic management based on deterministic guarantees is expected to lead to lower network resource utilization compared to that under stochastic guarantees. Nevertheless, such considerations are necessary when deterministic guarantees are required by the applications and can provide insight and methodology for the consideration of stochastic guarantees.

Under the GPS scheduling discipline traffic is treated as an infinitely divisible fluid. A GPS server that serves N sessions is characterized by N positive real numbers $\phi_1, ..., \phi_N$, referred to as weights. These weights affect the amount of service provided to the sessions (or, their bandwidth shares). More specifically, if $W_i(\tau, t)$ denotes the amount of session i traffic served in a time interval $(\tau, t]$ then $W_i(\tau, t)/W_j(\tau, t) \geq \phi_i/\phi_j$, $j = 1, 2, ...N$ will hold for any session i that is continuously backlogged in the interval $(\tau, t]$; session i is considered to be backlogged at time t if a positive amount of that session traffic is queued at t.

---

* This work has been supported in part by the G.S.R.D. of Greece under grant PENED99(99ED 92) and the IST Program of the EU IST-1999-10160.

L. Wolf, D. Hutchison, and R. Steinmetz (Eds.): IWQoS 2001, LNCS 2092, pp. 419–434, 2001.

The GPS scheduling discipline has been introduced in [1],[2] where bounds on the induced delay have been derived for single node and multiple nodes systems, respectively. These (loose) delay bounds have a simple form allowing for the solution of the inverse problem, that is, the determination of the weight assignment for sessions demanding specific delay bounds, which is central to the Call Admission Control (CAC) problem. Tighter delay bounds have been derived in [7] and in [8] (also reported in [10]). These efforts have exploited the dependencies among the sessions –due to the complex bandwidth sharing mechanism of the GPS discipline– to derive tighter performance bounds. Such bounds could lead to a more effective CAC and better resource utilization. The inverse problem in these cases, though, is more difficult to solve. For example, tighter delay bounds have been presented in [3] in conjunction with a CAC algorithm for the single node case. The CAC procedure employs an exhaustive search having performance bound calculations as an intermediate step. In this paper the problem of optimal CAC in a GPS scheduling environment is investigated by following a different philosophy than in [3].

A major contribution of this paper is an algorithm which determines the optimal weights ϕ directly from the QoS requirements of the sessions, rather than through a recursive computation of the induced delay bounds and weight re-assignment. The major results are derived by considering a mixed traffic environment in which the bandwidth resource controlled by the GPS server is assumed to be shared by a number of QoS sensitive streams and best effort traffic. This system will be referred to as a Best Effort Traffic Aware Generalized Processor Sharing (BETA-GPS) system. The developed algorithm determines the minimum ϕ assignments for the QoS sensitive streams which are just sufficient to meet their QoS and, consequently, maximizes the (remaining) ϕ assignment to the best effort traffic. Based on the main results an optimal CAC scheme is proposed in this paper for a decoupled system of GPS-controlled QoS sensitive traffic and best effort traffic (referred to as pure QoS system, see section 4). The formulation of the pure QoS system facilitates the derivation of the minimum required GPS scheduler capacity to support N QoS sensitive streams, which is, in itself, an interesting problem.

2   Definitions and Description of the BETA-GPS System

The QoS sensitive sessions will be assumed to be leaky bucket constrained. That is, the amount of session i traffic arriving at the GPS server over any interval $(\tau, t]$ –referred to as the (assumed to be left continuous as in [1]) session i arrival function $A_i(\tau, t)$– will be bounded as follows: $A_i(\tau, t) \leq \sigma_i + \rho_i(t-\tau), \forall t \geq \tau \geq 0$; σ_i and ρ_i represent the burstiness and long term maximum mean arrival rate of session i. A session i is characterized as greedy starting at time τ, if the aforementioned bound is achieved, that is if $A_i(\tau, t) = \sigma_i + \rho_i(t-\tau), \forall t \geq \tau$. A GPS system busy period is defined to be the maximal time interval such that at least one session is backlogged at any time instant in the interval.

An all-greedy GPS system is defined as a system in which all the sessions are greedy starting at time 0, the beginning of a system busy period. The significance of the all-greedy system follows from [1] (Theorem 3): If the input link speed of any session i exceeds the GPS service rate, then for every session i, the maximum delay D_i^* and the maximum backlog Q_i^* are achieved (not necessarily at the same time) when every session is greedy starting at time zero, the beginning of a system busy period. This implies that if the server can guarantee an upper bound on a session' s delay under the all greedy system assumption this bound would be valid under any (leaky bucket constrained) arrival pattern. In view of the previous observation and by examining only all greedy systems, the CAC problem for a GPS system is simplified.

Let $t = 0$ denote the beginning of a system busy period in an all greedy system. For each session i the arrival function takes the form $A_i(0,t) = \sigma_i + \rho_i \cdot t, \forall t \geq 0$. If $Q_i(t)$ denotes the amount of session i traffic queued in the server at time t, then $Q_i(t) = A_i(0,t) - W_i(0,t)$ and $Q_i(t) = 0$ for all $t \leq 0$ by assumption. Let e_i denote the backlog clearing time of session i, then:

$$e_i = max\{t > 0 : Q_i(t) > 0\} \tag{1}$$

and $B_i = (0, e_i]$, corresponds to the session i busy period.

The QoS sensitive sessions will be assumed to have a stringent delay requirement, denoted by D_i for session i. Thus, a QoS sensitive session will be characterized by the triplet (σ_i, ρ_i, D_i). To ensure that the delay constraint for the QoS sensitive session i is met, a minimum amount of service $N_i(0,t)$ must be provided by the GPS server to session i over the interval $(0, t]$, where

$$N_i(0,t) = \begin{cases} \sigma_i + \rho_i(t - D_i) & t \geq D_i \\ 0 & t < D_i \end{cases} \tag{2}$$

That is the actual amount of service (work) $W_i(0,t)$ provided by the GPS server to session i over the interval $(0, t]$ must satisfy $W_i(0,t) \geq N_i(0,t), \forall t \geq 0$. The function $N_i(0,t)$ is referred to as session i requirements.

The Best Effort Traffic Aware (BETA) GPS system is depicted in figure 3. The BETA-GPS server capacity C_G is assumed to be shared by N QoS sensitive greedy sessions with descriptors (σ_i, ρ_i, D_i), $i = 1, \ldots, N$ and best effort traffic represented by an additional session. Each session is provided a buffer and the input links are considered to have infinite capacity.

Quantities associated with a QoS sensitive session (best effort session) will be identified by a subscript i (be), $i = 1, \ldots, N$. To avoid degenerate cases and be consistent with the GPS definitions it is assumed that $\sigma_i \rho_i D_i \neq 0, i = 1, \ldots, N$ and that the ϕ assignment of the BETA-GPS scheduler to a session can not be zero ($\phi_i > 0, \quad i = 1, \ldots, N, be$).

Generally, the task of CAC is to determine whether the network can accept a new session without causing QoS requirement violations. In the case of a GPS scheduler it should also provide the server with the weight assignment which will be used in the actual service of the admitted calls. A CAC scheme for a

GPS server is considered to be optimal if its incapability to admit a specific set of sessions implies that no ϕ assignment exists under which the server could serve this set of sessions (without causing QoS requirement violation even under the worst case arrival scenario (all greedy system)). In addition, an optimal CAC scheme for the BETA-GPS system should seek to maximize the amount of service provided to the (traffic unlimited) best effort session under any arrival scenario and over any time horizon, while satisfying the QoS requirement of the (traffic limited) QoS sensitive sessions. That is, it should seek to maximize the normalized weight assigned to the best effort traffic (ϕ_{be}), while satisfying the QoS requirement of QoS sensitive sessions. In view of this discussion, the following definition may be provided.

Definition 1. *(a) The optimal CAC scheme for the BETA-GPS system is the one that is based on the optimal ϕ assignment for the BETA-GPS system. (b) The optimal ϕ assignment for the BETA-GPS system is the one that allows the QoS sensitive sessions to meet their QoS requirements - provided that it is possible - and achieves:* $\max\{\phi_{be}\} = \max\{1 - \sum_{i=1}^{N} \phi_i\}$ *or, equivalently,* $\min\{\sum_{i=1}^{N} \phi_i\}$, *where* $\phi_i \in \mathbb{R}_+^*$, $i = 1, \ldots, N$, *be according to the definition of GPS.*

3   Optimal CAC for the BETA-GPS System

Because of the all greedy system assumption, all QoS sensitive sessions are backlogged at time $t = 0^+$. Let $\mathcal{B}(t)$ denote the set of sessions that are backlogged in the interval $(0, t]$ and let $\mathcal{E}(t)$ denote the set of sessions which have emptied their backlog before time t, that is, $\mathcal{B}(t) = \{i : e_i \geq t, \quad i = 1, \ldots, N\}$ and $\mathcal{E}(t) = \{i : e_i < t, \quad i = 1, \ldots, N\}$ where e_i is defined in (1). Each session $k \in \mathcal{E}(t)$ requires a rate equal to ρ_k. Consequently, the bandwidth that can be considered to be available for allocation to the sessions i, $i \in \mathcal{B}(t)$ is equal to $(C_G - \sum_{k \in \mathcal{E}(t)} \rho_k)$. Session $i \in \mathcal{B}(t)$ will be allocated a share of that bandwidth equal to $\phi_i(1 - \sum_{k \in \mathcal{E}(t)} \phi_k)^{-1}$ and will be served with a rate $\phi_i(C_G - \sum_{k \in \mathcal{E}(t)} \rho_k)(1 - \sum_{k \in \mathcal{E}(t)} \phi_k)^{-1}$. Let

$$\hat{C}(t) \triangleq (C_G - \sum_{j \in \mathcal{E}(t)} \rho_j)(1 - \sum_{j \in \mathcal{E}(t)} \phi_j)^{-1} \tag{3}$$

be referred to as the Normalized Backlogged Sessions Allocated (NBSA) bandwidth. Clearly $\hat{C}(t)$ changes value each time a session empties its backlog and remains constant between two consecutive backlog clearing times. Thus, $\hat{C}(t)$ is a piecewise constant function with the discontinuity points coinciding with the backlog clearing times of the sessions.

Let $\{b_i\}_{i=1}^{L}$, $L \leq N$ denote the ordered set of distinct backlog clearing times and let $b_0 = 0$ be the beginning of the system busy period. For two consecutive backlog clearing times b_{j-1} and b_j $\hat{C}(b_{j-1}^+) = \hat{C}(b_j^-)$. Treating the NBSA bandwidth as a left continuous function implies that $\hat{C}(b_j) = \hat{C}(b_j^-)$ and

$$\hat{C}(t) = (C_G - \sum_{k \in \mathcal{E}(b_j)} \rho_k)(1 - \sum_{k \in \mathcal{E}(b_j)} \phi_k)^{-1} \quad \forall t \in (b_{j-1}, b_j] \tag{4}$$

$\hat{C}(t)$ is an increasing function of time since it preserves a constant value between two consecutive backlog clearing times and $\hat{C}(b_j) < \hat{C}(b_j^+)$ for a backlog clearing time b_j (the proof may be found in [11]). The amount of scheduler's work that is shared among the backlogged sessions $i \in \mathcal{B}(b_j)$ over the time interval $(b_{j-1}, b_j]$ is equal to $(C_G - \sum_{i \in \mathcal{E}(b_j)} \rho_i)(b_j - b_{j-1})$. Let

$$\hat{W}(b_{j-1}, b_j) \triangleq \hat{C}(b_{j-1}^+)(b_j - b_{j-1}) \tag{5}$$

be referred to as the Normalized Backlogged Sessions Allocated (NBSA) work. Then – in view of (4) and (5) –, session $i \in \mathcal{B}(b_j)$ is allocated an amount of work equal to $\phi_i \hat{W}(b_{j-1}, b_j)$ over $(b_{j-1}, b_j]$.

3.1   Optimizing an Acceptable ϕ Assignment

In this section a process is developed that converts an acceptable ϕ assignment into a more efficient acceptable one. **An acceptable ϕ assignment** is one which is feasible (that is $\sum_{i=1}^{N} \phi_i < 1^1$) and delivers the required QoS to each of the supported QoS sensitive sessions. **A ϕ assignment is more efficient than another** if the sum of ϕ's $\sum_{i=1}^{N} \phi_i$ under the former assignment is smaller than that under the later. The aforementioned process will be referred to as the XMF (eXpand Minimum busy period First) process. According to the XMF process each QoS sensitive session's busy period is expanded as much as its QoS would permit, starting from the set of QoS sensitive sessions that empty their backlog first in order. A very important property of the XMF process is that it converts any acceptable ϕ assignment into the optimal one.

Let Π denote the set of acceptable policies (or equivalently, ϕ assignments) and let $\pi_a \in \Pi$. The application of the XMF process to π_a results in an acceptable policy $\pi_o = XMF(\pi_a)$, which is not less efficient than π_a; $XMF(\pi)$ denotes a policy that is generated by applying the XMF process to π. In particular, it will be shown that π_o is unique and more efficient than π_a, except for the case in which $\pi_a = \pi_o$. Let $^a\mathcal{I}_k$ denote the set of QoS sensitive sessions that empty their backlog k-th in order under π_a and let ^ab_k be the time instant when this happens. Let $^a\mathcal{I}_k^P = \bigcup_{s=1}^{k-1} {}^a\mathcal{I}_s$ and $^a\mathcal{I}_k^F = \bigcup_{s \geq k+1} {}^a\mathcal{I}_s$ denote the sets of sessions that empty their backlog before (past) and after (future) ^ab_k, respectively. The following definitions will be needed:

Definition 2. *(a) A session i is compressed (decompressed) in ϕ- space if its weight is decreased (increased). (b) A session i is decompressed in t- space, or its busy period is expanded, if its backlog clearing time is increased. (c) Sessions in $^a\mathcal{I}_k$ are uniformly decompressed in t- space, or their busy periods are uniformly expanded, if their backlog clearing times are equally increased. (d) A session i preserves its position in ϕ-(t-) space if its weight (backlog clearing time) remains unchanged. (e) A set $A \subseteq {}^a\mathcal{I}_k$ is compressible in ϕ- space if $\forall i \in A \, d\phi_i > 0$*

---

[1] Strict inequality is assumed to avoid the degenerate case which under equality would not leave any remaining ϕ to be assigned to the best effort traffic.

exists, such that sessions $i \in A$ do not violate their delay bounds when they are assigned a weight $\phi_i - d\phi_i$, under the conditions: a) sessions in $\mathcal{T}_k^P \cup \{\mathcal{T}_k \setminus A\}$ preserve their position in ϕ- space; b) sessions which emptied their backlog after sessions in A still empty their backlog after sessions in A.

The XMF process applied to an acceptable policy $\pi_a \in \Pi$ is described in figure 1. At this point the following should be noted. XMF is a conceptual process which is not directly applicable at a computational level. The weights assigned to sessions and their busy periods change in a continuous way under this process. In order to keep the presentation clear and simple, each time that the process modifies the policy it is applied to, the process is presented to be reapplied in its entirety to the modified policy, although not necessary. The rationale for this approach is that since XMF is a conceptual process it is not a concern how many times it will be applied, as long as it terminates (generates results) after a finite number of steps.

The XMF process forces all sessions to empty their backlog as late as possible. The only parameters that impose an upper limit on the expansion of the busy periods are the sessions' delay bounds, which do not depend on π_a. Thus, one could expect π_o not to depend on π_a. The following propositions hold. Their proofs may be found in [11].

Proposition 1. *The (intermediate) policy π_b that is defined at the end of steps (II.1.a), (II.1.b) or (II.2.a) is acceptable and more efficient than π_a.*

Proposition 2. *The final policy that results when the application of the XMF process to an arbitrary original acceptable policy is terminated, is acceptable and does not depend on the original policy. That is, $\forall \pi_{a1}, \pi_{a2} \in \Pi$, $XMF(\pi_{a1}) = XMF(\pi_{a2}) = \pi_o, \pi_o \in \Pi$.*

From Propositions 1 and 2 it is easily concluded that a) π_o is the only policy that remains unchanged under the XMF process; and b) for any $\pi_a \in \Pi$, $XMF(\pi_a) = \pi_o$ is more efficient than π_a, except for the case where $\pi_a = \pi_o$. In view of the above the following proposition is self-evident.

Proposition 3. *Policy π_o, $\pi_o = XMF(\pi)$ for any $\pi \in \Pi$ is optimal and unique.*

3.2   Properties of the Optimal ϕ Assignment (Policy π_o)

In this section some properties of the optimal policy π_o are provided. These properties help establish in the next section that the proposed CAC algorithm is optimal. It is shown that in order to determine the optimal policy it is sufficient to observe the all greedy system at certain time instances, which coincide with either the delay bound or the backlog clearing time of some session. For this reason the notion of the checkpoints is introduced.

Definition 3. *Let $\tau_0 = 0$, that is τ_0 coincides with the beginning of the system busy period of an all greedy system. Let $\{\tau_m\}_{m=1}^M$, $M \leq 2N$ denote the ordered set of distinct time instances which coincide with either the delay bound or the backlog clearing time of some session. The time instant τ_m, $m = 0, \ldots, M$ will be referred to as the mth ordered checkpoint.*

(0) Initially[a] , $k = 0$.

(I) $k = k + 1$, If ${}^{a}I_k = \emptyset$ goto (III).

(II) Sessions in ${}^{a}I_k$ are considered. Sessions in ${}^{a}I_k^P$ preserve their position in ϕ- space. This implies that the sessions in ${}^{a}I_k^P$ preserve their position in t- space as well[b].

(II.1) If ${}^{a}I_k$ is compressible the sessions in ${}^{a}I_k$ are compressed in ϕ- space in such a way that their busy periods are uniformly expanded. At the same time the sessions in ${}^{a}I_k^F$ are decompressed in ϕ- space in such a way (under the condition) that they receive the same amount of work up to the end of the (modified) busy periods of sessions in ${}^{a}I_k$ as they did under π_a. This "conditional exchange of weights" is possible [b] and does not alter the backlog clearing times of the sessions in ${}^{a}I_k^F$, that is the sessions in ${}^{a}I_k^F$ preserve their position in t- space. The "conditional exchange of weights" between sessions in ${}^{a}I_k$ and sessions in ${}^{a}I_k^F$ takes place continuously until one of the following happen:

(II.1.a) The backlog clearing time of sessions in ${}^{a}I_k$ becomes equal to the backlog clearing time of sessions which empty their backlog k+1-th in order under π_a (${}^{a}I_{k+1}$), or

(II.1.b) sessions in ${}^{a}I_k$ can not be compressed any further in ϕ-space (their busy periods be uniformly expanded), because some session will miss its delay bound.

At the end of step (II.1) new ϕ's will have been assigned to (all) streams in ${}^{a}I_k$ and ${}^{a}I_k^F$ while streams in ${}^{a}I_k^P$ will have maintained their original ϕ's under π_a. Thus, a new policy π_b is defined in terms of the new ϕ's which is shown [b] to be acceptable and more efficient than π_a. At the end of step (II.1), the XMF process is applied to policy π_b (modified π_a) from the beginning (from step (0)).

(II.2) If ${}^{a}I_k$ is not compressible then a uniform expansion of the busy period of sessions in ${}^{a}I_k$ is not feasible. In this case, ${}^{a}I_k$ is divided into two subsets, that is ${}^{a}I_k = {}^{a}I_k^C \cup {}^{a}I_k^{NC}$, where: ${}^{a}I_k^C$ is the maximum subset of ${}^{a}I_k$ which is compressible in ϕ- space and ${}^{a}I_k^{NC} = {}^{a}I_k \setminus {}^{a}I_k^C$.

(II.2.a) If ${}^{a}I_k^C \neq \emptyset$ then the two sets ${}^{a}I_k^C$ and ${}^{a}I_k^{NC}$ are separated by performing an infinitesimal uniform expansion of the busy periods of sessions in ${}^{a}I_k^C$ (at the same time the weights of sessions in ${}^{a}I_k^F$ are increased in such a way that they receive the same amount of work up to the end of the (modified) busy periods of sessions in ${}^{a}I_k^C$ as they did under π_a (as in step (II.1))), resulting in a new policy π_b. If step (II.2.a) is followed, the XMF process is applied to policy π_b (modified π_a) from the beginning (from step (0)).

(II.2.b) If ${}^{a}I_k^C = \emptyset$ then no stream in ${}^{a}I_k$ may be compressed in ϕ- space any further and the next set ${}^{a}I_{k+1}$ needs to be considered. Thus the XMF process continues from step (I).

(III) End of the XMF process. At this step the unique optimal policy πo (see Proposition 3) has been determined, that is the original acceptable policy π_a has been optimized. Under the resulting policy π_o the QoS sensitive sessions are assigned some weights $\phi_i^{\pi_o}, i = 1, \ldots, N$ and the best effort traffic is assigned weight $\phi_{be}^{\pi_o} = 1 - \sum_{i=1}^{N} \phi_i^{\pi_o}$.

---

[a] Throughout the description only the treatment of the QoS sensitive sessions is considered, and this is sufficient since the weight assigned to best effort traffic is given by $1 - \sum_{i=1}^{N} \phi_i^{\pi}$ under a policy π.

[b] See proof of Proposition 1 in [11].

Figure 1. Description of the XMF Process.

Definition 4. *Let $d_i = \{m : \tau_m = D_i\}$. That is, checkpoint τ_{d_i} coincides with the time instant at which the deadline of session i expires. Then the following quantities are defined for session $i, i = 1, \ldots, N$ at all checkpoints τ_m such that $D_i \leq \tau_m < e_i$ (that is, checkpoints at which session i is still backlogged and requires nonzero service to meet its QoS).*

$$\phi_i^-(\tau_m) = N_i(0, \tau_m)/\hat{W}(0, \tau_m) \tag{6}$$

$$\phi_i^+(\tau_m) = \rho_i/\hat{C}(\tau_m^+) \tag{7}$$

where $\hat{W}(0, \tau_m) = \sum_{k=1}^{m} \hat{W}(\tau_{k-1}, \tau_k)$. The quantity $\phi_i^-(\tau_m)$ represents the fraction of the total NBSA work that must be assigned to session i in order for session i to be assigned work exactly equal to $N_i(0, \tau_m)$ up to time τ_m, given that session i has not emptied its backlog before time τ_m. In particular, the denominator of the right hand side of equation (6) is the total amount of NBSA work which is assigned to backlogged sessions up to time τ_m, including sessions which cleared their backlog earlier and are no longer backlogged at time τ_m. Each session i, which is still backlogged at τ_m gets a percent of this work equal to ϕ_i, that is, it is assigned work equal to $\phi_i \cdot \left(\sum_{k=1}^{m} \hat{W}(\tau_{k-1}, \tau_k) \right)$ over $(0, \tau_m)$. The quantity $\phi_i^+(\tau_m)$ represents the percent of the NBSA bandwidth just after τ_m that must be assigned to session i in order for session i to be served with rate equal to ρ_i. It is easily seen that this is sufficient to ensure that its requirements are satisfied for $t > \tau_m$, if session i is assigned work at least equal to $N_i(0, \tau_m)$ up to time τ_m. The usefulness of these quantities follows from Proposition 5 (All proofs may be found in [11].).

Proposition 4. *Under π_o, $e_i > D_i$, $\forall i \in QoS$ ($QoS \triangleq \{1, 2, \ldots, N\}$). That is, each QoS sensitive session empties its backlog after checkpoint $\tau_{d_i} = D_i$.*

Proposition 4 follows directly from the fact that no QoS sensitive session is compressible in ϕ- space under the optimal policy. Since the requirements ($N_i(0, t)$) of a QoS sensitive session i have zero value for $t < D_i$, session i would be compressible in ϕ- space if it emptied its backlog before its delay bound.

Proposition 5. *Under π_o, QoS sensitive session i is assigned weight:*

$$\phi_i^{\pi_o} = \begin{cases} \phi_i^-(\tau_k) \text{ if } \exists k \text{ such that } k = \min\{m : \phi_i^-(\tau_m) \geq \phi_i^+(\tau_m)\} \\ \phi_i^+(\tau_{l_0}), k = \max\{n : \tau_n < \infty\} \text{ , otherwise} \end{cases} \tag{8}$$

The inequality $\phi_i^-(\tau_m) \geq \phi_i^+(\tau_m)$ (in Proposition 5) implies that the service rate of the QoS sensitive session i would be greater than or equal to ρ_i for $t > \tau_m$, if session i were assigned weight equal to $\phi_i^-(\tau_m)$ (its requirements would be met for $t > \tau_m$). On the other hand, if $\phi_i^-(\tau_m) < \phi_i^+(\tau_m)$ session i requirements would be satisfied if session i were assigned weight equal to $\phi_i^+(\tau_m)$, but session i would be compressible in ϕ- space, except for the case where τ_m were the last checkpoint with finite value. Proposition 5 combines these two facts.

Proposition 6. *Assume that* $\tau_k, \hat{C}(\tau_k^+), \forall k \leq j-1$, *are known.* τ_j *is given by*[2]:

$$\tau_j = \min(\min_{i \in Nex_{j-1}} D_i, \min_{i \in Phi_{j-1}} vct_i(\tau_{j-1})) \tag{9}$$

where Nex_{j-1} *is the set of sessions with delay bound greater than* τ_{j-1}, Phi_{j-1} *is the set of sessions for which* $\exists k \leq j-1 : \phi_i^-(\tau_k) \geq \phi_i^+(\tau_k)$ *and have not cleared their backlog up to time* τ_{j-1} *and*

$$vct_i(\tau_{j-1}) = \tau_{j-1} + (A_i(0,\tau_{j-1}) - W_i(0,\tau_{j-1}))(\phi_i^{\pi_o} \cdot \hat{C}(\tau_{j-1}^+) - \rho_i)^{-1} \tag{10}$$

is the virtual clearing time of session i *at* τ_{j-1}, *that is the backlog clearing time of session* i *assuming that no other session is going to empty its backlog before session* i *does.*

By definition τ_j coincides with either the delay bound or the backlog clearing time of some session. Sessions which could empty their backlog at τ_j are sessions i which are served with a rate greater than or equal to ρ_i for $t > \tau_{j-1}$ (these are the sessions in Phi_{j-1}). Proposition 6 states that τ_j is the minimum of the minimum of the delay bounds which are greater than τ_{j-1} and the minimum of the virtual clearing times of sessions in Phi_{j-1}.

3.3   Optimal Call Admission Control Algorithm

The CAC algorithm presented in this section determines progressively the optimal policy π_o, based on Propositions 5 and 6. It includes two conceptually distinct functions. One which (based on Proposition 5) examines whether the optimal weight of the QoS sensitive sessions can be determined at a specific checkpoint and another which (based on Proposition 6) determines the next checkpoint.

The initial condition of the all greedy system and the first checkpoint are known ($\tau_0 = 0$ and $C(0^+) = C_G$). Assume that τ_k, $\hat{C}(\tau_k^+)$, $\forall k \leq j-1$, are known (later it will become clear how the algorithm determines these quantities). Proposition 6 provides a mechanism to determine the next checkpoint[3]. Once the next checkpoint (τ_j) is determined, the phase of the determination of the optimal ϕ's starts. Proposition 5, in conjunction with Definition 4, indicates that the interval $[D_i, e_i)$ –which will be referred to as examination interval for session i– is critical for the determination of the weight assigned to session i under the

---

[2] In order to avoid unnecessary complexity it is assumed that $\min(\emptyset) = \infty$.

[3] The weights of sessions in Phi_j, which are needed in order to determine the next checkpoint, have already been determined using Proposition 5 at some previous (or the current) checkpoint. In particular, Phi_j contains sessions which fulfill the first condition of Proposition 5 at some previous (or the current) checkpoint. All other quantities needed to determine the next checkpoint are known or can be computed. In particular, $A_i(0,\tau_{j-1}) = \sigma_i + \rho_i \tau_{j-1}$, $W_i(0,\tau_{j-1}) = \phi_i^{\pi_o} \hat{W}(0,\tau_{j-1})$, $\hat{W}(0,\tau_{j-1}) = \sum_{k=1}^{j-1} \hat{W}(\tau_{k-1},\tau_k)$, $\hat{W}(\tau_{k-1},\tau_k) = \hat{C}(\tau_{k-1}^+)(\tau_k - \tau_{k-1})$.

optimal policy π_o. Notice that the beginning of this interval is provided while its termination is shaped by the optimal policy.

Sessions i for which the current checkpoint τ_j satisfies $\tau_j < D_i$ need not to be examined at τ_j. The weight assigned to such sessions under π_o can not be determined at τ_j, since Proposition 5 is not applicable, and it is not needed yet, since Proposition 4 indicates that such sessions are still backlogged at τ_j^+ (and at τ_{j+1}, since τ_{j+1} is at most equal to the delay bound of such sessions), implying that the value of $\hat{C}(\tau_j^+)$ and the next checkpoint (τ_{j+1}) can be determined without knowing the optimal weight of such sessions.

There are two kinds of sessions which are examined by the algorithm at τ_j. Sessions which are examined for the first time at τ_j (sessions for which $\tau_j = D_i$, that is τ_j is the beginning of their examination interval) and sessions which have been examined at some previous checkpoint but the algorithm could not determine the optimal weight for them. According to Proposition 5, if τ_j is the first checkpoint, within the examination interval of session i, at which $\phi_i^-(\tau_j)$ $\geq \phi_i^+(\tau_j)$ then $\phi_i^{\pi_o} = \phi_i^-(\tau_j)$; that is, the weight assigned to session i under the optimal policy is determined. In case where the optimal weight can not be determined at the specific checkpoint, the session will be examined again at the next checkpoint. Sessions for which the above criterion does not hold for any checkpoint with a finite value (of time), are assigned weight at the last checkpoint with a finite value, as the second case of Proposition 5 indicates. Such sessions (whose optimal weight is determined according to the second case of Proposition 5) have a backlog clearing time which tends to infinity.

Sessions i for which the weight assigned by the optimal policy has been determined continue to be taken into consideration by the algorithm until their backlog has been emptied. This is necessary in order to specify future checkpoints and keep track of the value of the NBSA bandwidth. Sessions i which have emptied their backlog at the current or a previous checkpoint need not to be considered any more. It is obvious that the sessions which empty their backlog at τ_j are the sessions whose virtual clearing time at τ_{j-1} is equal to τ_j. Having determined the sessions which empty their backlog at the current checkpoint τ_j (and since the sessions which empty their backlog at some previous checkpoint have been determined at some previous checkpoint) it is straightforward to compute $\hat{C}(\tau_j^+)$ using equation (3). (In particular, $\hat{C}(\tau_j^+) = (C_G - \sum_{i \in Empty_j} \rho_i)(1 - \sum_{i \in Empty_j} \phi_i)^{-1}$ (see the definition of $Empty_j$).). At τ_j the following sets of sessions are defined:

- Nex_j: contains sessions i which have not been examined yet, that is sessions whose delay bound (D_i) is greater than τ_j. Each of those sessions will be examined for the first time at $\tau_k, k > j : \tau_k = D_i$.
- $Empty_j$: contains sessions which have emptied their backlog at the current or a previous checkpoint. Their weights have been determined at a previous checkpoint.
- Phi_j: contains sessions whose weights have been determined, that is the condition of the first case of Proposition 5 holds at $\tau_k, k \leq j$, but have not emptied their backlog yet.

Determine_$\pi_o(C_G, Nex_0)$

/* C_G is the bandwidth controlled by the GPS scheduler, Nex_0 is the set of QoS
sensitive sessions under investigation. */

A. $j=0, Trans_0=Phi_0=Empty_0=\emptyset$, $Nex_0 = QoS$, $\tau_0 = 0$

/* Initialization of the algorithm. At the beginning all sets are empty except
Nex_0 (the set of not examined sessions). The initial value of time is equal to 0,
the beginning of the system busy period of the all greedy system. */

B. repeat (B1.-B6.) until ($\phi_i = \phi_i^{\pi_o}$ $\forall i \in Nex_0$)

/* (B1.-B6.) is executed until the ϕ's of all QoS sensitive sessions are determined.*/

B1. $j = j + 1$, $\tau_j = \min\{\min_{i \in Nex_{j-1}} D_i$, $\min_{i \in Phi_{j-1}} vct_i(\tau_{j-1})\}$

/* Using Proposition 6 the value of the current checkpoint is computed. */

B2. If ($\tau_j = \infty$) then $\forall i \in Trans_{j-1}$

$\{\phi_i = \phi_i^{\pi_o} = \phi_i^+(\tau_{j-1})$, if $\sum_{i \in QoS} \phi_i \geq 1^a$ then error$\}$, goto **C**.

/* If $\tau_j = \infty$ the previous checkpoint was the one with the maximum finite
value of time and according to the second case of Proposition 5 the weights
of sessions whose weight has not been determined yet must be determined
at τ_{j-1}. If the sum of the weights assigned to sessions is greater or equal to
1 then the sessions are not schedulable and the algorithm terminates.*/

B3. $PE_j = \{i \in Phi_{j-1} : vct_i(\tau_{j-1}) = \tau_j\}$, $Empty_j = Empty_{j-1} \cup PE_j$

/* Sessions which empty their backlog at τ_j, that is sessions whose virtual
clearing time is equal to τ_j, are moved to set $Empty_j$. */

B4. $TP_j = \{i \in Trans_{j-1} : \phi_i^-(\tau_j) \geq \phi_i^+(\tau_j)\}$, $NP_j = \{i \in Nex_{j-1} : D_i = \tau_j, \phi_i^-(\tau_j) \geq \phi_i^+(\tau_j)\}$, $NT_j = \{i \in Nex_{j-1} : D_i = \tau_j, \phi_i^-(\tau_j) < \phi_i^+(\tau_j)\}$

*/ The rest temporary sets are determined (see their definitions). */

B5. $Nex_j = Nex_{j-1} \backslash (NP_j \cup NT_j)$, $Trans_j = (Trans_{j-1} \backslash TP_j) \cup NT_j$

/* Main sets $Nex_j, Trans_j$ are updated. */

B6. $\forall i \in NP_j \cup TP_j$ $\{\phi_i = \phi_i^{\pi_o} = \phi_i^-(\tau_j)$, if $\sum_{i \in QoS} \phi_i \geq 1$ then error$\}$

$Phi_j = (Phi_{j-1} \backslash PE_j) \cup NP_j \cup TP_j$

/* Weight assignment and update of set Phi_j. If the sum of the weights
assigned to sessions is greater or equal to 1 then the sessions are not schedulable and the algorithm terminates. */

C. $\phi_{be} = 1 - \sum_{i \in QoS} \phi_i$ /* Computation of $\phi_{be}^{\pi_o}$. */

---

^{a} All ϕ's are considered to be initially undefined. $\sum_{i \in QoS} \phi_i$ denotes the summation
over all sessions that have been assigned weight by the algorithm.

Figure 2. Description of the Optimal CAC Algorithm for the BETA-GPS
System.

– $Trans_j$: contains sessions which have been examined, at a previous checkpoint, but their weights have not been determined yet, that is the condition
of the first case of Proposition 5 does not hold for any $\tau_k, k \leq j$.

These sets are updated at each checkpoint. During the examination of the sessions at each checkpoint (τ_j) some temporary subsets are defined:

- $NT_j \subseteq Nex_{j-1}$ contains sessions which are examined for the first time at the current checkpoint (τ_j) but their weight is not determined at τ_j.
- $NP_j \subseteq Nex_{j-1}$ contains sessions which are examined for the first time at τ_j and their weight is determined at τ_j, that is, the condition of the first case of Proposition 5 is fulfilled at τ_{d_i}.
- $TP_j \subseteq Trans_{j-1}$ contains sessions which have been examined at a previous checkpoint, that is at $\tau_k, k < j$, but their weight is determined at τ_j.
- $PE_j \subseteq Phi_{j-1}$ contains sessions which have been examined at a previous checkpoint, that is at $\tau_k, k < j$, their weight is determined at a previous checkpoint and empty their backlog at τ_j.

The optimal Call Admission Control algorithm for the BETA-GPS system is described in figure 2. In the description the only constraint concerning the best effort traffic is that it must be assigned a nonzero weight. The CAC algorithm can be modified in order to apply a different requirement for the service received by the best effort traffic. For example, the algorithm can guarantee to the best effort traffic a minimum weight assignment $\phi_{be(min)}$ (and a minimum service rate equal to $\phi_{be(min)}C_G$) if the condition ... "$\sum_{i=1}^{N} \phi_i \geq 1$ then error"... is replaced by ... "$\sum_{i=1}^{N} \phi_i > 1 - \phi_{be(min)}$ then error"....

4   Pure QoS System

In a system where only QoS sensitive sessions are present the existence of an extra session, denoted as "dummy", may be assumed and the presented algorithm be applied. Nevertheless, a slight modification must be made to the algorithm since the "dummy" session can be assigned a weight equal to zero. Specifically, the condition $\sum_{i=1}^{N} \phi_i < 1, \phi_i \in \mathbb{R}_+^*$, which must hold for the QoS sensitive sessions in the BETA-GPS system, must be modified to $\sum_{i=1}^{N} \phi_i \leq 1, \phi_i \in \mathbb{R}_+^*$ for the pure QoS system. The modified algorithm for the pure QoS system is referred to as Modified Optimal CAC Algorithm (MOCA) [4].

The input of the MOCA is a traffic mix consisting only of QoS sensitive sessions $i, i = 1, \ldots, N$. If the MOCA finds a solution it returns as output a ϕ assignment ($\phi_1, \phi_2, ..., \phi_N, \phi_d$). Obviously the QoS sensitive sessions can be admitted being assigned weights ($\phi_1, \phi_2, ..., \phi_N$)(or normalized to sum to one ($\phi_1, \phi_2, ..., \phi_N$) $\cdot (1 - \phi_d)^{-1}$). If the MOCA does not find a solution then this implies that a solution does not exist, since the MOCA minimizes $\sum_{i=1}^{N} \phi_i$. In this sense the MOCA can be considered as optimal for the pure QoS system.

Another capability that could be required by a CAC scheme for the pure QoS system is to be able to compute the minimum capacity of the GPS server

---

[4] MOCA is exactly the same as the optimal CAC algorithm for the BETA-GPS system except the check of the sum of the weights assigned to QoS sensitive sessions in steps B.2 and B.6, which should be replaced by ..."if ($\sum_{i \in QoS} \phi_i > 1$ or $\phi_i = 0$(according to the definition of GPS)) then error"... . In addition, in step C. ϕ_{be} should be replaced by ϕ_d, d for "dummy".

$C_{G(min)}$ required to support the N QoS sensitive sessions (and the appropriate ϕ assignment). In this case the following proposition is useful (the proof may be found in [11]).

Proposition 7. *Suppose that an acceptable policy is computed by the MOCA in a pure QoS system where the GPS scheduler controls capacity C_G. QoS sensitive sessions, assigned the computed weights and being served by a GPS scheduler (of a pure QoS system) controlling capacity $C_G(1 - \phi_d)$, do meet their QoS requirements.*

This indicates that if it is desirable to compute the minimum capacity ($C_{G(min)}$) required to support the N QoS sensitive sessions a recursive process can be followed, which amounts to cutting slices of size $\phi_{d(n)}C_n$ from the capacity C_n (with $C_1 = C_G$) controlled by the GPS scheduler at the n-th iteration, until the last cutted slice becomes smaller than an arbitrary predefined small quantity ϵ. where Nex_0 is the traffic mix (consisting only of QoS sensitive sessions) and ϵ is

$compute\_C_{G(min)}(Nex_0, C_G)$
$n = 1, C_{(1)} = C_G, for(;;)$
$\{\phi_{d(n)} = \phi_d(MOCA(Nex_0, C_{(n)})),\ \ C_{(n+1)} = C_{(n)}(1 - \phi_{d(n)})$
$if\ (\phi_{d(n)}C_{(n)} < \epsilon)\ then\ \{C_{G(min)} = C_{(n+1)},\ exit\}$
$n = n + 1\}$

an arbitrary small positive number. $\phi_{d(n)}$ is the weight assigned to the "dummy" session by the MOCA, assuming that the GPS scheduler controls capacity $C_{(n)}$. The process stops when $\phi_{d(n)}C_{(n)}$ becomes less than a predefined quantity ϵ. The feasibility of this process and the fact that it can approximate as closely as desired the minimum GPS capacity required to support the N QoS sensitive sessions can easily be concluded (see [11]).

5   Numerical Results

Deterministic effective bandwidth ([9]) can be used in a straightforward way to give a simple and elegant CAC scheme. A similar approach is followed in [4] for the deterministic part of their analysis. The deterministic effective bandwidth of a (σ_i, ρ_i, D_i) session is given by $w_i^{\text{eff}} = \max\{\rho_i, \frac{\sigma_i}{D_i}\}$. It is easy to see that the requirements of the QoS sensitive sessions are satisfied if they are assigned weights such that $\frac{\phi_i}{\phi_j} = \frac{w_i^{\text{eff}}}{w_j^{\text{eff}}}, \forall i, j \in QoS$. In this section the presented algorithm is compared with the effective bandwidth based CAC scheme.

Since the GPS is a work conserving scheduler, $C_G \cdot t = \sum_{i \in TF} W_i(0, t)$,where $TF = QoS \cup be$ in the best effort aware system and $TF = QoS$ in the pure QoS system, holds for arbitrary t in the system busy period. This implies that $C_G = \sum_{i \in TF} \overline{W}_i(0, t)$ where $\overline{W}_i(0, t) = W_i(0, t)/t$ is the mean work assigned to session i in $(0, t]$. The mean requirements of session i in the interval $(0, t]$ are $\overline{N}_i(0, t) = \frac{\sigma_i + \rho_i(t - D_i)}{t}$ for $t \geq D_i$ and $N_i(0, t) = 0$ for $t < D_i$ which implies $\frac{\partial \overline{N}_i(0, t)}{\partial t} = \frac{\rho_i D_i - \sigma_i}{t^2}$, $t \geq D_i$. Although session requirements are always an increasing function of time, the mean requirements of a session are an increasing (decreasing) function of time (for $t \geq D_i$) if $\rho_i D_i > \sigma_i$ ($\rho_i D_i < \sigma_i$). As it will be demonstrated this determines whether the bandwidth utilization can be improved by the algorithm.

Although the algorithm can support an arbitrary number of delay classes the numerical investigation is limited to the case of three delay classes. Two cases are investigated.

Case 1: The traffic mix consists only of QoS sensitive sessions whose mean requirements are a decreasing function of time for $t \geq D_i, \forall i$. The sessions under investigation for this case are shown in Table 1. All quantities are considered normalized with respect to the link capacity C. In order to compare the presented algorithm with the effective bandwidth based CAC scheme the following scenario is considered. The effective bandwidth based CAC scheme admits the maximum number of sessions under the constraint that a nonzero weight remains to be assigned to best effort traffic. From Table 1 it can 'be seen that the effective bandwidth of each QoS sensitive sessions is 1/25 of the server 's capacity (which is considered to be equal to the link capacity ($C_G = C$)), implying that for the BETA-GPS system at most 24 QoS sensitive sessions can be admitted under the effective bandwidth based CAC scheme. This means that $N_1 + N_2 + N_3 = 24$ must hold and that the best effort traffic is assigned weight equal to 0.04 for each such triplet (N_1, N_2 and N_3 denote the number of admitted sessions of type s_1, s_2 and s_3 respectively). For each triplet (N_1, N_2, N_3), $N_1 + N_2 + N_3 = 24$ the weight assigned to the best effort traffic by the optimal BETA-GPS CAC scheme is computed. The results are illustrated in figure 4. It is easily seen that the improvement achieved by the optimal algorithm depends on the diversity of the traffic mix. For heterogeneous traffic mixes a significant improvement is achieved. On the other hand for pure homogeneous traffic mixes (only one type of sessions) the optimal algorithm can not result in any improvement.

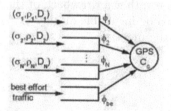

Figure 3. The BETA-GPS System.

Table 1. Sessions under Investigation.

Case 1 s_1	s_2		s_3
Case 2 s_1		s_2	s_3
σ_i	0.04	0.16 0.04	0.64
ρ_i	0.01	0.01 0.04	0.01
D_i	1	4 4	16
w_i^{eff}	0.04	0.04 0.04	0.04

Case 2: The traffic mix consists of both types of QoS sensitive sessions (with increasing and decreasing mean requirements). To demonstrate this case session s_2 is replaced by a session with the same effective bandwidth but with mean requirements which are an increasing function of time (see Table 1). The same scenario as in *Case 1* is followed. The results are illustrated in figure 4. The

achieved improvement is less than in *Case 1*, in particular when sessions of type s_1 are a minor part of the traffic mix.

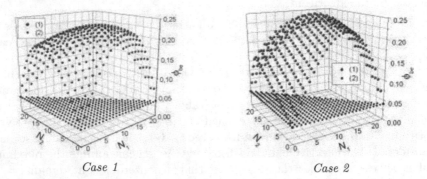

Case 1   *Case 2*

Figure 4. Weight assigned to the best effort traffic according to the (1) optimal BETA-GPS CAC (2) effective bandwidth based CAC scheme, both under the constraint $N_1 + N_2 + N_3 = 24$. The minimum guaranteed rate to the best effort traffic is $\phi_{be}C_G$.

6   Conclusions

In this paper a systematic treatment of the optimal choice of GPS weights has been attempted and the possibility to exploit the dependencies among the sessions in order to achieve better bandwidth utilization has been demonstrated. The presented optimal CAC scheme suffers from the drawback that the ϕ assignment must be modified each time that the traffic mix changes (in order to remain optimal). Nevertheless, this seems to be rather a drawback of the GPS discipline itself, in the sense that GPS seems to be incapable of fully exploiting the available bandwidth under a "static" ϕ assignment.

References

1. A. Parekh and R. Gallager. A generalized processor sharing approach to flow control in integrated services networks: The single-node case. *IEEE/ACM Transactions on Networking*, 1(3):344–357, June 1993.
2. A. Parekh and R. Gallager. A generalized processor sharing approach to flow control in integrated services networks: The multiple node case. *IEEE/ACM Transactions on Networking*, 2(2):137–150, April 1994.
3. R. Szabó, P. Barta, F.Németh, and J.Biró. Call admission control in generalized processor sharing (gps) schedulers using non-rate proportional weighting of sessions. *Proceedings of IEEE INFOCOM*, 3:1243–1252, March 2000.

4. A. Elwalid and D. Mitra. Design of generalized processor sharing schedulers which statistically multiplex heterogeneous qos classes. *In Proceedings of IEEE INFO-COM*, pages 1220–1230, March 1999.
5. Z. Zhang, D. Towsley, and J. Kurose. Statistical analysis of the generalized processor sharing scheduling discipline. *Proc. of ACM SIGCOMM*, September 1994.
6. Z. Zhang, Z. Liu, J. Kurose, D. Towsley. Call admission control schemes under the generalized processor sharing scheduling. *The Journal of Telecommunication Systems, Modeling, Analysis, Design, and Management*, 7(1), July 1997.
7. Z. Zhang, Z. Liu, and D. Towsley. Closed-form deterministic end-to-end performance bounds for the generalized processor sharing scheduling discipline. *Journal of Combinatorial Optimization (Special Issue on Scheduling)*, 1(4), 1998.
8. R. Szabo, P. Barta, J.Biró, F.Németh, and C-G. Perntz. Non-rate proportional weighting of genaralized processor sharring schedulers. *Proceedings of GLOBE-COM*, 2:1334–1339, December 1999.
9. J.-Y. Le Boudec. Application of network calculus to guaranteed service networks. *IEEE Transactions on Information Theory*, 44(3):1087–1096, May 1998.
10. L. Georgiadis, R. Guérin, V. Peris, and K. Sivarajan. Efficient network qos provisioning based on per node traffic shaping. *IEEE/ACM Transactions on Networking*, 4(4):482–501, August 1996.
11. A. Panagakis and I. Stavrakakis. Optimal call admission control under generalized processor sharing scheduling. *Technical report*, available at http://www.di.uoa.gr/~istavrak/pub/gps.ps, 2001.

Author Index